ENOLOGICAL CHEMISTRY

Wine is the friend of the wise man and the enemy of the drunkard. It is bitter and useful like the advice of the philosopher; it is permitted to the gentleman and forbidden to the idiot. It lures the stupid into the fog and guides the wise toward God.

Avicenna (980–1037)

ENOLOGICAL CHEMISTRY

Written and Edited by

JUAN MORENO AND RAFAEL PEINADO

Dpto. Química Agrícola y Edafología, Universidad de Córdoba, Córdoba, Spain

Translated by

ANNE MURRAY AND IAIN PATTEN

AMSTERDAM • BOSTON • HEIDELBERG • LONDON • NEW YORK • OXFORD PARIS
SAN DIEGO • SAN FRANCISCO • SINGAPORE • SYDNEY • TOKYO
Academic Press is an imprint of Elsevier

Academic Press is an imprint of Elsevier
32 Jamestown Road, London NW1 7BY, UK
225 Wyman Street, Waltham, MA 02451, USA
525 B Street, Suite 1800, San Diego, CA 92101-4495, USA

First edition 2012

Originally published in Spanish under the title *Química Enológica*, by Juan Moreno and Rafael Peinado. A. Madrid Vicente, Ediciones, 2010.

Notice
No responsibility is assumed by the publisher for any injury and/or damage to persons or property as a matter of products liability, negligence or otherwise, or from any use or operation of any methods, products, instructions or ideas contained in the material herein. Because of rapid advances in the medical sciences, in particular, independent verification of diagnoses and drug dosages should be made

British Library Cataloguing-in-Publication Data
A catalogue record for this book is available from the British Library

Library of Congress Cataloging-in-Publication Data
A catalog record for this book is available from the Library of Congress

ISBN : 978-0-12-388438-1

For information on all Academic Press publications visit our website at www.elsevierdirect.com

Typeset by TNQ Books and Journals

Printed and bound by CPI Group (UK) Ltd, Croydon, CR0 4YY

Transferred to digital print 2012

Working together to grow
libraries in developing countries

www.elsevier.com | www.bookaid.org | www.sabre.org

ELSEVIER   BOOK AID International   Sabre Foundation

Dedication

To our families and loved ones, for the time we could not dedicate to them whilst writing this book. Thank you for your endless encouragement, patience, and understanding.

The authors.

Contents

Preface

Enology has traditionally been considered both a science and an art. Progress in the field has been linked to the work of a core group of professionals who have produced empirical evidence essentially by trial and error. These insights, however, have often failed to reach an audience sufficiently wide to drive innovation in those countries with a strong winemaking tradition. Nowadays, a wealth of scientific evidence is available to support continued innovation and development, and these developments will be essential to successfully overcoming the challenges faced by the wine industry.

Modern enology is now inconceivable without specialist training to cover the breadth of knowledge that has been accumulated in the field. To this end, enology can now be studied as a university degree. This volume is the product of more than 10 years' experience in teaching the chemistry of wine to students on the degree course in enology at the University of Córdoba in Spain. The authors draw on material from lectures and laboratory sessions to provide an up-to-date review of the subject that will be of use to students, professionals, and all those who want to increase their knowledge of wine science. The content is designed to be accessible to anyone with a solid foundation in chemistry.

About the Authors

Juan Moreno was born in Córdoba, Spain in 1955. He holds a degree in chemistry and a PhD in science from the University of Córdoba, where he is currently a tenured professor. Prof. Moreno has conducted extensive enological research since 1983 and has published over 86 articles, 64 of which have appeared in international journals; he has also presented over 100 papers at conferences worldwide. His research interests include the chemistry of grape ripening, the characterization and fermentation of musts, the biological aging of Andalusian wines, and the production of sweet wines from raisined grapes. As part of his work at the University of Córdoba, he has taught general chemistry to students on the degree course in chemistry since 1996, and since 1999, he has taught wine chemistry to students on the degree course in enology and general chemistry, and to forest engineering students.

Rafael Peinado was born in Córdoba, Spain, in 1975. He holds degrees in chemistry (1998) and enology (2001), and he completed his doctoral studies at the University of Córdoba in 2004. His early research interests were the biological deacidification of must and wine and the development of a new cell immobilization system. He is currently working on the characterization of musts made from different varieties of sun-dried grapes and studying the antioxidant properties of these musts and vinification sub-products. He has published more than 35 articles in international journals and made numerous presentations at national and international congresses. He combines his research activities with lecturing on food and technology science, enology, and forest engineering.

1

The Vine

1. BIOLOGICAL CYCLES OF THE VINE

Vines are herbaceous or sarmentose shrubs. Their leaves are simple and more or less palmate or lobulate in shape, and they have tendrils that grow in the opposite direction. Their flowers have five petals joined at the apex which are grouped in narrow panicles. Commonly, the flowers are unisexual. The fruit is an oligospermic berry with a spherical or ovoid shape that contains pear-shaped seeds and a soft pulp.

The vine belongs to the *Vitaceae* family, which includes a dozen genera. Among these are *Ampelopsis* and *Parthenocissus*, which include wild vines, and *Vitis*, which is responsible for all table and wine grape varieties. The genus *Vitis* contains around 40 species, the most important being *Vitis vinifera* or the European species, which is used in the production of high-quality wines, and the American species *Vitis rupestris, riparia, berlandieri, labrusca,* etc., which have been used as rootstocks and direct-producing hybrids. Within each species, there are varieties that only conserve their characteristics by vegetative propagation, and can therefore be considered as clones. There are 6800 known varieties of *V. vinifera* alone, although no more than 100 are used to produce the most recognizable wines worldwide.

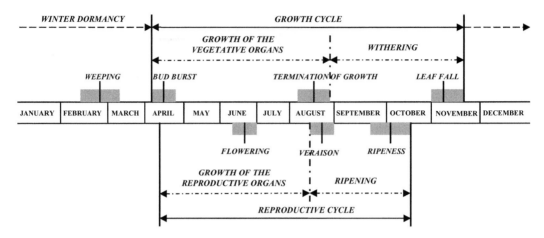

FIGURE 1.1   Growth and reproductive cycle of the vine.

As vines are perennial plants, they undergo characteristic morphological changes with the seasons. Throughout each yearly cycle, the vine undergoes:

- The growth and development of the vegetative organs (shoots, leaves, tendrils, and roots), its survival through the accumulation of reserves (withering), and the acquisition of latency by the buds. This is the growth cycle.
- The growth and development of reproductive organs (inflorescences, flowers, and berries) and their maturation. This is the reproductive cycle.

The morphological cycle of the vine has particular characteristics known as phenological events. In chronological order these are weeping, bud burst, flowering, fruit set, ripening, and leaf fall.

There are two clearly distinguishable periods in the yearly cycle of the vine, namely winter dormancy, during which no morphological changes occur and which extends from leaf fall until bud burst, and active growth, which begins with bud burst and ends with leaf fall. During the period of active growth, the organs are constructed, the seeds and berries form, and the materials necessary for survival accumulate in the living parts of the plant. The period of active life contains two very different phases: growth, which occurs from bud burst until fruit set, and ripening, which occurs from veraison until leaf fall.

1.1. The Growth Cycle

Prior to the onset of vegetative growth at the end of the winter, liquid oozes from the wounds created by pruning. This is known as weeping and can last for a number of days. It is caused by the activation of the root system as a result of the increasing temperature of the soil, and is halted by the development of bacteria that form a viscous mass within the liquid, that ultimately obstructs the xylem vessels.

Bud burst is the process by which the protective scabs covering the buds break open. Not all buds break open and the process is not simultaneous for all buds on the vine, a phenomenon known as acrotonic budding. Buds that burst in the spring do so because a latent bud will have formed during the previous growth cycle.

Foliation involves the appearance and development of the leaves. This phenomenon cannot be separated from the growth of the shoots.

Withering begins at the end of the veraison and continues until leaf fall. The most important event that occurs during this process is the accumulation of reserves, particularly of starch, in the trunk and shoots.

At the end of the period of active life, the leaves lose their green color, photosynthesis ceases, and the leaves fall. At this point the plant can be considered to have entered a dormant phase, although the translocation of reserve substances continues for a few days after leaf fall.

1.2. The Reproductive Cycle

The reproductive cycle begins during the period of growth and continues during part of the withering period. It comprises two successive cycles, since the flower cluster exists in embryonic form in the fertile buds formed during the previous growth cycle.

Flowering involves the opening of the corolla of the flower and is linked to fertilization. It is difficult to separate these two processes in time since the same vine will carry flowers that have yet to open and others that are already fertilized. After flowering, the inflorescence is termed a raceme or cluster. It is made up of a principal axis, together with secondary axes formed by the stalks or stems that support the fruit or berries. The structure of the raceme and the number and volume of berries are determined by the inflorescence; the cluster can be loose, intermediate, or compact.

The berries begin to develop at fertilization. The development process can be divided into four periods: herbaceous growth, veraison, ripening, and over-ripening.

In varieties containing seeds, the fruit begins to develop after fertilization of the ovary. At this stage the fruit is said to be set. Under favorable conditions for production, this type of growth and formation of the fruit generates berries of the maximum size for each variety.

The herbaceous period extends from fruit set until veraison. Normally not all of the flowers are fertilized and therefore do not form berries. The grape cluster behaves as a green organ during the herbaceous period, and as it contains chlorophyll, it contributes to photosynthesis. The stalks also reach their final size during this period and the berries increase in volume but remain hard and green. Their sugar content is low, but acids begin to accumulate and reach their maximum concentration when the grapes are close to veraison.

Veraison is characterized by a change in the color of the grapes, leading to the development of the color typical of the variety. Not all grapes change color at the same time and the process takes approximately two weeks.

During veraison, the berries become softer and more elastic due to changes in the cell walls. They lose chlorophyll and change color due to the formation of pigments; white grapes become translucent and some develop a yellowish color, whereas red grapes begin to develop their characteristic color in a series of increasingly strong red tones. By the end of veraison, the seeds or pips of the grapes are perfectly formed and able to reproduce the plant; they have thus achieved physiological maturity much earlier than the fruit.

During veraison, the pulp rapidly begins to accumulate sugars while the acidity is considerably reduced. At this point, the grape berries enter the next stage of their development, namely ripening. During this stage, the composition of the berry is modified extensively by the accumulation of substances derived from other organs and by the transformation of those that are already present. Once ripeness has been achieved, the grapes enter a phase of over-ripening. During this phase, external physical factors have a greater influence than the function of the plant, and the grapes become increasingly fragile. The grape receives almost no contribution from the plant and there is a partial evaporation of water from the pulp that leads to concentration of the sugars. In parallel, respiratory combustion of some acids continues and the grapes begin to change their consistency. In other words, over-ripening reduces the yield of fruit juice but increases the richness of the juice in terms of sugars and reduces its acidity. Over-ripening is essential to obtain wines with high alcohol content, since the concentration of alcohol in the final product is proportional to the sugar content of the grapes from which it was produced.

2. MORPHOLOGY OF THE GRAPE CLUSTERS

The grape clusters comprise two elements, namely the stalk and the actual fruit. The different elements of the inflorescence progressively achieve their final dimensions while the ovaries of the flowers are transformed into fruits and the ovules into seeds. All these elements together form the raceme or cluster.

2.1. The Stem or Stalk

The stalk of the grape clusters comprises a basal stem or peduncle, which is the part that joins the shoot, and all of its ramifications. The longest arm forms the main axis or backbone of the cluster. The finest branches, or pedicels, end in a swelling or receptacle into which the grape berry is inserted. This swelling is where the vascular bundles that transport nutrients to the inside of the berry travel. One part of these bundles, the brush, remains connected to the receptacle when the ripe berry is removed.

The stalk reaches its final dimensions during veraison, and during ripening of the grapes the peduncle becomes woody while the rest of the stalk remains herbaceous.

The texture of the cluster depends upon the length of the pedicels. If these are long and thin, the berries remain separate and the clusters are loose. If, on the other hand, the pedicels are short, the clusters will be compact and the berries are pressed against each other. Whereas loose clusters are desirable in table grapes, most varieties used in winemaking require tightly packed clusters.

Under normal conditions in which no accidents such as fruit shatter or failure to seed have occurred, the stalk accounts for between 3% and 7% of the weight of a ripe grape cluster. The proportion of stalk in the clusters is determined by the grape variety and the type of cluster (simple, branched, conical, winged), and even clusters belonging to the same grape variety can vary according to the type of care they receive, the weather conditions, diseases, etc. Consequently, it is difficult to establish an accurate average. In varieties with very tight clusters and fine stalks, and in climates and years with a wet summer, the stalk may account for only 2% of the total weight of the cluster. In contrast, it can account for 7% in fruits from plots with clusters that are spread out, either naturally or due to fruit shattering, in years with dry or very dry summers, and in poor soils with little additional management during the growing season. In general, an acceptable percentage is 4% for most grape varieties grown in Spain and 2% for Pedro Ximénez grapes grown in the Montilla-Moriles region.

2.2. The Grape Berry

Grapes are fleshy berries. Their shape can vary substantially between varieties but is consistent within the same variety. They can be lobular, elongated, flattened, ellipsoid, ovoid, etc. Prior to veraison, they are green, contain chlorophyll, and can perform photosynthesis. Following veraison, the berries of white grape varieties acquire a yellowish color, whereas berries from red varieties acquire a reddish-violet color.

The grapes are always very hard until veraison and after that their consistency depends upon the variety. The size of the berries depends on a number of factors, principally the soil, cultivation method, development of the seeds, and also the number in the cluster. Morphologically, the grapes comprise skin, pulp, xylem and phloem vessels, and seeds or pips.

The skin has a heterogeneous structure comprising cuticle, epidermis, and hypodermis. On the surface of the skin at the other end from the pedicel there is a small, darker spot, the navel, which is clearly visible in white grapes and corresponds to the remnant of the stigma. The cuticle is very thin in varieties of *V. vinifera* and is covered by the bloom, a layer with a waxy appearance that comes off to the touch. This layer is important as it captures microorganisms present in the air. Most notable among these microorganisms are the yeasts, which are responsible for spontaneous fermentation of must. The bloom also acts as an impermeable layer that blocks the evaporation of water from inside the berry.

The epidermis and hypodermis, which lie below the cuticle, contain layers of cells of different sizes that contain pigments, aromatic substances, and tannins. While these substances are only weakly soluble in cold water, they are soluble in alcohol, and consequently diffuse during fermentation when the sugary juice becomes enriched in alcohol. The odorant substances mainly comprise monoterpene alcohols and heterosides, and they endow the wine with floral and fruity notes corresponding to the varietal aroma. Tannins are more abundant in red than white grapes.

The proportion of the ripe grape made up by the skin depends mainly on the variety and the climate (due to its influence on transpiration). In varieties grown in Spain, the proportion is 7% to 8%, in France it is 15% to 20%, and in California, 5% to 12%.

The pulp is made up of large cells containing vacuoles, the structures which contain the fluid that will form the must. The grape contains 25 to 30 layers of cells from the epidermis to the endocarp. Since the same number is present in the ovary, it is apparent that the swelling of the grape berry is caused by an increase in the volume of the cells rather than by cell division.

The vascular system of the berry, which carries sugars from the leaves and minerals from the roots, comprises 10 to 12 bundles that are left attached to the receptacle when the berry is removed and form the so-called brush. The bundles branch off within the pulp to form a network.

The proportion of pulp in the mature berry varies between different varieties of *V. vinifera*, but the differences are not substantial and the pulp accounts for 85% of the mass on average. In the case of Pedro Ximénez grapes, the percentage is between 90% and 92%.

The seeds acquire definitive characteristics at the end of veraison, when they achieve physiological maturity. They therefore remain relatively unaffected by the chemical changes that occur in the berry between veraison and full ripeness.

Although the berry should normally contain four seeds, derived from the four ovules in the ovary, there are nearly always fewer due to the absence or abortion of one or more of these ovules. If fertilization is defective, the berries contain no seeds, or the seeds are smaller than normal and hollow. When this occurs, the berries will remain very small, although they may ripen and even develop high concentrations of sugars. The mass of the berry, its sugar content, and often its acid content are related to the number of seeds. An absence of seeds may also be typical of the vine, which is desirable when producing table grapes or raisins.

The seeds contain two enveloping woody layers in the form of a skin. These are known as the testa and tegmen (rich in tannins), and they surround the endosperm (rich in fatty acids). Inside the endosperm, towards the tip of the seed, the germ or embryo of the new plant is found.

The proportion of the mass formed by the seeds depends upon the variety and, in particular, the number of seeds within the berry. The average found in varieties grown in Spain is 4%. In grapes belonging to the Pedro Ximénez variety, the seeds account for between 3% and 4% of the mass.

The grape berry increases steadily in volume and mass from the time the fruit is set until it is ripe. In addition to the effect of the variety and the number of seeds, the weather conditions in a given year also affect the mass of the grapes.

From mid-veraison to ripeness, the mass of the berries increases by 50%. The volume and mass of the ripe berry depend mainly on the rainfall after veraison and the water reserves in the soil. The volume and mass of the berry can be affected by various diseases. The maximum weight is achieved a few days before the harvest, and a slight loss of mass can be observed in the week before harvesting due to loss of water from the berries. This can reach up to 10% of the total mass.

3. CHEMICAL COMPOSITION OF THE FRUIT

3.1. Composition of the Stalk

The chemical composition of the stalk is similar to that of the leaves and tendrils, although it is particularly rich in polyphenols. It has a low sugar content and intermediate concentrations of acid salts due to the abundance of minerals, and its cell content has a high pH (>4). Contrary to popular belief, maceration of the stalks during vinification of red grapes without destemming leads to a reduction rather than an increase in acidity, with a slight increase in pH.

The ash from the stalks accounts for 5% to 6% of the dry weight and comprises approximately 50% potassium salts. After potassium, the most abundant cations are calcium and magnesium, followed by sodium, iron, copper, manganese, and zinc in much lower proportions.

The stalks are rich in phenolic compounds (particularly in red grape varieties), and the concentrations of these compounds in wine is therefore increased when vinification is carried out with the stalks remaining present. The polyphenols present in the stalks have a bitter flavor, however, and therefore reduce the quality of the wine.

TABLE 1.1 Chemical Composition of the Stalk (Milliequivalents per kg of Stalk)

Sugars (g/kg)	< 10
pH	4.1 − 4.5
Free acids	60 − 90
Acid salts	102 − 140
Tartaric acid	30 − 90
Malic acid	80 − 150
Citric acid	4 − 10
Total anions	170 − 183
Total cations	160 − 205
Soluble polyphenols (g/kg)	5.4 − 15.2

Although the stalks account for only around 4.5% of the weight of the cluster, they contribute around 20% of the total phenolic compounds, 15% of the tannins, 26% of the leucoanthocyans (constituents of condensed tannins and therefore linked to astringency), 15% of the catechins, 16% of the gallic acid, and 9% of the total caffeic acid.

3.2. Composition of the Seeds

The outer layers (woody parts) of the seeds are rich in tannins, containing, depending on the crop, between 22% and 56% of the total polyphenols of the grape. These include the procyanidins (67% to 86%) and a substantial proportion of the total gallic and caffeic acid. The woody part (testa and tegmen) is surrounded by a thin film that is also rich in tannins.

The endosperm contains a lipid fraction that comprises on average 50% linoleic acid, 30% oleic acid, 10% saturated fatty acids, and 1% unsaponifiable residue. This oil is commonly extracted from the flour obtained upon pressing the grape seeds using an appropriate solvent, and up to half a liter of oil can be extracted per hectoliter of wine.

Whereas the substances contained in the seed coat can be beneficial (phenolic compounds, nitrogenated substances, and phosphates that are partially dissolved during the production of red wines), those present on the inside of the seeds would have a negative effect on the quality of the wine if they were to dissolve, hence rupture of the seeds during pressing should be avoided.

When the seed reaches physiological maturity, it begins to lose up to a fifth of its nitrogen content in the form of ammonium cations. Nevertheless, the seeds remain richer in nitrogen than the remaining solid parts of the grape cluster.

The minerals contained in the seeds account for 4% to 5% of their weight and the distribution of cations differs from that of the other parts of the cluster, since calcium tends to be the most abundant (particularly in chalky soils) followed by potassium, magnesium, and sodium, and then much lower levels of iron, manganese, zinc, and copper, in that order.

TABLE 1.2   Chemical Composition of the Seeds
(Percentage of the Total Mass)

Water	25 − 45
Sugars	34 − 36
Oils	13 − 20
Tannins	4 − 6
Nitrogenous compounds	4 − 6.5
Minerals	2 − 4
Free fatty acids	1

3.3. Composition of the Skin

The skin of the grapes has an important role to play in winemaking, since the type of wine (white or red) is defined by the way in which the different parts of the grapes are used in vinification. The skins contain most of the substances responsible for the color and aroma of the grapes and make a substantial contribution to the color, aroma, and flavor of musts and wines.

The bloom is made up of two thirds oleanoic acid and the remainder comprises hundreds of different compounds such as alcohols, esters, fatty acids, and aldehydes. Grape skins contain appreciable quantities of malic acid, but its concentration declines during ripening, and the skins of ripe grapes contain mainly tartaric, malic, and citric acids, in that order. The most characteristic substances in the skins of ripe grapes are yellow and red pigments and aromatic substances. The typical color of the grape variety begins to appear at the veraison and peaks when the grape is ripe.

TABLE 1.3   Chemical Composition of the Skin
(Milliequivalents per 100 g of Skin)

Sugars (g/1000 berries)	0.7 − 3
pH	3.8 − 4.3
Free acids	55 − 94
Acid salts	65 − 148
Tartaric acid	64 − 99
Malic acid	40 − 132
Citric acid	3 − 9
Total anions	123 − 240
Total cations	120 − 242
Soluble polyphenols (g/kg)	26 − 68

The amounts of phenolic compounds in the skins are highly variable, and depend mainly on the grape variety. The skin contains between 12% and 61% of the total polyphenol content of the fruit, between 14% and 50% of the tannins, 17% to 47% of the procyanidins, and almost all of the anthocyans in red grape varieties. They are rich in cellulose, insoluble pectins, and proteins. Chlorophyll, xanthophyll, and carotenoids are present in appreciable quantities when the grapes are green, but their concentrations are lower in the ripe grape.

The minerals in the skins have an almost identical distribution to that in the stalks, with potassium accounting for more than 30% of the total mineral content. In decreasing order of concentration, potassium is followed by calcium and magnesium, and then at much lower concentrations, by sodium, iron, copper, manganese, and zinc.

3.4. Composition of the Pulp

The pulp contains those components that predominate in the grape juice or must. The solid part of the pulp is made up of cell walls and vascular bundles, and accounts for no more than 0.5% of its mass. It is this that forms the sediment or deposit that remains in the tanks after the must is decanted.

The sugars in the pulp are mainly glucose and fructose. During veraison, the glucose content is twice that of fructose, whereas in ripe grapes the two sugars are present in almost equal proportions. Sucrose is only present in wine grapes in trace amounts, since, although it is the main sugar synthesized in the leaves, it is hydrolyzed during translocation to the fruit. In addition to glucose, fructose, and sucrose, other sugars such as arabinose, xylose, rhamnose, maltose, and raffinose have been identified in grapes.

Sugars are not uniformly distributed in the grape berry, and the part at the opposite side to the pedicel is richer in sugars than that closest to it. Similarly, if the pulp is divided into three parts, one closest to the skin, one surrounding the seeds, and one in the region in between, it is this last intermediate region that is richest in sugars. This distribution has consequences for the winemaking techniques used, particularly for the production of white wines, since free-run juice is richer in sugars than subsequent press fractions.

TABLE 1.4 Chemical Composition of the Pulp
(Milliequivalents per kg of Pulp)

Sugars (g/kg)	180 − 240
pH	3.2 − 4.0
Free acids	98 − 125
Acid salts	43 − 58
Tartaric acid	45 − 90
Malic acid	70 − 90
Citric acid	1.5 − 2.9
Total anions	130 − 170
Total cations	150 − 170

TABLE 1.5 Chemical Composition of Grape Must (g/L)

Water	800 − 860
Sugars	120 − 250
Organic acids	6 − 14
Minerals	2.5 − 3.5
Nitrogenous compounds	0.5 − 1.0
Other substances	< 1

TABLE 1.6 Physical and Chemical Composition of Different Fractions of Grape Berries (Percentage Fresh Weight)

Components	Stalk	Berry		
		94−97		
Cluster % Weight	3−6	Skin	Seeds	Pulp
		7−12	0−6	83−91
pH	4−4.5	3.8−4.3		3−4.5
Water	78−80	78−80	25−45	70−85
Sugars	0.5−1.5		34−36	14−26
Organic acids	0.5−1.6	0.8−1.6		0.6−2.7
Lipids			13−20	
Free fatty acids			1	
Polysaccharides				0.3−0.5
Polyphenols				0.05
Anthocyans		0−0.5		
Tannins	2−7	0.4−3	4−10	
Aromatic substances		<< 1		< 0.01
Waxes		< 1		
Minerals	2−2.5	1.5−2	2−4	0.08−0.28
Nitrogenous compounds	1−1.5	1.5−2	4−6.5	0.4−0.7
Vitamins				0.02−0.08

Other major components of the pulp are organic acids, mainly tartaric and malic acid. Citric acid is also present, although at lower levels.

3.5. Composition of the Must

The liquid component of the pulp, which is obtained when the grapes are crushed, makes up what is commonly referred to as must. The must is characteristically made up of sugars, acids, and other substances in proportions that are very similar to those of the pulp from which it is derived, and its composition depends upon various factors, such as the grape variety, the location of the winery, the composition of the soil, and the weather conditions during the growth and ripening of the grapes. Diseases of the vine and the processes used to obtain the must and begin fermentation can also influence its composition.

Logically, the compounds that are present at the highest concentrations are important because they form the main raw materials for the transformation of must into wine. On the other hand, a number of minor components are necessary for the growth and survival of yeast, and others contribute sensory qualities. These last components include polyphenols, which give color to the wine, and monoterpene alcohols, which are important due to their delicate aromas, responsible for the varietal aroma of the wine. In the following chapters, we will therefore explore each of these groups of compounds in detail before considering how they change during the ripening of the grapes.

Composition of Grape Must

1. GRAPE MUST

Grape must is the liquid obtained by the gentle crushing or pressing of grapes. Pressing takes place once the grapes (either destemmed or still in clusters) have been gently crushed. Even within the same winemaking region, must composition varies according to several factors, including:

- The type and variety of grapes used,
- The ripeness and health of the grapes (ripeness depends on a range of factors, such as the climate during the growing season, the type of soil, and the fertilizers used),
- The pressure exerted on the grapes.

Musts are classified as free-run must (or juice), which is obtained by the simple crushing of grapes, or press-fraction must, which is obtained by subjecting the grapes to increasing levels of pressure. There are therefore many types of must.

TABLE 2.1   Composition of Musts Obtained at Increasing Press Pressures

	Liquid Volume (%)	Dry Extract (g/100 mL)	Sugars (g/L)	Acidity (g/L)	Ash (g/L)	Alkalinity of Ash (meq/L)	Tartaric Acid (g/L)	Malic Acid (g/L)
Free run	60	21	194	7.5	3.4	32	5.6	3.8
First press	25	22	192	7.7	3.4	34	5.8	4.0
Second press	10	22	191	6.5	3.8	34	4.4	4.2
Third press	4	25	187	5.4	4.9	40	3.6	4.3
Fourth press	1	31	176	5.1	5.6	46	3.7	4.3
Total	100	22	193	7.3	3.6	34	5.3	3.9

Adapted from De Rosa, 1988.

Quality white wines are made only from free-run must (known in Spain as *mosto flor* or *mosto yema*) or first-press fractions. Subsequent fractions can be used to make other products, such as the more intense, deeper-colored press wines.

As the press fractions, and logically, the pressure exerted on the grapes increase to improve the yield, the resulting juice becomes increasingly rich in substances derived from the solid parts of the grape, such as the stems (when the grapes are crushed in clusters), the skins, and the pips.

Figure 2.1 shows variations in must pH and the concentration of several compounds derived from grape solids over successive press fractions. As can be seen, polyphenol and potassium levels increase after several presses, as the increased pressure on the solid parts of the grape extracts a greater proportion of these compounds. The pH of successive press fractions is related to the amount of free acids and acid salts in the different parts of the grape and is therefore also a reflection of the pH of the tissue of the solid parts of the grape (which is higher than that of the pulp). pH is also influenced by potassium levels, as potassium ions neutralize most of the acids in the berry. Iron levels are also directly related to the number of presses, but it should be noted that pressing equipment, which is generally made of iron or stainless steel, can also contribute to these levels.

It is thus clear that there is no such thing as a *single* must, and that to understand must composition we must take into account the different treatments that both the grape and its juice undergo in order to obtain a reliable raw material for fermentation.

The following substances or groups of substances, shown in order of abundance, are present in must:

Water
Sugars
Organic acids
Nitrogen compounds
Minerals
Polyphenols
Vitamins
Aromatic compounds

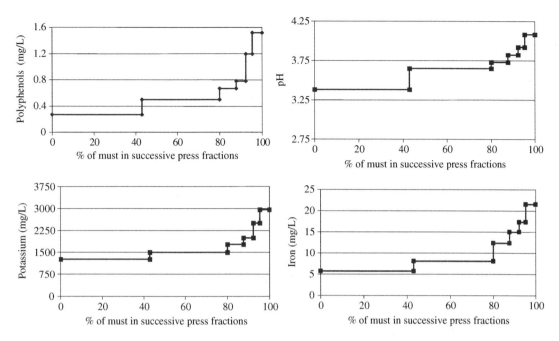

FIGURE 2.1 **Must composition according to number of presses.**

2. CHEMICAL FAMILIES PRESENT IN MUST

All chemical families are characterized by a functional group, which is a collection of one or more atoms within a molecule that provide the molecule with a unique chemical function, specific to the functional group.

2.1. Sugars

The sugars present in must are polyalcohols with a carbonyl group. These are organic compounds that contain several −OH groups along with an aldehyde or ketone group. According to the established nomenclature, the suffix -ose should be used to refer to these compounds, which can be aldoses or ketoses. The D and L prefixes used in sugars are related to the (+) or (−) enantiomer of the glyceraldehyde from which they are derived. Accordingly, monosaccharides in which the chiral center that is furthest from the aldehyde or ketone group has the same configuration as the corresponding D-glyceraldehyde are members of the D series, while those with the opposite configuration are members of the L series. In a Fischer projection, the −OH group is to the right of the chiral center of D-glyceraldehyde. The D or L configuration is not related to the ability of the enantiomers to rotate polarized light to the right or left.

The most abundant sugars found in must are monosaccharides with six carbon atoms:

- Glucose: 6-carbon aldose (dextrose)
- Fructose: 6-carbon ketose (levulose)

The chemical structures for D-glucose and D-fructose are shown below:

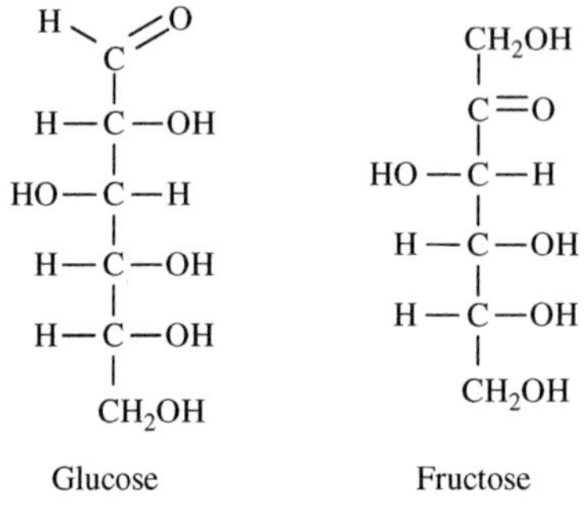

Glucose   Fructose

The most relevant 5-carbon monosaccharides are the D-aldoses:

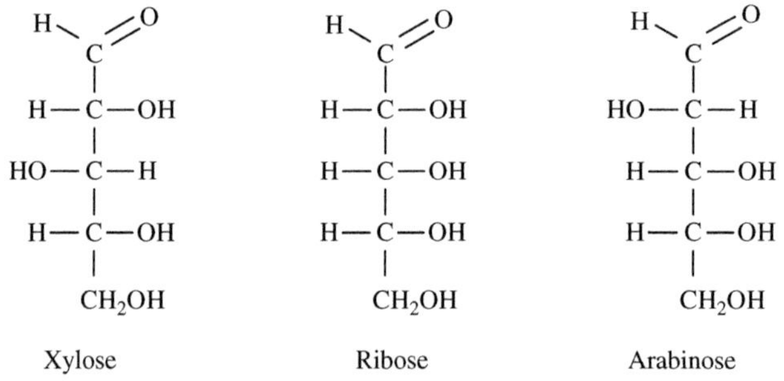

Xylose   Ribose   Arabinose

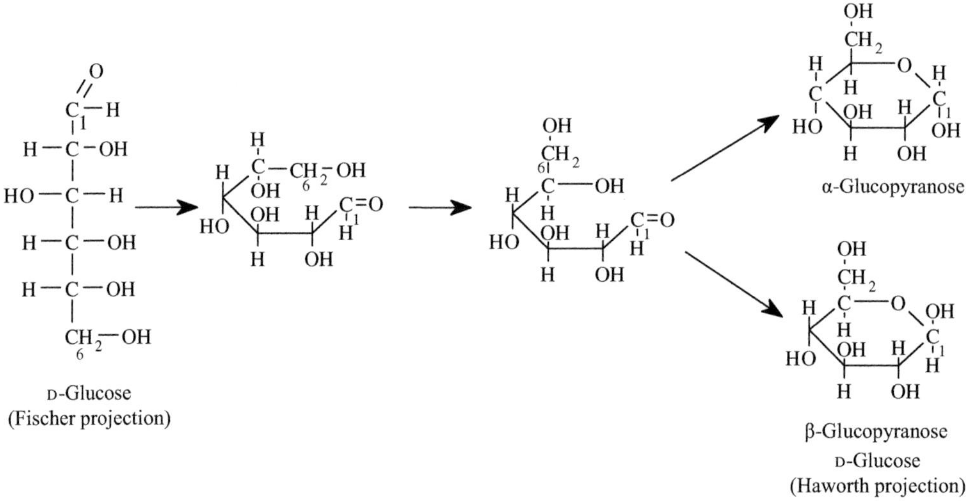

D-Glucose
(Fischer projection)

α-Glucopyranose

β-Glucopyranose
D-Glucose
(Haworth projection)

FIGURE 2.2 Haworth projection of glucose.

Disaccharides

Disaccharides are formed by joining pairs of various monosaccharides via α- or β-glycosidic bonds. A hemiacetal hydroxyl group formed from the oxygen of the carbonyl group ($-C=O$) always participates in the formation of these bonds. In certain cases, all the carbonyl groups in the molecule are used. This means that the resulting product (e.g., sucrose) lacks reducing power, because the two monosaccharide units are linked by the hemiacetal hydroxyls of the two molecules, hence both are blocked.

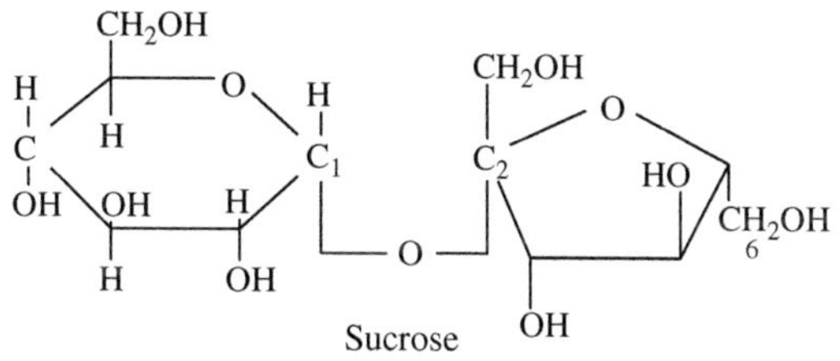

CH$_2$OH ... Sucrose

In the case of maltose, which is a disaccharide formed by two glucose molecules, only one hemiacetal hydroxyl group is blocked, meaning that the second one retains the reducing properties characteristic of aldehydes.

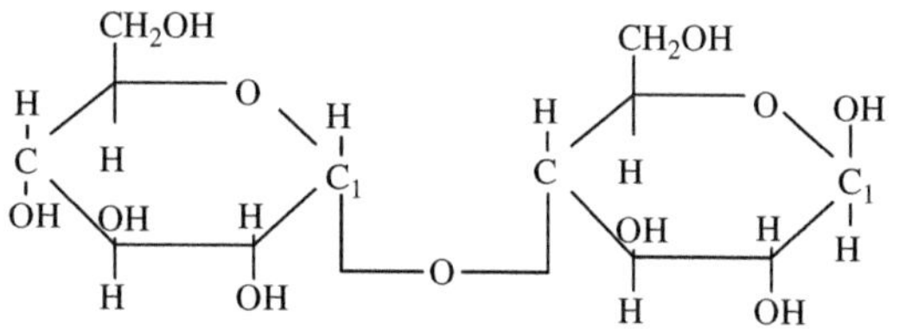

Maltose

Glycosidic bonding of these disaccharides to other monosaccharides gives rise to polysaccharides.

2.2. Organic Acids

Organic acids possess a $-COOH$ functional group. In must, however, acids also possess other groups, such as the $-OH$ group of alcohols.

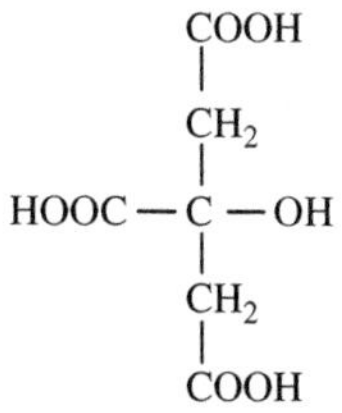

Citric acid 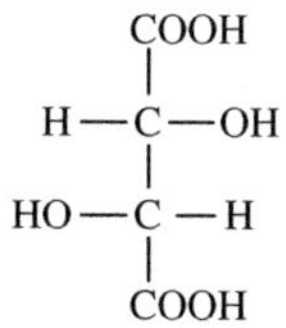   Tartaric acid 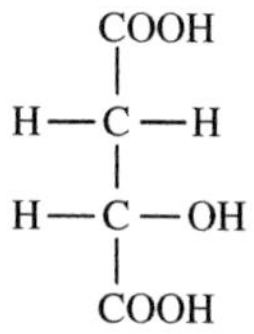   Malic acid

Other sugar-related compounds are uronic acids, which are the result of the oxidation of the $-OH$ group on carbon 6 of a sugar to form a carboxyl group, and aldonic acids, which are formed by the oxidation of the aldehyde group of an aldose. Other acids of interest are galacturonic acid, which is the main component of grape pectins, and gluconic acid, which is highly abundant in rotten grapes. Must produced from rotten grapes can also contain large quantities of acetic acid.

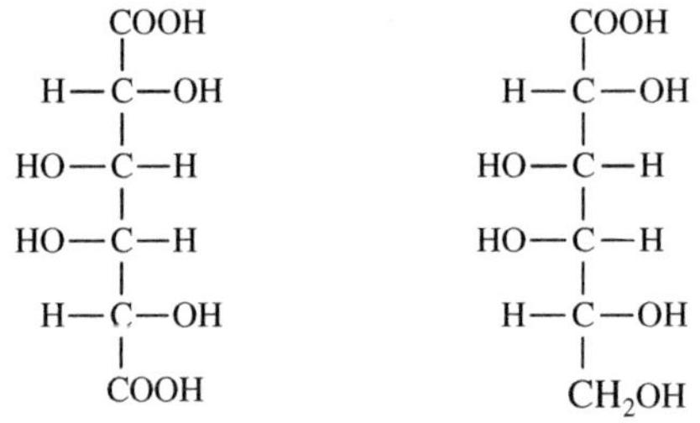

$$
\begin{array}{cc}
\text{COOH} & \text{COOH} \\
| & | \\
\text{H}-\text{C}-\text{OH} & \text{H}-\text{C}-\text{OH} \\
| & | \\
\text{HO}-\text{C}-\text{H} & \text{HO}-\text{C}-\text{H} \\
| & | \\
\text{HO}-\text{C}-\text{H} & \text{HO}-\text{C}-\text{H} \\
| & | \\
\text{H}-\text{C}-\text{OH} & \text{H}-\text{C}-\text{OH} \\
| & | \\
\text{COOH} & \text{CH}_2\text{OH}
\end{array}
$$

Galacturonic acid   Gluconic acid

2.3. Nitrogen Compounds

The main nitrogen compounds found in must are amino acids, either in free form or as polypeptides or proteins.

The characteristic functional group of amino acids is shown below:

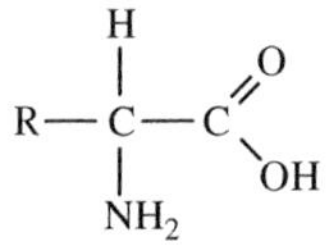

$$
\begin{array}{c}
\text{H} \\
| \quad\quad \text{O} \\
\text{R}-\text{C}-\text{C} \diagup\diagup \\
| \quad\quad \diagdown \text{OH} \\
\text{NH}_2
\end{array}
$$

All naturally occurring amino acids are α-amino acids, because the amine group (NH_2) is bound to the carbon immediately adjacent to the one bearing the acid group (which is preferentially used to name these compounds). Amino acids are joined by peptide bonds to form peptides and proteins.

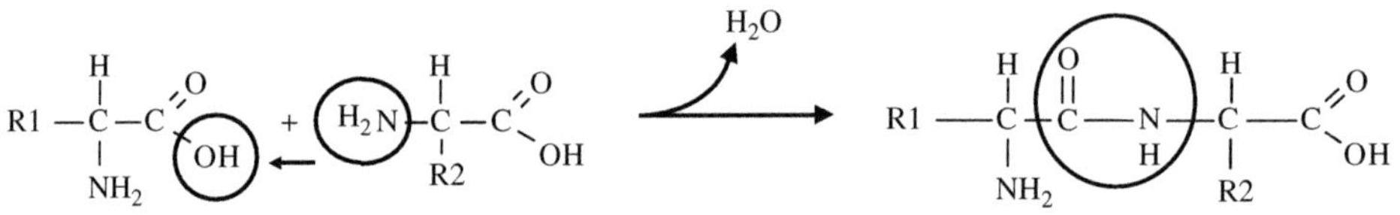

Peptide bond

Nitrogen in the form of ammonium ions is the most assimilable form of nitrogen for yeasts, and its deficit in must can cause stuck fermentation.

2.4. Minerals

The mineral content of must and wine refers to the cations and elements that these contain. Musts contain many mineral substances, which can be classified according to their electric charge and abundance.

- Cations
  Abundant: K^+, Ca^{2+}, Mg^{2+}, Na^+, and Si^{4+} (plant macronutrients)
  Less abundant: Fe^{3+}, Mn^{2+}, Zn^{2+}, Al^{3+}, Cu^{2+}, Ni^{2+}, Li^+, Mo^{4+}, Co^{2+}, and V^{3+}
  (micronutrients)
  Trace levels: Pb^{2+}, As^{3+}, Cd^{2+}, Se^{4+}, Hg^{2+}, and Pt^{2+} (ppb)
- Anions
  Abundant: PO_4^{3-}, SO_4^{2-}, Cl^-
  Less abundant: Br^-, I^-

2.5. Polyphenols

Grapes acquire their color from different compounds in the berries. The most noteworthy of these are:

- Chlorophyll
- Carotenoids
- Betalains
- Polyphenols
  Anthocyans $\Rightarrow$ red
  Yellow flavonoids $\Rightarrow$ yellow
  Tannins $\Rightarrow$ brown

Polyphenolic compounds play an essential role in both grapes and wine, as they are responsible for a range of sensory properties, such as appearance (color), taste (astringency, bitterness), and aroma (volatile phenols). They can be classified as follows:

- Simple (non-flavonoid) polyphenols
- Flavonoids
- Tannins
- Others (stilbenes)

Simple phenol   Flavonoid phenol

2.6. Vitamins

Grapes contain approximately 90 mg of vitamin C (ascorbic acid) per kilogram, as well as small quantities of B-group vitamins (of which there are 10). Vitamins are particularly important in wine making as they are very useful to yeasts and therefore essential for successful alcoholic fermentation.

TABLE 2.2   Vitamin Content of Grapes and Must

	Grapes (μg/1000 grapes)	Must (μg/L)
Thiamin	253	160–450
Riboflavin	3.6	3–60
Pantothenic acid	660	0.5–1.4
Nicotinamide	700	0.68–2.6
Pyridoxine	260	0.16–.50
Biotin	2.2	1.5–4.2
Myo-inositol	297	380–710
Aminobenzoic acid	14	15–92
Folic acid	1.3	0–1.8
Choline	24	19–45
Cyanocobalamin		0–0.2
Ascorbic acid		30,000–50,000

2.7. Aromatic Compounds

Two types of compounds confer aroma to must: grape-derived compounds (terpenes, carotenoids, and pyrazines) and compounds that arise during flavor extraction and pre-fermentation treatments (alcohols and C_6-aldehydes).

Terpenes

Terpenes are derived from isoprene units (2-methyl butadiene).

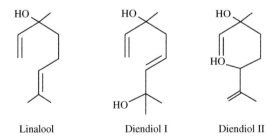

Linalool   Diendiol I   Diendiol II

Carotenoids

The main carotenoids found in must are β-carotene and lutein. When the grape berry bursts, they can break down into compounds with 9, 10, 11, or 13 carbon atoms that are more powerful odorants than their precursors. Of particular note are the C_{13}-norisoprenoid derivatives. These are divided into two groups: megastigmane forms and non-megastigmane forms.

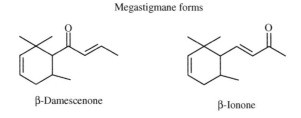

Megastigmane forms

β-Damescenone

β-Ionone

Non-megastigmane forms

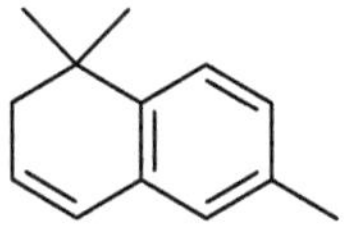

Trimethyldihydronaphthalene 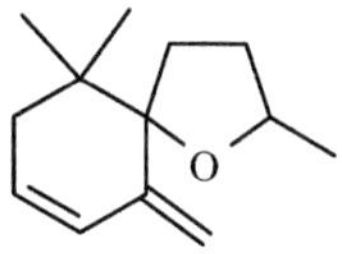   Vitispirane 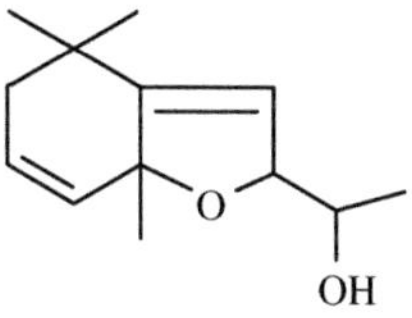   Actinidol

Pyrazines

Methoxypyrazines are nitrogen heterocyclic compounds with the following general structure:

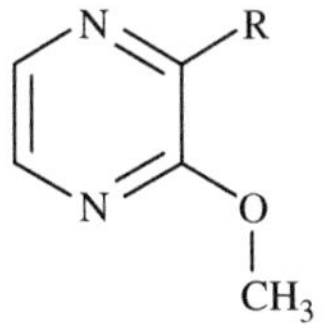

They are responsible for the vegetal aroma of certain grape varieties, such as Cabernet Sauvignon, Sauvignon Blanc, and Merlot.

Alcohols and Aldehydes

The most important aromatic alcohols and aldehydes are those with 6 carbon atoms (saturated and unsaturated). These compounds originate enzymatically during prefermentation treatments via the aerobic oxidation of linoleic and linolenic acid ($C_{18:2}$ and $C_{18:3}$). The following compounds have been identified: hexanal, (E)-2-hexenal, (Z)-3-hexenal, hexanol-1, (E)-2-hexen-1-ol, and (Z)-3-hexen-1-ol.

Hexanol-1	(Z)-3-hexen-1-ol	(E)-2-hexen-1-ol

The list below shows the average quantitative composition of must, shown by groups of compounds:

pH	3–4.5
Water	700–850 g/L
Sugars*	140–250 g/L
Organic acids	4–17 g/L
Nitrogen compounds	4–7 g/L
Polysaccharides	3–5 g/L
Minerals	0.8–2.8 g/L
Polyphenols	0.5 g/L
Vitamins	0.25–0.8 g/L
Aromatic compounds	<0.5 g/L

* These levels can be much higher in certain musts, such as those made from raisined grapes or grapes with noble rot.

Other compounds of interest in musts are the wax and oleanolic acid present in the grape bloom. This bloom retains traces of products used during the wine-growing process, such as pesticides and other mineral compounds, e.g., copper sulfate. The bloom also contains yeasts that participate in alcoholic fermentation.

Must Aromas

1. INTRODUCTION

Substances that contribute to the aroma of musts fall into two main categories: those that are already present in the grapes (terpenes, carotenoids, and pyrazines) and those that are generated during must extraction and as a result of the treatments applied prior to fermentation. The latter group encompasses C_6-alcohols and aldehydes, which are generated through the activity of lipoxygenase enzymes present in the grapes. These enzymes come into contact with linoleic and linolenic ($C_{18:2}$ and $C_{18:3}$) fatty acids during pressing or crushing and cause them to break down.

2. TERPENES

Terpenes are usually found at concentrations below 1 mg/L, and are more abundant in aromatic grape varieties such as Muscat. These compounds constitute the primary aroma of the wine and are largely responsible for the varietal character. Terpenes can be present as free or glycosidically bound forms. Free terpenes are volatile and largely responsible for the aroma of grapes and must, whereas the bound terpenes are non-volatile and constitute what is known as the "hidden aroma" of the grapes.

In general, terpenes are isolated from so-called essential oils, which are in turn mixtures of volatile compounds responsible for the fragrance of plants. These essential oils have been investigated since the 16th century due to their applications in perfumery. Initially, the term terpene was applied to mixtures of $C_{10}H_{16}$ hydrocarbon isomers found in *trementina* and other essential oils, but nowadays it refers to all oxygenated and/or unsaturated compounds of plant origin with the molecular formula $(C_5H_8)_n$. In other words, they are secondary metabolites made up of isoprene (2-methyl-1,3-butadiene) subunits bonded head-to-tail (though tail-to-tail bonding does occur). In recent years, many thousands of terpene compounds have been isolated and identified in numerous species of higher plants and also in microorganisms.

2.1. Chemical Description

Terpenes are classified according to the number of isoprene units they contain. Thus, naturally occurring terpenes are classified as monoterpenes (containing two isoprene subunits), sesquiterpenes (three subunits), diterpenes (four subunits), triterpenes (six subunits), tetraterpenes (eight subunits), and finally polyterpenes (n isoprene subunits). No naturally occurring terpenes containing five or seven isoprene subunits have been isolated.

2.2. Biosynthesis

Terpenes are synthesized from isopentenyl pyrophosphate (IPP), which is the active form of isoprene. IPP is obtained via the mevalonic acid pathway. Mevalonic acid is a key intermediate in the synthesis of terpenes and is generated by condensation of acetoacetyl coenzyme A (CoA) and acetyl CoA, followed by reduction by the reduced form of nicotinamide adenine dinucleotide phosphate (NADPH), as shown in Figure 3.1.

Mevalonic acid (I) is then converted to mevalonic acid-5-phosphate in a reaction involving conversion of adenine triphosphate (ATP) into adenine diphosphate (ADP), after which, the donation of another phosphate group by ATP generates mevalonic acid-5-pyrophosphate. Decarboxylation and dehydration of mevalonic acid-5-pyrophosphate, which involves the conversion of another molecule of ATP into ADP and inorganic phosphate, generates IPP, which exists in equilibrium with the isomer β,β-dimethylallyl pyrophosphate (DMAPP). IPP and DMAPP are the active forms of isoprene involved in the biosynthesis of terpenes (Figure 3.2).

Hemiterpenes are generated from IPP or DMAPP via condensation to produce geranyl pyrophosphate, which is used as the substrate for formation of the different monoterpenes.

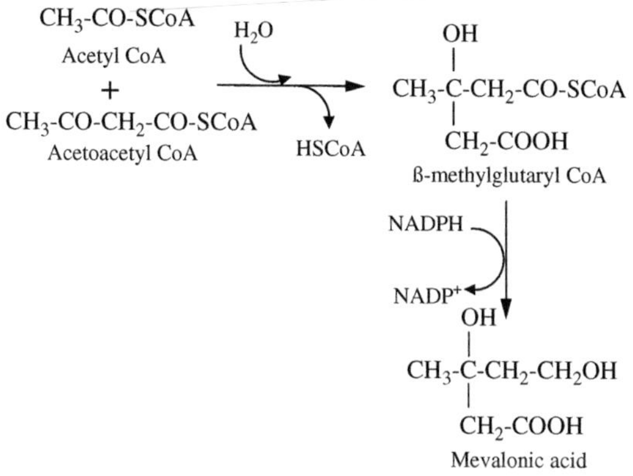

CH$_3$-CO-SCoA
Acetyl CoA
+
CH$_3$-CO-CH$_2$-CO-SCoA
Acetoacetyl CoA

H$_2$O

HSCoA

OH
|
CH$_3$-C-CH$_2$-CO-SCoA
|
CH$_2$-COOH
ß-methylglutaryl CoA

NADPH

NADP$^+$

OH
|
CH$_3$-C-CH$_2$-CH$_2$OH
|
CH$_2$-COOH
Mevalonic acid

FIGURE 3.1 Biosynthesis of mevalonic acid. CoA = Coenzyme A; NADPH = reduced nicotinamide adenine dinucleotide phosphate.

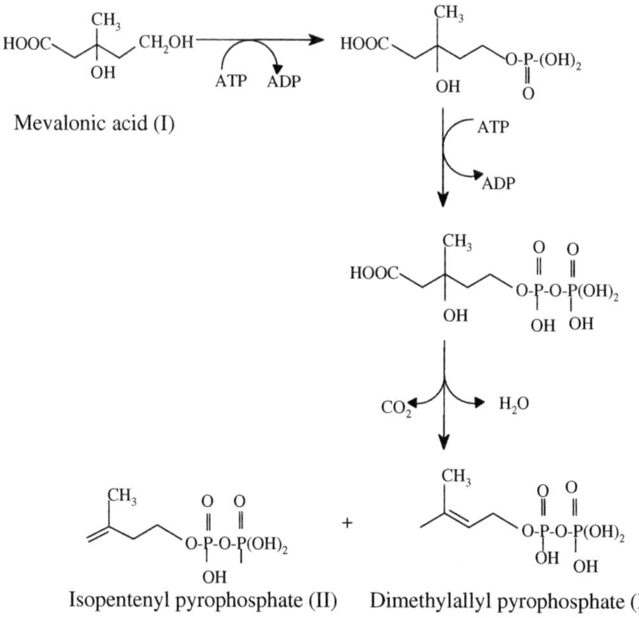

Mevalonic acid (I)

ATP   ADP

ATP

ADP

CO$_2$ H$_2$O

Isopentenyl pyrophosphate (II)   Dimethylallyl pyrophosphate (III)

FIGURE 3.2 Biosynthesis of the active forms of isoprene. ADP = adenosine diphosphate; ATP = adenosine triphosphate.

The addition of two or more molecules of IPP to geranyl pyrophosphate followed by cyclization and/or condensation gives rise to all of the terpenes or terpenoids.

There are four different structural types of monoterpene: acyclic monoterpenes, such as nerol, geraniol, and citral; monocyclic monoterpenes, such as α-terpineol and limonene; bicyclic monoterpenes; and tricyclic monoterpenes.

Generic pathways exist for the biosynthesis of terpenes in the plant kingdom. As shown in Figure 3.3, monoterpenes are synthesized by the enzymatic condensation of β,β-dimethyl-allyl pyrophosphate (III), and isopentenyl pyrophosphate (II) generates the pyrophosphate form of the monoterpene geraniol (IV). Steric hindrance prevents the ion (V) derived from this last species from being cyclized, but it can form acyclic monoterpenes. In the majority of cases, (V) is isomerized to (VI), which is a derivative of neryl pyrophosphate, and this, in turn, can be cyclized to form (VII), a derivative of α-terpineol. This can form menthane derivatives by functionalization, carane derivatives by 1,3-elimination, and bornane and pinane derivatives by internal addition. The ions (V) and (VI) can also be derived from linalool, although there is no evidence that this is an obligatory intermediate.

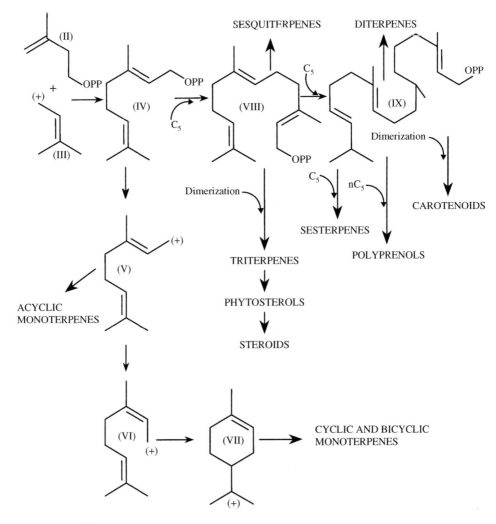

FIGURE 3.3   Biosynthesis of terpenes from the active forms of isoprene.

Alternatively, as shown in the scheme, condensation of geranyl pyrophosphate (IV) with another molecule of isopentenyl pyrophosphate leads to the generation of farnesyl pyrophosphate (VIII) and geranylgeranyl pyrophosphate (IX), from which the entire family of terpenes can be produced via cyclization and/or dimerization.

2.3. Odorant Characteristics

Around 70 terpene compounds have been identified in grapes and wine. Although most of these are monoterpenes, some sesquiterpenes have also been found, along with the corresponding alcohols and aldehydes, which are the most interesting compounds from the point of view of aroma. We can distinguish between the following groups of terpenes:

- Monoterpene hydrocarbons, which are not particularly aromatic and include limonene and *p*-cymene.
- Monoterpene alcohols, which are generally more odorant and have characteristic floral aromas reminiscent of rose, camomile, or lavender, but also honey and beeswax. These include α-terpineol, terpinen-4-ol, linalool, nerol, β-citronellol, geraniol, and hotrienol. The most aromatic are citronellol and linalool, which also act synergistically. Monoterpene polyols are less odorant but can nevertheless convert to more odorant monoterpenes such as hotrienol in acidic media.
- Sesquiterpene alcohols, such as farnesol, which contribute rose aromas and oily notes.

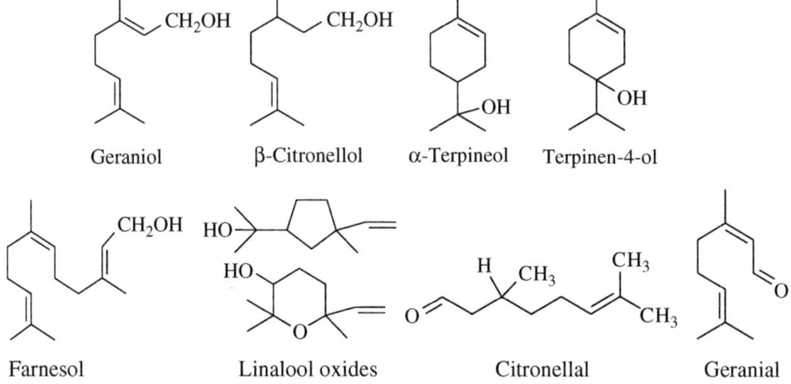

FIGURE 3.4   Terpene groups.

- Linalool oxides, which contribute floral notes, but are generally less odorant than the preceding terpenols.
- Aldehydes, such as geranial and citronellal, which are highly odorant but aromatically more aggressive than the corresponding alcohols. Their aroma is suggestive of lemon or citrus fruits.

Classification of Grape Varieties

In recent years, studies linking the concentration of monoterpenes to the aromatic properties of grapes have resulted in the following classification of grape varieties:

a. Varieties with intense aromatic properties, such as Muscat, have a total concentration of free and bound terpenes of between 4 and 6 mg/L. This category includes Muscat of Alexandria and Muscat Blanc à Petits Grains.
b. Highly aromatic, non-Muscat varieties have a total monoterpene concentration of between 1 and 4 mg/L. Examples include Riesling, Sylvaner, and Gewürztraminer.
c. Aromatic varieties have terpene concentrations below 1 mg/L. These include Cabernet Sauvignon, Chardonnay, Merlot, and Verdejo.
d. Neutral varieties have concentrations of monoterpenes well below 1 mg/L and their odorant characteristics are not dependent upon the monoterpene compounds they contain. Examples include Airen, Palomino, Garnacha, and Pedro Ximénez.

Terpenes and Key Compounds

Studies analyzing the concentration of monoterpenes and certain short-chain acids and esters in the Ruländer, Scheurebe, Morio-Muscat, Gewürztraminer, Riesling, and Müller-Thurgau varieties during the 1969−1973 harvests showed that monoterpenes were the compounds that most clearly distinguished between the different varieties. During the same period, studies of the composition of Riesling musts showed that certain compounds have a specific varietal aroma, which can vary according to the geographic origin of the grapes but is never masked. These compounds − mainly linalool, citronellol, and geraniol − are the so-called "key" compounds or substances, and their levels differ in other varieties such as Morio-Muscat, Sylvaner, and Bachus.

Currently, certain furanic and pyranic derivatives of terpene compounds are used to characterize grape varieties. Thus, high concentrations of terpene alcohols such as linalool, hotrienol, geraniol, nerol, and α-terpineol have been found in the most aromatic varieties of *Vitis vinifera*, along with the so-called terpene oxides, such as nerol oxide, rose oxide isomers, and *trans*-geranic acid.

2.4. Distribution of Free and Bound Terpenes in the Grape

Current consumer preference for wines with a high aromatic potential means that winemakers may want to produce wines with a high concentration of terpene compounds. This can be achieved through the use of specific vinification techniques. However, approaches must be based on an understanding of terpene biosynthesis and, in particular, the distribution of these compounds within the grapes.

Free Terpenes

There is an unequal distribution of certain free monoterpenes in the skin, pulp, and must of Muscat grapes. Most of the geraniol and nerol is found in the skin, whereas free linalool and diendiol I are more uniformly distributed between the skin and pulp.

The high levels of geraniol found in the skin (90% of the total content) suggest that geraniol biosynthesis and/or storage takes place in the hypodermal cells of the fruit. Linalool is evenly distributed and acts as a substrate for the generation of compounds with a higher oxidation state, such as diendiol I and II.

Approximately 46% of the total concentration of terpene compounds is located in the solid parts (skin and pulp) of the grape. Consequently, processes that facilitate exchange between the solid and liquid components favor the extraction of these compounds.

Linalool   Diendiol I   Diendiol II

Bound Terpenes

The addition of sugars to free terpenes is known as glycosylation. A glycosylated compound therefore contains two components: a non-sugar part (in this case a terpene), referred to generically as the aglycone, and a sugar part, which may be a mono- or disaccharide.

Glycosylated monoterpenes can function as intermediates in terpene biosynthesis. They mainly act as *in vivo* reagents, and this accounts for the presence of carbonium ions, which are intermediates in the biosynthesis pathway described by Ruzicka (isoprene rule) and cannot exist in their free form. Glycosylation generates compounds that are highly soluble in aqueous media. They are more soluble than free terpenes and can cross cellular barriers more easily and transport terpenes to storage areas until they are used or transformed. Their high solubility also facilitates the distribution of terpenes among the different fractions of the grape. Glycosylation can also be considered as a means of terpene storage.

Bound terpenes are responsible for what has come to be known as the hidden aroma of the grape. This is because, although they are less volatile than free terpenes, their characteristic aroma can be revealed by enzymatic or acid hydrolysis.

The most common glycosylated derivatives are those formed from monosaccharides such as glucose, and disaccharides such as rutinose (made up of rhamnose and glucose). Apioglycosides (apiose and glucose) and arabinoglycosides (arabinose and glucose) have also been described.

FIGURE 3.5   Structure of the disaccharides responsible for terpene glycosylation.

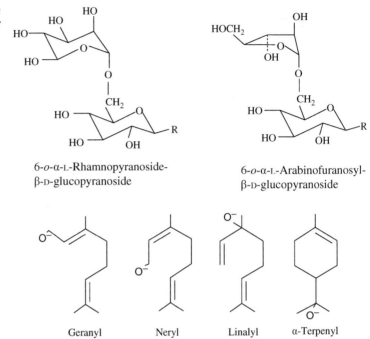

6-*o*-α-L-Rhamnopyranoside-β-D-glucopyranoside

6-*o*-α-L-Arabinofuranosyl-β-D-glucopyranoside

Geranyl   Neryl   Linalyl   α-Terpenyl

2.5. Terpene Profiles During Grape Ripening

The concentration of terpenes increases during ripening, and ripe grapes can therefore contain terpene concentrations that are 5 or 6 times higher than those found in unripe fruits. Consequently, terpenes can sometimes be used as indicators of optimal ripening.

Beginning at veraison, there is an increase in the concentration of free terpenes, in particular diendiol I, which is found at higher concentrations than other terpenes. The concentrations of geraniol, nerol, and α-terpineol do not change substantially.

The concentration of bound terpenes begins to increase at veraison and exceeds that of free forms (except for diendiol I). Veraison is therefore a stage involving a high level of glycoside biosynthesis.

Exploiting the Glycosylated Aromatic Potential of the Grape

The localization and distribution of monoterpenes in the grape (skin, pulp, and seeds) can be exploited to extract and conserve the characteristic aromas of the grape variety by increasing the length of contact with the skins, altering pressing conditions, etc. The must has a high aromatic potential that is mainly related to the high concentration of bound terpenes. Although grapes contain β-glycosidases that are capable of releasing certain terpene alcohols, under winemaking conditions the presence of these enzymes is limited for various reasons:

- The optimal pH for the enzymes is 5.
- They lack specificity for the glycosides (terpene-sugar combinations).
- The clarification of the must limits glycosidase activity.

TABLE 3.1 Distribution (%) of Free Terpene Alcohols in Different Parts of the Muscat Grape after Hot Maceration of the Crop

	Linalool	Nerol	Geraniol
Skin	26	95.6	94.2
Pulp	24	2.7	3.3
Juice	50	1.7	2.5

TABLE 3.2 Concentrations of Monoterpene Alcohols (mg/L) in Natural Sweet Wines from Two Muscat Grape Varieties

Monoterpenes	Muscat Variety	Free	Bound	Total
Linalool	Muscat à petit grains	600	200	800
	Muscat of Alexandria	600	200	800
Nerol	Muscat à petit grains	100	650	750
	Muscat of Alexandria	75	250	325
Geraniol	Muscat à petit grains	100	400	500
	Muscat of Alexandria	150	500	650
Total	Muscat à petit grains	800	1250	2050
	Muscat of Alexandria	825	950	1775

Yeasts also contain glycosidases (β-glycosidases, α-arabinosidases, α-rhamnosidases) but their activity is substantially limited because, as in the grape, the optimal pH for these enzymes is 5.

Some commercial pectolytic enzymes have residual β-glycosidase activity caused by impurities. Although the use of enzyme preparations from *Aspergillus niger* cultures has been proposed, these preparations are only effective in dry wines, since the glycosidases are inhibited by glucose. Furthermore, not all aromatic precursors are glycosidic compounds and even glycosidic compounds can contain non-terpene aglycones. On the other hand, it is not always desirable to give wines a terpene base aroma. Research is currently focused on improving enzyme preparations and employing wine yeasts with a high glycosidase activity.

The ratio of total geraniol to total nerol differs between wines. For wines made from Muscat of Alexandria grapes, for example, it is $650/325 = 2.0$, whereas for wines made from Muscat à petit grains it is $500/750 = 0.7$.

3. CAROTENOIDS

Many grape varieties that are not especially aromatic typically produce wines that develop their aromatic character during aging, and so generate high-quality wines. For instance, Riesling wines have a distinctive character that is derived from carotenoids, which are non-aromatic precursors present in the grapes.

CAROTENOIDS

The carotenoids belong to the family of C_{40}-terpenes (tetraterpenes) and their concentration in the berry can range from 15 to 2000 µg/kg depending on the grape variety, the region in which it is cultivated, and the weather conditions during ripening. Although the concentrations may appear low, they are sufficient to generate a unique character in the wines, since they are precursors of highly odorant compounds. The main carotenoids found in grapes are β-carotene and lutein, a dehydroxylated derivative of α-carotene. They have very little aqueous solubility and are more soluble in lipids. They are present in the pulp and at higher concentrations in the skin. As a result, they are not generally found in those musts obtained without maceration. These substances are light sensitive and can be degraded by oxidases from the grape (polyphenol oxidases and lipoxygenases) through a process of coupled oxidation that gives rise to C_9, C_{10}, C_{11}, and C_{13} compounds that are more soluble in the must, more volatile, and more odorant than their precursors.

3.1. C_{13}-Norisoprenoid Derivatives

Among the compounds derived from carotenoids, the C_{13}-norisoprenoids are notable for their aromatic importance and their contribution to the definitive character of certain wines. In Chardonnay, for instance, 70% of the volatile compounds which have been identified are C_{13}-norisoprenoid derivatives. Two groups can be distinguished chemically: the megastigmane and non-megastigmane forms.

Megastigmane forms

β- Damascenone   β- Ionone

Non-megastigmane forms

Trimethyldihydronaphthalene   Vitispirane   Actinidol
(TDN)

The differences between megastigmane and non-megastigmane forms are both structural and sensory, since each compound has a characteristic aroma. β-Damascenone has complex aromas with floral tones and aromas of exotic fruit and stewed apple. It is probably present in all wine varieties, but with a very variable concentration and at somewhat higher levels in red than in white wines. β-Ionone produces violet aromas. Like β-damascenone, it may be present in all varieties but its concentrations are even more variable. It makes little contribution to the aroma of white wines but can have a substantial impact on that of reds.

Among the non-megastigmane forms, trimethyldihydronaphthalene (TDN) is notable for the petrol notes it contributes to aged Riesling wines. Actinidols and vitispirane contribute camphor and eucalyptus aromas. C_{13}-norisoprenoid derivatives can be found in both free and glycosylated forms, although non-volatile precursors (glycosylated forms and carotenoids) are more abundant in the grape.

A reduction in the concentration of carotenoids together with an increase in that of their derivatives has been observed during the ripening of grapes, although it is not known whether free or glycosylated forms are involved.

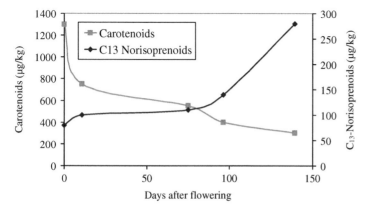

FIGURE 3.6 Temporal profile of carotenoid and norisoprenoid concentrations during grape ripening.

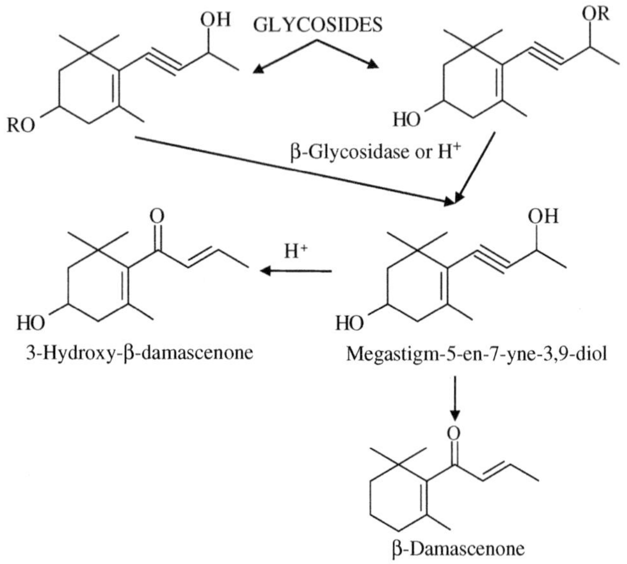

FIGURE 3.7   Formation of β-damascenone from glycosylated derivatives.

3-Hydroxy-β-damascenone

Megastigm-5-en-7-yne-3,9-diol

β-Damascenone

The odorant effect of the C_{13} derivatives is greater in more extensively aged wines. This may be explained by the formation of β-damascenone from glycosylated precursors in the acidic medium of the wine (Figure 3.7).

4. PYRAZINES

Cabernet Sauvignon, Sauvignon Blanc, and Merlot wines, among others, often have a "vegetal" or "green" aroma that has been attributed to methoxypyrazines. These molecules have been detected in musts and wines from the Chardonnay, Riesling, Pinot Noir, and Gewürztraminer grape varieties, although usually at levels below the perception threshold.

The methoxypyrazines are nitrogenous heterocyclic compounds that have the following general structure:

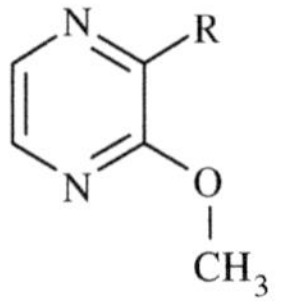

Methoxypyrazines

Essentially three methoxypyrazines have been described: 2-methoxy-3-isobutylpyrazine, 2-methoxy-3-isopropylpyrazine, and 2-methoxy-3-secbutylpyrazine.

$$-CH_2-CH\begin{smallmatrix}CH_3\\\\CH_3\end{smallmatrix} \qquad -CH\begin{smallmatrix}CH_3\\\\CH_3\end{smallmatrix} \qquad -CH-CH_2-CH_3 \;|\; CH_3$$

Isobutyl   Isopropyl   Secbutyl

The most abundant methoxypyrazine is isobutylmethoxypyrazine, which has a very low perception threshold (2 ng/L). Secbutylpyrazine has an even lower perception threshold (1 ng/L).

Given the very low concentrations of these compounds in grapes and wine (a few nanograms per liter), they are very difficult to detect and quantify. The mean concentration in Cabernet Sauvignon wines is highly variable, and analysis of wines from different regions of the world has revealed values ranging from 5.6 to 42.8 ng/L. The concentration of pyrazines in grapes generally decreases between veraison and ripeness, and higher concentrations are observed when the weather has been poor during ripening (low temperatures and little sunlight). However, these compounds are light sensitive and therefore defoliation favors a reduction in their concentration. A correlation between isobutylmethoxypyrazine and malic acid concentrations has been observed during ripening. Since this acid can be easily quantified, it can be used as an indicator of vegetal aroma and flavor in musts. Finally, under similar climatic conditions, the soil can have some influence on the final concentration of pyrazines.

Little information is available on the distribution of pyrazines in grapes; the few studies available have revealed a substantial fraction of 2-methoxy-3-isobutylpyrazine in the skins of Cabernet Sauvignon grapes (50% of the total), indicating that pressed wines or those obtained following long periods of maceration have higher concentrations of methoxypyrazines than those derived from free-run juice. The pulp and the must of this variety contain 10% and 40%, respectively, of the concentration of 2-methoxy-3-isobutylpyrazine.

5. SUBSTANCES DERIVED FROM TREATMENTS PRIOR TO FERMENTATION

Enzymes that catalyze the aerobic oxidation of linoleic and linolenic acids ($C_{18:2}$ and $C_{18:3}$) generate saturated and unsaturated C_6-alcohols and aldehydes. While this process does occur in the grape, it is enhanced by the mechanical processes that are used at different points between harvesting and alcoholic fermentation. These mechanical processes rupture the berries and cause release of the juice, and as a result, the fatty-acid substrates come into contact with the relevant enzymes and give rise to the corresponding C_6 compounds. These fatty acids form part of the cell membranes and are released by the enzyme acyl hydrolase.

The presence of these compounds in grapes and must has been widely studied, since they can potentiate the vegetal aroma and flavor that contributes undesirable sensory characteristics to the wine, despite being below the detection threshold. Hexanal, (E)-2-hexenal, (Z)-3-hexenal, hexanol-1, (E)-2-hexen-1-ol, and (Z)-3-hexen-1-ol have all been detected.

5.1. Synthesis

The synthesis of C_6 compounds involves four steps. The first is mediated by acyl hydrolase, which releases fatty acids from the plasma membrane of cells in the grape. The next step consists of oxygen fixation by lipoxygenase. Although this generates both C_9- and C_{13}-hydroperoxide isomers, lipoxygenase preferentially generates the C_{13}-hydroperoxide from linoleic acid (80%). The next step involves cleavage of the peroxides. If lipoxygenase acts on the hydroperoxide at position 9, it will give rise to C_9-aldehydes, whereas if it acts at position 13, it will give rise to C_6-aldehydes. Due to the predominance of C_{13}-hydroperoxide formation, C_6-aldehydes predominate. The final step involves reduction of the aldehydes to form C_6-alcohols in a reaction mediated by alcohol dehydrogenase (ADH).

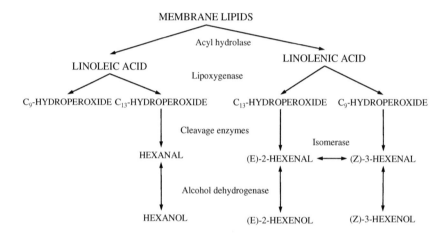

The enzyme acts at C_{13} of the hydroperoxides of linoleic and linolenic acid to generate hexanal and (Z)-3-hexenal, respectively. Hexanal is then reduced by ADH to form hexanol, whereas (Z)-3-hexenal can isomerize to form (E)-2-hexenal, and each of these gives rise to the corresponding alcohol by ADH-mediated reduction. The diagram above shows a schematic summary of the formation of C_6-aldehydes and alcohols from linoleic and linolenic acid.

The concentrations of C_6-alcohols and aldehydes in must depend on a number of factors, including the variety and ripeness of the grapes, treatments prior to fermentation, and temperature and duration of contact with the skins. Oxygen plays a fundamental role in the production of these C_6 compounds. Consequently, must from harvests that are subjected to intense mechanical processing, such as destemming, have higher concentrations than must from harvests that have not undergone such a process. Crushing also increases the levels of C_6 compounds found in musts. Nevertheless, few C_6-aldehydes have been identified as components of the must aroma, probably due to their reduction by ADH present in the grape. At the end of alcoholic fermentation, the concentration of C_6-aldehydes is almost zero, suggesting that enzymes from *Saccharomyces cerevisiae* catalyze the production of hexanol or hexenols from the corresponding aldehydes.

Hexanol contributes an aroma that is described in terms such as "green", "peanut", and "spicy", and it appears in wines at concentrations of between 0.3 and 12 mg/L. Since its perception threshold is 8 mg/L, it is generally not considered to make an important contribution to aroma. Nevertheless, it can be found at higher concentrations in some wines. These compounds can potentiate vegetal aromas, however, and their concentrations should be minimized in order to prevent a negative effect on wine quality. Consistent with this potentiation effect, statistical analyses of samples from young Riesling wines have shown a negative correlation between the concentration of hexanol and the bouquet of the wine.

To prevent the formation of these compounds, clarification of the must should be performed as quickly as possible, enzymatic activity and air exposure minimized, and low temperatures maintained. Alternatively, the enzymes can be denatured by heat. It has been shown that carbonic anaerobiosis of Muscat grapes generates notably lower concentrations of these compounds compared with control musts. It is also possible to reduce the concentration of these compounds by removing stems and leaves during harvesting. In contrast, maceration leads to an appreciable increase in their concentration in the must.

6. SUBSTANCES RELEASED DURING FERMENTATION: MERCAPTANS

Sulfur compounds containing thiol groups (mercaptans) are generally associated with olfactory defects, but they have a well-established contribution to fruity aromas such as grapefruit, kiwi, and guava. The aromatic characteristics of wines from Sauvignon grapes have been linked to the presence of certain thiols.

The first thiol to be detected in grapes was 4-mercapto-4-methylpentan-2-one, and although its concentration does not exceed 40 ng/L, it has a perception threshold of 0.8 ng/L and can therefore make a notable contribution to wine aroma. Other thiols found at relevant concentrations are 3-mercaptohexan-1-ol (3–13 μg/L) and its acetate, which have perception thresholds of 60 and 4 ng/L, respectively. These compounds, when present, can therefore have a significant effect on aroma. Other mercaptans, such as 4-mercapto-4-methylpentan-1-ol and 3-mercapto-3-methylbutan-1-ol have also been detected, but they have less influence on aroma since the former rarely exceeds its perception threshold (55 ng/L) and the latter never exceeds it (1500 ng/L).

Although mercaptans have been detected in wines produced from grape varieties such as Gewürztraminer, Merlot, Chenin Blanc, and Sauvignon, they are not detected in the musts. This is because they are present as cysteine S-conjugates and are not odorant in this form. The corresponding aromas are unmasked during fermentation possibly in a reaction mediated by a β-lyase. The yeast strain plays an important role in this process, but the exact mechanism of action of the enzyme on the odorant precursors remains unclear. The release of these precursors has also been observed in the presence of ascorbic acid. The possibility that enzymes in the mouth also cause release of sulfated derivatives during tasting cannot be ruled out.

3-Mercaptohexan-1-ol   3-Mercaptohexan-1-ol acetate   4-Mercapto-4-methyl-pentan-2-one

In terms of the influence of technological factors, skin maceration can lead to the release of sulfated derivatives, whereas musts without added sulfites can be subject to substantial loss of aroma. This loss also occurs in the presence of copper, which is mainly introduced as part of phytosanitary measures.

7. THE IMPORTANCE OF VOLATILE COMPOUNDS IN AROMA

Although all of the senses contribute to the enjoyment of wine, the greatest role is probably played by smell. The qualitative analysis of volatile compounds in wine has shown that around 1000 compounds belonging to various chemical families can be found over a wide range of concentrations, from between 14% and 16% right down to ultratrace compounds at concentrations below 0.5 ng/L.

In order to produce a sensation of smell, a substance must reach the olfactory bulb in sufficient quantities to generate a response that can be transmitted to the brain. This is difficult to achieve if the substance is not very volatile, hence only volatile compounds are responsible for wine aroma. In order to determine the sensory importance of a substance, it is essential to know, in addition to its concentration, its threshold for olfactory perception in the matrix of interest; in this case the must or the wine. The perception threshold of a compound is defined as the minimum concentration at which at least 50% of a tasting panel can detect a difference between solutions with and without the compound. The recognition threshold is the minimum concentration at which at least 50% of the panel can identify its aroma. Based on these concepts, we can define the odorant activity value (OAV) as the ratio of the concentration of a compound to its perception threshold. In principle, compounds with a concentration at or above the perception threshold (OAV $\geq$ 1) will make an active contribution to the aroma of must and wine.

TABLE 3.3 The Influence of Maceration Time in Pedro Ximénez Grapes on the Concentration of Monoterpene and C_6-Alcohols

	Time				
	0 Hours	4 Hours	16 Hours	24 Hours	48 Hours
Linalool + Nerol + α-Terpineol	15 µg/L	18 µg/L	100 µg/L	108 µg/L	84 µg/L
Hexanol + (E)-2-hexenol + (Z)-3-hexenol	184 µg/L	197 µg/L	234 µg/L	214 µg/L	293 µg/L

In the case of Muscat varieties, the aroma of the wine is determined by the presence of terpene compounds. Linalool, geraniol, and nerol are the main terpenes present; they also have a low perception threshold (0.5–1 mg/L). These three terpenes are found in numerous grape varieties, but all three need to be present to generate the characteristic Muscat aroma. As with other compounds, these terpenes have a synergistic effect, since a mixture of the terpenes has a lower perception threshold than any of them individually. In addition, their concentration is also a determining factor in the typical aroma of Muscat, which is actually not perceived if the sum of the concentrations of the three terpenes is less than 650 μg/L, or indeed if the terpene concentration exceeds 1400 μg/L. The optimal concentration is around 1000 μg/L.

Other varieties such as Cabernet Sauvignon and related strains are characterized by the presence of methoxypyrazines, among other compounds. In the case of Sauvignon Blanc, the characteristic aroma is produced by certain sulfur compounds (mercaptans).

The ripeness of the grapes and the processes used prior to fermentation (maceration, clarification, crushing, pressing, addition of enzymes, etc.) should be taken into account, particularly in weakly aromatic grape varieties, since the degree of aromatic complexity that results will depend on these factors. Enologists must therefore optimize these factors in order to obtain the desired wine.

Composition of Wine

1. THE TRANSFORMATION OF MUST INTO WINE

The transformation of must into wine through alcoholic fermentation considerably alters the chemical composition of the medium. Fermentation is a catabolic process that essentially consists of the degradation of sugars to obtain energy, with the production of ethanol and carbon dioxide as waste products.

Fermentation:

$$C_6H_{12}O_6 \rightarrow 2\ CH_3CH_2OH + 2\ CO_2 + 40\ Kcal/mol\ (2\ ATP + 23.5\ Kcal)$$

where ATP is adenosine triphosphate.

The energy yield of fermentation is lower than that of the catabolic process of respiration.

Respiration:

$$C_6H_{12}O_6 + 6\ O_2 \rightarrow 6\ CO_2 + 6\ H_2O + 686\ Kcal/mol\ (38\ ATP + 386\ Kcal)$$

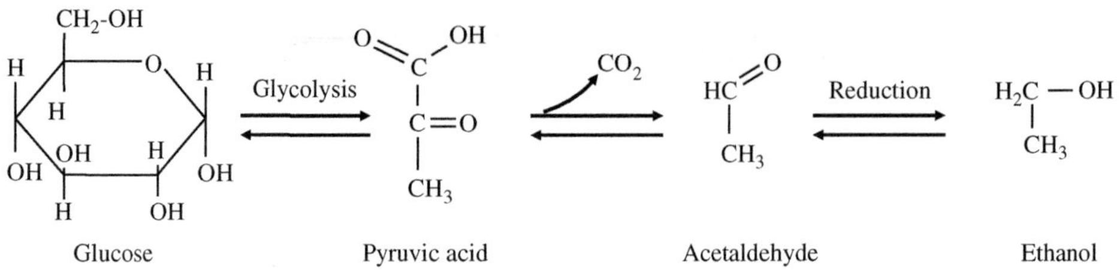

FIGURE 4.1 Catabolism of glucose in alcoholic fermentation.

Both conversion processes are performed by yeasts. In the initial phases, the yeasts grow under aerobic conditions, consuming the sugars in the must and proliferating until practically no dissolved oxygen remains. No ethanol is produced during this stage. The greatest conversion of glucose and fructose into ethanol occurs under anaerobic conditions.

Alcoholic fermentation can be divided into two phases: the production of pyruvic acid from glucose (this also occurs during respiration) and the conversion of pyruvic acid to acetaldehyde and, ultimately, ethanol.

In this next section, we will perform a mass balance and calculate the percentage of glucose that can theoretically be converted to ethanol.

According to the following equation, if all the glucose and fructose in the must were converted to ethanol, 100 g of sugar would yield 51.1 g of ethanol and 48.9 g of carbon dioxide.

$$C_6H_{12}O_6 \rightarrow 2\ CH_3CH_2OH + 2\ CO_2 \quad 180\ \text{g/mol}\ 2 \times 46 = 92\ \text{g}\ 2 \times 44 = 88\ \text{g}$$

$$\%EtOH = \frac{92}{180} \times 100 = 51.1\ \text{g EtoH/100 g sugar}$$

$$\%CO_2 = \frac{88}{180} \times 100 = 48.9\ \text{g CO}_2\text{/100 g sugar}$$

Considering the density of ethanol (0.791 g/mL), the percentage of ethanol (%EtOH) can be expressed as follows:

$$\%EtOH\ (vol/wt) = \frac{51.1\ \text{g EtoH/100 g sugar}}{0.791\ \text{g/mL}} = 64.6\ \text{mL EtOH/100 g sugar}$$

This is the equivalent of saying that for every 15.5 g of sugar in 1 L of must, we would obtain 10 mL of pure ethanol (1 unit of alcohol). In practice, however, the alcohol yield is lower (approximately 90–94% of the theoretical yield), as a small proportion of the fermentable sugars (approximately 1–2%) is used by the yeast to produce components of the cell, and another, larger, proportion is converted to by-products of alcoholic fermentation such as glycerol, acetic acid, succinic acid, and 2,3-butanediol. Glycerol is the most abundant of these by-products, and it is also the third most abundant component of wine, after water and ethanol.

1.1. Glyceropyruvic Fermentation

Glyceropyruvic fermentation produces glycerol and pyruvic acid from glucose. In this process, glycerol is formed by the reduction reaction required in order to subsequently oxidize pyruvic acid to ethanol, except that here pyruvic acid is converted to other metabolites.

$$C_6H_{12}O_6 \rightarrow CH_2OH\text{-}CHOH\text{-}CH_2OH + CH_3\text{-}CO\text{-}COOH$$

180 g/mol	92 g/mol	88 g/mol
Glucose	Glycerol	Pyruvic acid

Glycerol, which actually competes with the acetaldehyde formed during alcoholic fermentation as a hydrogen acceptor, is produced by the reduction of dihydroxyacetone-3-phosphate as pyruvic acid levels accumulate. Dihydroxyacetone-3-phosphate is in equilibrium with glyceraldehyde-3 phosphate; the two compounds are formed in the initial stages of glycolysis (they both occur in the fermentation and respiration of glucose), when glucose, which has 6 carbon atoms, splits into two molecules (glyceraldehyde and dihydroxyacetone), each with three carbon atoms.

| Glyceraldehyde-3-phosphate | Dihydoxyacetone-1-phosphate | Glycerol-3-phosphate |

To calculate the true percentage of glucose that is converted to glycerol, the percentage of sugar used for the formation of by-products can be estimated from the average content of glycerol in wine (8 g/L).

Let us assume that wine contains 12% (vol/vol) ethanol and 8 g/L of glycerol. One mole of glycerol is obtained from the fermentation of 1 mole of sugar.

Hence, let us calculate the equivalent of 8 g of glycerol in terms of moles of glycerol and glucose:

$$\text{Moles of glycerol} = \frac{8}{92} \approx 0.09 \text{ moles} \equiv 0.09 \text{ moles of glucose}$$

Using the same process, we can calculate the number of moles of glucose required to obtain 12% (vol/vol) ethanol via alcoholic fermentation:

$$C_6H_{12}O_6 \rightarrow 2\ CH_3\text{-}CH_2OH + 2\ CO_2$$

180 g/mol 2 × 46 = 92 g/mol 2 × 44 = 88 g/mol
Glucose Ethanol Carbon dioxide

ENOLOGICAL CHEMISTRY

$$\text{Moles of ethanol} = \frac{120 \text{ mL/L} \times 0.791 \text{ g/mL}}{46} \approx 2.06 \text{ moles} \equiv 1.03 \text{ moles of glucose}$$

In other words, to obtain a wine with 12% (vol/vol) ethanol and 8 g/L of glycerol, we would need:

$$0.09 \text{ moles} + 1.03 \text{ moles} = 1.12 \text{ moles of } C_6H_{12}O_6$$

According to the above equation, the molar percentage would be:

$$\text{Molar \% of glucose converted to ethanol} = 100 \times 1.03/1.12 = 91.96\%$$
$$\text{Molar \% of glucose converted to glycerol} = 100 \times 0.09/1.12 = 8.04\%$$

The next step is to calculate the corresponding weight percent:
It is first necessary to transform the moles of glucose that are converted to ethanol and glycerol into grams:

$$1.12 \text{ moles} \times 180 \text{ g/mol} = 201.6 \text{ g glucose}$$

Considering that approximately 1.5% of the sugar present in the must is used by yeasts to form their cell walls, the above quantities actually correspond to 98.5% of the total sugar. The initial content will thus be:

$$201.6 \text{ g} \times 100/98.5 = 204.7 \text{ g of total glucose}$$

Let us now calculate the weight %:

$$\text{\% sugar converted to ethanol} : 100 \times \frac{120 \text{ mL} \times 0.791}{204.7} = 46.4\%$$

$$\text{\% sugar converted to glycerol} : 100 \times \frac{8}{204.7} = 3.9\%$$

% sugar converted to cell mass : 1.5%

$$\text{\% sugar converted to CO}_2 \text{ (2.06 moles} \times 44 \text{ g/mol)} : 100 \times \frac{90.64}{204.7} = 44.5\%$$

TOTAL : 96.3%

Using a similar procedure to that used to calculate the mass balance in the chemical transformation of fermentable sugars, the number of grams of sugar needed to obtain 1 unit of alcohol can be calculated:

$$\%\text{EtOH (vol/wt)} = \frac{46.4 \text{ g EtOH/100 g sugar}}{0.791 \text{ g/mL}} = 58.66 \text{ mL EtOH/100 g sugar}$$

In other words, in practice, 10 mL of ethanol (1 unit of alcohol) is obtained for every 17 g of sugar per liter of must.

In addition to glycerol, glyceropyruvic fermentation generates pyruvic acid from the remaining "fermentable" sugar initially present in the wine, i.e. 3.7% (100% − 96.3%) fermentation. This is then transformed into other metabolites, the most abundant of which include acetic acid, succinic acid, and butanediol.

Glyceropyruvic fermentation mostly occurs in the early stages of alcoholic fermentation; at these stages, the yeasts grow in the presence of oxygen, and pyruvate decarboxylase and alcohol dehydrogenases, the enzymes responsible for the conversion of pyruvic acid to ethanol, are weakly expressed. Under these conditions, pyruvic acid levels increase. This reaction also occurs when acetaldehyde binds to other products such as SO_2, and hence becomes blocked.

The concentration of pyruvic acid in wine is low (around 80 mg/L). This acid is the building block for metabolites that are essential in wine, such as:

- 2,3-butanediol (CH_3-CHOH-CHOH-CH_3)
- Succinic acid (COOH-CH_2-CH_2-COOH)
- Acetic acid (CH_3-COOH)
- Acetoin (acetyl-methylcarbinol) (CH_3-CO-CHOH-CH_3)
- Diacetyl (CH_3-CO-CO-CH_3)
- Butyric acid (CH_3-CH_2-CH_2-COOH)
- Lactic acid (CH_3-CHOH-COOH)
- Acetoacetic acid (CH_3-CO-CH_2-COOH) and acetone (CH_3-CO-CH_3)

2. ALCOHOLIC FERMENTATION AND THE COMPOSITION OF WINE

The ethanol produced by the fermentation of sugars in the must (glucose and fructose) causes both qualitative and quantitative changes in its chemical composition. The main quantitative changes obviously affect the sugars being fermented the most, as these practically disappear. Next, ethanol, which is practically nonexistent in the must, becomes the most abundant component, after water, in wine. Glycerol is the third most abundant component.

Other compounds affected, albeit indirectly, are the acids, and particularly the acid salts initially present in the must. These are affected by the influence of ethanol on the medium and the dielectric constant of the medium. This change in dielectric constant in turn influences the solubility of the ionic and polar compounds ($E_{water} = 78.5$ q^2/Nm^2; $E_{ethanol} = 24.3$ q^2/Nm^2; both at 25°C). The salt content of wine is thus lower than that of the original must due to the formation of ethanol.

Another effect is the increase in phenolic compounds. This is particularly evident in red wines, which are fermented in the presence of grape skins. Phenolic compounds found in these skins are not very polar and are therefore extracted more efficiently when the ethanol content of the medium increases. Alcohol is also responsible for extracting the polyphenols which are present in the barrels in which wine is aged.

It would therefore be useful to study changes in the concentrations of key groups of compounds during the transformation of must into wine, as these will be closely linked to the characteristics of the product.

2.1. Sugars

The concentration of reducing sugars in must, which is initially around 150 to 250 g/L, falls to < 5 g/L (generally 1.2–3 g/L) in dry wines and to between 60 and 90 g/L in semi-dry or sweet wines. Yeasts preferentially consume glucose, explaining why wine contains more fructose than glucose. The glucose to fructose ratio is considerably less than 1 (generally < 0.50). Sucrose is not normally found in wine, as the small amounts initially present in the grapes are hydrolyzed to glucose and fructose during fermentation. The presence of sucrose indicates that the wine has been adulterated by the addition of sugar.

Wines can be classified by their sugar content. In Spain, sweet wines (*vinos dulces*), for example, have a residual sugar content of over 50 g/L while dry wines (*vinos secos*) have less than 5 g/L. Other categories are *abocado* (5–15 g/L), *semiseco* (15–30 g/L), and *semidulce* (30–50 g/L).

2.2. Acids

A range of acids derived from yeast metabolism appear during vinification. The main metabolites are succinic acid and acetic acid (the main component of volatile acidity), but small amounts of lactic acid are also produced. All of these acids are by-products of the metabolism of pyruvic acid. Grapes also contain acids (mostly tartaric acid and malic acid), but levels decrease during vinification. Tartaric acid levels decrease with increasing ethanol concentrations due to the decrease in the solubility of its salt, potassium bitartrate (also known as cream of tartar).

Storage temperature is key to ensuring the stability and clarity of wine, as a decrease in temperature reduces the solubility of tartaric acid salts and other sparingly soluble compounds, resulting in wine haze. Wines undergo a physical stabilization process. This involves subjecting the wines to temperatures of under $0°C$ according to the equation (%vol/vol of EtOH/2) $- 1 = °C$ to prevent haze.

Tartaric acid is highly resistant to attack by bacteria and other common wine microorganisms. Accordingly, changes in tartaric acid levels are generally due to physical or chemical phenomena. Only a few bacteria are capable of degrading tartaric acid and causing what is known as *tourné* or tartaric spoilage. The metabolites of tartaric acid are lactic acid, acetic acid, and succinic acid.

Malic acid is prone to attack by microorganisms, and several strains of yeast are capable of consuming up to 45% of all malic acid present in must. The degradation of malic acid by yeast produces ethanol. In malolactic fermentation, however, all the malic acid is consumed by lactic acid bacteria, which convert it to L-lactic acid. Yeasts of the genus *Schizosaccharomyces*

TABLE 4.1   Solubility of Tartaric Acid in Water and in Alcohol Solution

	Solubility (g/L) at 20°C
H$_2$O	4.91
EtOH 12%	2.76

are capable of rapidly metabolizing malic acid as they have a malic acid transporter. In *Saccharomyces* yeasts, however, uptake occurs via simple diffusion.

Citric acid is present in musts at a level of between 0.1 and 1 g/L. It is not metabolized by yeast, although it is by bacteria during malolactic fermentation, resulting in a slight increase in volatile acidity. While its addition as an antioxidant is permitted at levels not exceeding 1 g/L, this is not advisable as it can cause microbiological instability if broken down by undesirable microorganisms.

Wine Acidity: Forms of Expression

The term *total acidity* refers to the total amount of acid present in a wine. A more correct term, however, is *titratable acidity*, as the acidity of wine is usually determined by simple testing (titration) of pH to an endpoint of 7, according to the International Organization of Vine and Wine (OIV), or of 8.2 according to the American Society for Enology and Viticulture (ASEV). This term is more correct, firstly because a weak acid cannot be fully quantified at a pH of 7, and secondly, because part of the acid content of must and wine will have been neutralized by cations absorbed by the roots of the vine.

The *volatile acidity* of wines is measured by steam distillation, a process in which volatile acetic acid is separated from the other, non-volatile acids present in the wine. The volatile acidity (whose main component (95–99%) is acetic acid), can then be measured by reacting the distillate with a base. The difference between titratable acidity and volatile acidity is known as *fixed acidity*. Logically, both types of acidity should be expressed in the same units and hence they are expressed in chemical units such as mmol/L, also known as chemical equivalents (meq/L).

2.3. Polyphenols

Polyphenols have different properties, but the most interesting ones from a winemaking perspective are:

- Bactericidal and antifungal activity (note that phenol, which has a characteristic medicinal smell, was the first antiseptic to be used)
- Sensory properties
    Color
    Flavor (bitter)
    Mouthfeel (astringent)
    Aroma (volatile phenols)

Although polyphenols are present in must, their levels increase considerably during vinification. These compounds play an important role in red wines as these are fermented on grape skins, which contain the highest concentrations of polyphenols. Grape skins are particularly rich in anthocyans, which confer color.

Classification of Polyphenolic Compounds

1. Simple phenols, containing a single benzene ring
- C_6-C_1 hydroxybenzoic compounds
- C_6-C_3 cinnamic compounds

Benzoic acids   Cinnamic acids

2. Flavonoids (C_6-C_3-C_6)
These have a common structure, with two benzene rings, but they differ according to the substituents
- Flavonols
- Flavanonols
- Flavanols
- Anthocyans

3. Polymerized phenolic compounds: tannins, which can interact with proteins (during the maceration of skins)
4. Minor phenolic compounds: e.g., stilbenes

2.4. Mineral Substances

Successful alcoholic fermentation depends on, among other things, the presence of potassium, magnesium, copper, iron, calcium, cobalt, and zinc. These minerals are usually present in sufficient quantities in the must.

Potassium and calcium levels generally decrease as the ethanol content of the medium increases and potassium bitartrate and neutral calcium tartrate sink to the bottom of the fermentation tanks. Here, these salts settle and form a crystalline precipitate known as cream of tartar. Nonetheless, levels of Ca^{2+} ions can increase in wines that come into contact with materials such as cement or that are treated with $CaSO_4$ (calcium sulfate). The use of these materials is permitted in some regions.

The levels of other metal cations such as iron and copper can also increase due to contact between the must or wine and iron, stainless steel, or bronze. These cations can cloud the wine, causing what is known as metal haze. They are also involved in numerous redox reactions.

Other key mineral substances found in wine, apart from the organic anions associated with the main acids in wines, are inorganic acid anions. The most relevant of these are phosphates and sulfates. Phosphates can cause blue casse (similar to a blue cloud) due to the formation of colloidal $(FePO_4)_n$. The levels of this substance in wines can vary greatly (from 70 to 1000 mg/L), depending on whether or not ammonium sulfate is added to activate fermentation. Given the risk of blue casse, however, the addition of this ammonium salt is recommended. Sulfates are also added in certain parts of Spain, such as Jerez and Montilla-Moriles, to reduce the pH of the must, although it is important to ensure that the levels of $CaSO_4$ in the wine do not exceed the legally permitted level of 1 g/L.

$$CaSO_4 + 2\,(C_2H_4O_6)HK \rightarrow K_2SO_4 + (C_2H_4O_6)H_2 + (C_2H_4O_6)Ca$$

| Calcium sulfate | Potassium bitartrate | | Tartaric acid | Calcium tartrate |

Sulfur dioxide (SO_2), which is added to must for its bactericidal, antiseptic, and antioxidant properties, can give rise to other important anions, namely SO_3^{2-} and HSO_3^{-}. Finally, the levels of other anions, namely Cl^{-} (= 0.4 g/L), Br^{-} (=3 mg/L), I^{-} (<0.3 mg/L), and F^{-} (<5mg/L) should never exceed the levels shown in brackets. Higher levels may be harmful and in some cases are even banned (e.g., the addition of fluoride as an antiseptic agent).

2.5. Nitrogen Compounds

The amino acid content of must varies greatly from one year to the next, even within the same wine region and for the same grape variety. Nitrogenous fertilizers and climate both have a significant influence on the qualitative and quantitative composition of the amino acids in must. Amino acid usage by yeasts during fermentation largely depends on the composition of the must. Nitrogen in the form of ammonium is the most rapidly assimilable source of nitrogen for yeast, and its levels must exceed 50 mg/L for fermentation to proceed smoothly.

Yeasts consume all the nitrogen from ammonium that is present in the medium during fermentation, leading to qualitative and quantitative changes in amino acid content, as several amino acids are consumed while others are released into the liquid. Nevertheless, the term *easily assimilable nitrogen* (EAN) refers to all the ammonia and amino acids — except proline — present in the fermentation medium. The level of EAN decreases as fermentation progresses.

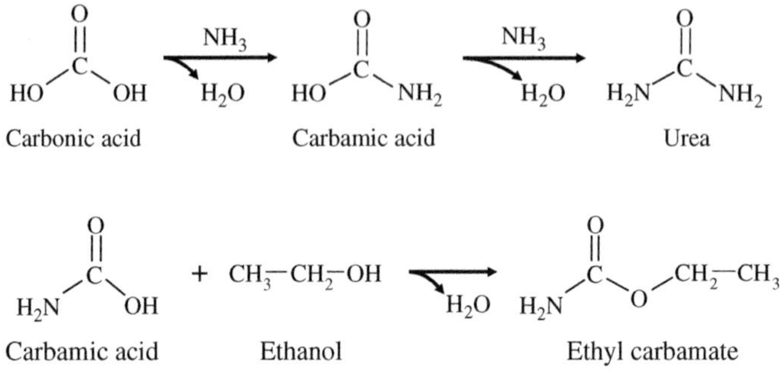

FIGURE 4.2   **Formation of urea and ethyl carbamate.**

The amino acid used most by yeast is arginine, followed by lysine, serine, threonine, leucine, aspartic acid, and glutamic acid. The least used amino acids are glycine, tyrosine, tryptophan, and alanine. Generally speaking, proline is the most abundant amino acid in both must and wine, as it is not used by yeast under anaerobic conditions. Certain yeasts, such as those that form a *velo de flor* (literally, flower veil) under which special wines such as Jerez and Montilla-Moriles wines undergo biological aging, consume proline in their aerobic development phase.

Must also contains polypeptides and proteins that are not used by yeast during fermentation, since these microorganisms are incapable of hydrolyzing them. They do, however, appear to be able to take up peptides formed by chains of two to five amino acids.

Other important nitrogen compounds formed during alcoholic fermentation are urea and ethyl carbamate. Urea is derived from the metabolism of amino acids, and ethyl carbamate is a carcinogen that is mostly derived from urea and ethanol. These compounds generally form during poorly controlled malolactic fermentation, in which undesirable strains of lactic acid bacteria are allowed to develop. The formation of ethyl carbamate has also been associated with excess levels of ammonium.

2.6. Vitamins

Vitamins can be classified into two main groups: water-soluble vitamins and fat-soluble vitamins (K, A, D, and E). Only vitamins from the first group are found in must and wine. They include vitamins B1, B2, B3, B5, B6, B8, B9, and B12, vitamin C (ascorbic acid), and vitamin P (flavonoids). *myo*-Inositol, which used to be referred to as vitamin B7, is also present. The essential fatty acids linoleic acid and linolenic acid (vitamin F) are found in must but they oxidize in the presence of lipoxygenase, giving rise to C_6-alcohols.

Ascorbic acid oxidizes rapidly when it comes in contact with oxygen in the atmosphere and therefore disappears almost completely during must extraction. Its use as an antioxidant in musts and wines is permitted but its levels are regulated. Between 75% and 90% of thiamin is used by yeasts, and riboflavin, which appears during alcoholic fermentation, is a growth factor for bacteria (but not for yeast), and therefore plays an essential role in malolactic fermentation. Thiamin and riboflavin levels increase when fermented must is left to rest on the lees, as the yeasts transfer practically all their vitamins to the liquid; this phenomenon has also been observed in biologically aged wines. Because riboflavin is photosensitive, its levels in wine can become quickly depleted. Pyridoxine is a growth factor for yeasts. The

TABLE 4.2   Vitamin Levels in Musts and Wines (μg/L)

Vitamins	Musts	White Wine	Red Wine
Thiamin (B1)	160–450	2–58	103–245
Riboflavin (B2)	3–60	8–133	0.47–1.9
Pantothenic acid (B5)	0.5–1.4	0.55–1.2	0.13–0.68
Pyridoxine (B6)	0.16–0.5	0.12–0.67	0.13–0.68
Biotin (B8)	1.5–4.2	1–3.6	0.6–4.6
Folic acid (B9)	0.0–1.8	0.4–4.5	0.4–4.5
Cobalamin (B12)	0.0–0.20	0.0–0.16	0.04–0.1
Ascorbic acid (C)	30–50 mg/L	1–5 mg/L	1–5 mg/L

remaining vitamins undergo few changes throughout fermentation and are found in similar concentrations in musts and wines.

Of the different clarifying agents used, such as bentonite, gelatin, kaolin and even potassium ferrocyanide, only bentonite appears to negatively affect B-vitamin content.

In brief, vitamins play an essential role in fermentation as they are growth factors for the yeast and bacteria responsible for alcoholic fermentation and malolactic fermentation, respectively. Although thiamin disappears almost entirely and ascorbic acid and pyridoxine levels decrease, wine is still a valuable potential source of vitamins.

2.7. Volatile Compounds

Volatile compounds, which are produced by yeasts during the fermentation of must, have an important influence on the organoleptic properties of wine. These aroma-contributing compounds belong to a wide variety of chemical families. Their key characteristics are their volatility and their possession of an osmophore that stimulates receptors in the olfactory bulb. The aromas of the different volatile compounds in a wine combine to create a new aroma that makes each wine unique. The most important chemical families that originate during the fermentative metabolism of yeast are described below.

Higher Alcohols

Higher alcohols have an −OH functional group and more than two carbon atoms. They are the most abundant of the volatile compounds found in wine and can be classified into majority alcohols (present in concentrations of >10 mg/L) and minority alcohols (present in concentrations of ≤10 mg/L).

- Majority alcohols:
  - propanol-1
  - isobutanol (2-methyl-1-propanol)
  - isoamyl alcohols (2-methyl-1-propanol and 3-methyl-1-propanol)
  - 2-phenylethanol

- Minority alcohols:
  - butanol
  - pentanol (n-amylic alcohol)
  - hexanol, etc.

Carbonyl Compounds

Carbonyl compounds have a carbonyl functional group ($-C=O$) and, depending on their position, are either aldehydes (at the end of a carbon chain) or ketones (in the middle of a carbon chain).

- Aldehydes:
  - ethanal (acetaldehyde)
  - propanal, hexanal, benzaldehyde, etc.
- Ketones:
  - 2,3-butanodione (diacetyl)
  - 3-hydroxy-2-butanone (acetoin)

Esters

Esters contain an R-CO-O-R' functional group. Chemically, it is convenient to classify them into three groups depending on their R or R' side chain.

1. Alcohol acetates: in which the acid R-COO$^-$ is derived from acetic acid. These are the most volatile types of esters.
2. Ethyl esters derived from wine acids: when R' is ethanol. These are more or less volatile depending on the volatility of the acid from which they are derived. Key ethyl esters are
   - tartrate, malate, lactate, and monoethyl and diethyl succinate
   - Short- and medium-chain fatty acid ethyl esters (C_4–C_{10})
3. Other esters, including caffeoyl tartrate, pyruvates, lactates, tartrates, and succinates of alcohols other than ethanol.

1. INTRODUCTION

One of the most important properties that polyphenols contribute to foodstuffs is color. Assessing the color of foods such as fruits is a critical element in determining their quality, since color can function as an indicator of their development and ripeness, and also of spoilage due to microbiological, physiological, chemical, or biochemical factors. The different colors of plant-based foodstuffs are determined by the presence of chlorophylls, carotenoids, and phenolic compounds. The latter contribute, among others, yellow, red, and blue tones. Among the different families of phenolic compounds, flavonoids play a particularly important role, with anthocyans, for example, contributing pink, red, orange, violet, and blue coloration. Other phenols such as the chalcones, aurones, and yellow flavonoids contribute yellow tones.

Phenolic compounds are a heterogeneous group of substances that are found widely throughout the whole of the plant kingdom. They are metabolites present in higher plants and are known for being highly specific, and for playing an important role in the response and resistance of plants to infection by pathogenic microorganisms. They are also known to participate in important plant regulatory mechanisms. They are very important components of grapes and wine, as they contribute to sensory properties such as color, astringency, bitterness, and roughness. They are also involved in oxidation reactions, protein interactions, and wine-aging processes. In addition, they have important effects on human physiology and are thought to be responsible for the so-called French paradox, since they have been shown to protect against cardiovascular disease.

The structure of phenolic compounds comprises a benzene ring with at least one hydroxyl group attached to it. They can be divided into flavonoids and non-flavonoids based on their chemical structure. Flavonoids contain a backbone of two benzene rings linked by a chain containing three carbon atoms, which joins to form a heterocycle in which oxygen is the non-carbon atom.

The flavonoids are of particular interest in winemaking as they occur in a wide variety of structural forms. Among the non-flavonoid phenols, hydroxycinnamic acids are the largest group found in grapes and wine.

The following phenolic compounds are present at the highest concentrations:

- Tannins, which are responsible for the astringency of red wines and which also influence color
- Anthocyans, which are responsible for color in grapes and red wine but are absent from white grape varieties
- Phenolic acids, which must be taken into consideration in the production of white wines
- Flavonols and dihydroflavonols
- Stilbenes, which do not influence sensory characteristics but are nevertheless important from a health perspective

2. NON-FLAVONOID PHENOLS

The non-flavonoid phenols include phenolic acids, which are classified into two main groups: benzoic acids (C_6-C_1) and the cinnamic acids (C_6-C_3). This group also includes the stilbenes (C_6-C_2-C_6).

2.1. Phenolic Acids

Phenolic acids are present in the vacuoles of the cells in the pulp and skin of grapes, but are most abundant in the skin. They are colorless in aqueous alcohol solution, but can develop a yellow color when oxidized. While they do not have a particular aroma or flavor, they are precursors of some volatile phenols formed by microorganisms.

Benzoic Acids

Principal Benzoic Acids	R1	R2	
p-Hydroxybenzoic acid	H	H	
Gallic acid	OH	OH	
Syringic acid	OCH₃	OCH₃	
Vanillic acid	H	OCH₃	

Benzoic acids differ in the groups substituted on the benzene nucleus and are present in grapes in glycosylated forms or as esters. The principal benzoic acid is gallic acid, which is found at a mean concentration of 7 mg/L in white wines and 95 mg/L in red wines. p-Hydroxybenzoic, syringic, and vanillic acids have been found at concentrations of around 5 mg/L in red wines.

Cinnamic Acids

Cinnamic acids are mainly present as tartaric acid esters, although they may also be found in glycosylated forms. Their concentrations are always higher in the skins than in the pulp of grapes. The concentration in wines is around 10 mg/L and is higher in red wines than white.

The tartaric esters of cinnamic acids are highly oxidizable by the enzyme tyrosinase, which is found naturally in grapes, and also by the enzyme laccase found in *Botrytis cinerea*. Their oxidation leads to the darkening of white grape must and may be responsible for reduction of aromatic precursors.

Ferulic and *p*-coumaric acid can be transformed into 4-vinyl guaiacol and 4-vinylphenol, respectively, by the enzyme cinnamate decarboxylase of the yeast *Saccharomyces cerevisiae*.

Principal Cinnamic Acids	R1	R2	
p-Coumaric acid	H	H	
Caffeic acid	OH	H	
Ferulic acid	OCH₃	H	

These compounds can be generically classified as C_6-C_2 polyphenols and they have aromas of cloves and pharmaceuticals. They can produce a bad aroma and therefore reduce the quality of certain white wines. Cinnamate decarboxylase is an endocellular enzyme; in other words it only acts during alcoholic fermentation, so the concentration of vinyl phenols does not increase following contact with the lees. The enzyme is also inhibited by phenolic compounds and the concentration of vinyl phenols is therefore low in red wines.

4-Vinyl Guaiacol   4-Vinylphenol

Other yeasts, such as *Brettanomyces* sp., can act during the aging of certain red wines to produce volatile phenols such as 4-ethyl guaiacol and 4-ethyl phenol, although these can also be derived from the barrels in which the wine is aged. These compounds contribute aromas of stables, leather, and wood smoke, and, at certain concentrations, can cause deterioration of the sensory qualities of wine.

2.2. Stilbenes

Stilbenes contain two benzene rings joined by a molecule of ethanol or ethylene. Among these compounds, the *trans* isomer of resveratrol is produced by vines in response to fungal infections. It is mainly localized in the skin of the grapes and is therefore mainly extracted during the production of red wines; it is found at concentrations of between a few tenths of a milligram and a few milligrams per liter. Stilbenes do not make a notable contribution to the color or other sensory properties of wine, but they do appear to protect against cardiovascular disease.

Trihydroxy-3-5-4'-stilbene

3. FLAVONOID PHENOLS

Flavonoid phenols contain a backbone of 15 carbon atoms comprising two benzene rings (A and B) joined by a heterocycle (C). These compounds have the general structure C_6-C_3-C_6, and can be divided into three groups according to the structure of the heterocycle: flavonols, dihydroflavonols, and flavones; anthocyanidins; and flavanols.

The flavanols are further categorized into catechins and condensed tannins, which are themselves classified into procyanidins and delphinidins.

Structure of Flavonoids	Flavonoid Group	Heterocycle	Structure
	Flavonols and dihydroflavonols	Pyrone	
	Flavans and flavanols	Pyran	
	Anthocyanidins	Pyrrole	

3.1. Flavonols

The flavonols contain a pyrone heterocycle. They are yellow pigments and therefore make little contribution to the color of red wines. They are present in the skins of red and white grapes in glycosylated forms (at position 3). These forms are the most abundant in grapes, with eight monoglycosides and three diglycosides characterized to date. A large number of glucuronides can also be found. Other sugars that are found are galactose, xylose, and arabinose. The concentrations of flavonols in the grape vary from 10 to 100 mg/kg and the most abundant are quercetol derivatives. Myricetol derivatives and isorhamnetol glycoside appear to be specific to red grape varieties.

The heterosides are hydrolyzed during the vinification process to leave the free aglycone. In white wines, which are fermented in the absence of the solid elements of the grape cluster, the concentration is between 1 and 3 mg/L, whereas in red wines it can reach 100 mg/L.

Principal Flavonols	R1	R2
Kaempferol	H	H
Quercetin	OH	H
Myricetin	OH	OH
Isorhamnetin	OCH$_3$	H

3.2. Dihydroflavonols or Flavononols

Dihydroflavonols or flavononols differ structurally from flavonols in that they lack the double bond of the heterocycle. They are yellow pigments that have been identified in the skins of white grapes and are present as 3-glycosides, specifically 3-rhamnosides. Astilbin is the most abundant and can reach concentrations of 9 mg/kg in fresh material; engeletin is found at levels of around 1 mg/kg.

Principal Dihydroflavonols	R1	R2
Dihydrokaempferol (engeletin)	H	H
Dihydroquercetin (astilbin)	OH	H

3.3. Flavanols

Flavanols contain a pyran heterocycle. They form a large group made up of the different isomeric forms of catechin and its polymers.

The structure of catechin (R1 = H) includes two asymmetric carbons (C2 and C3), thereby making it possible to obtain four different isomers: (+/−) catechin and (+/−) epicatechin. In addition, if R1 is an OH group, we obtain four new isomers: (+/−) gallocatechin and (+/−) epigallocatechin. The hydroxyl group at position 3 may be esterified with gallic acid (galloylation). This is an important consideration, since higher levels of gallic acid in polymerized flavanols (tannins) will contribute greater astringency and bitterness.

Flavanol Monomers

If R1 is a hydrogen group, the compound is catechin/epicatechin. If the R1 group is a hydroxyl group, it is gallocatechin/epigallocatechin. These monomers are found in the skin and at higher concentrations in the seeds of grapes.

The flavanols tend to form oligomers and polymers that give rise to tannins. These are referred to as proanthocyanidins or procyanidins, since their hydrolysis in acidic media produces highly unstable carbocations that are transformed into grayish-brown condensation products and, in particular, cyanidin, which is red. (The term procyanidins replaces the older term leucoanthocyans.) If the polymer is formed of gallocatechin and/or epigallocatechin subunits, acid hydrolysis will generate delphinidin; in this case, the polymers are more correctly called prodelphinidins. However, the term procyanidins is generally used to refer to all condensed tannins. Given the tendency of these compounds to combine with each other, they are not found in glycosylated forms.

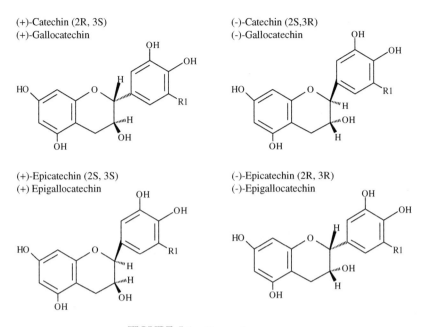

FIGURE 5.1 Flavanol monomers.

3.4. Tannins

The term *tannin* encompasses substances that are able to produce stable complexes with proteins and other plant polymers such as polysaccharides. Tannins cause astringency and bitterness by binding to proteins in saliva. They also cause flocculation during fining by binding to protein fining materials, and cause enzyme inhibition by binding to enzyme proteins.

Tannins are generated by the polymerization of flavanols, and must be sufficiently large in order to form stable complexes with proteins. The molecular mass required in order for tannins to complex with proteins is between 600 and 3500 kDa (between 2 and 10–12 flavanol subunits), hence flavanol monomers cannot be considered as tannins.

Grape-derived tannins are known as catechic or condensed tannins; during aging in barrels, however, the wood releases other tannins into the wine. These are known as hydrolyzable tannins, or ellagitannins, as their hydrolysis gives rise to ellagic acid. The addition of this type of tannins to wine is authorized (commercial tannin).

The procyanidin dimers can be classified into two categories according to the type of bonds between the monomers:

- Type-B procyanidins, which are dimers of two flavanol subunits linked by a C4-C8 or C4-C6 interflavan bond
- Type-A procyanidins, which, in addition to the interflavan bond, have an ether bond between carbons C2-C5 or C2-C7

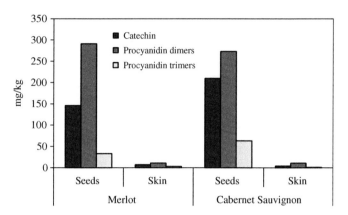

FIGURE 5.2 Catechin and procyanidin dimer and trimer concentrations in two grape varieties at the point of ripeness.

Trimeric procyanidins are classified into two categories: type-C procyanidins, linked by interflavan bonds, and type-D procyanidins, linked by an interflavan bond and another flavano-ether bond (as in type-A procyanidins).

Procyanidin oligomers contain between three and ten flavanol subunits, and the chiral centers present in the monomers allow for an almost infinite number of isomers. Condensed procyanidins are formed from more than ten flavanol subunits and have a molecular mass of more than 3000 kDa.

The procyanidins are responsible for bitter taste and astringency; they also contribute to yellow coloration and the aging capacity of the wine. Both bitter taste and astringency depend on the number of hydroxyl groups that can react with proteins in the saliva, and these factors therefore increase with the number of galloylated subunits. The bitterest procyanidins are formed from four flavanol subunits, whereas astringency is greatest when around ten subunits are present.

The total tannin content, degree of polymerization, and the proportion of constituent units vary from one grape variety to another and depend on the growing conditions and the degree of ripeness. In addition, in the grape varieties that have been analyzed, the tannins present in the skin can be distinguished from those in the seeds by the presence of gallocatechin and

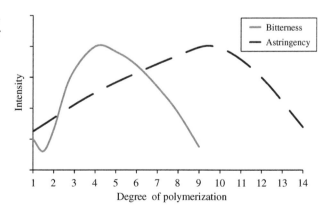

FIGURE 5.3 Relationship between the degree of polymerization and bitterness or astringency.

TABLE 5.1 Composition of Tannins and Degree of Polymerization in Different Grape Varieties

Tannins (mg catechin/kg)	Carignan		Cabernet Franc		Cabernet Sauvignon	
	Skin	Seeds	Skin	Seeds	Skin	Seeds
Galloylated procyanidols	40	415	30	706	60	1162
Procyanidols	300	996	120	2905	480	1660
Prodelphinidols	60		210		300	
Degree of polymerization	12	9	33	8.5	42	16

epigallocatechin subunits (prodelphinidols), a higher degree of polymerization, and a smaller proportion of galloylated subunits. The last two factors explain the greater astringency of the tannins found in seeds. It has also been observed that the concentration of flavanol monomers, dimers, and trimers in seeds is always greater than that in the skins.

3.5. Anthocyans, Anthocyanidins, and Anthocyanins

The term anthocyan is generic and covers both anthocyanidins and anthocyanins. The anthocyanidins are flavonoids that contain a pyrrole heterocycle and are responsible for the bluish-red color of the skin of red grapes and, therefore, for the color of red wine.

Anthocyanidins	R1	R2
Cyanidin	OH	H
Peonidin	OCH$_3$	H
Delphinidin	OH	OH
Petunidin	OCH$_3$	OH
Malvidin	OCH$_3$	OCH$_3$

Anthocyans are made up of an anthocyanidin bound to a sugar via a glycosidic bond. They are therefore glycosides in which only the aglycone, or non-sugar component, is a chromophore. These compounds are fairly soluble in water, and this favors their distribution in the plant and their transfer into the must and wine during vinification.

The anthocyans are mainly present in the skin of red grape varieties, although in grapes belonging to the Garnacha Tintorera variety they are also present in the pulp. They are much more stable in their heteroside form (anthocyanins) than as aglycones (anthocyanidins). In *Vitis vinifera* and in the corresponding wines, only monoglycosidic anthocyanins with a glycosidic bond at position 3 have been identified. The glucose may be esterified (acylated) at position 6 with caffeic, *p*-coumaric, or acetic acid.

Malvidin is the most abundant anthocyanidin in all grape varieties and can therefore be considered the main component responsible for the color of red grapes and red wine. The concentrations of other anthocyanidins and the different compounds they can form (with glucose and acylated forms) varies substantially with grape variety. This variability accounts for the different color tones of red grape varieties; although they all contain the same basic anthocyanidins, there are qualitative and quantitative differences in their composition.

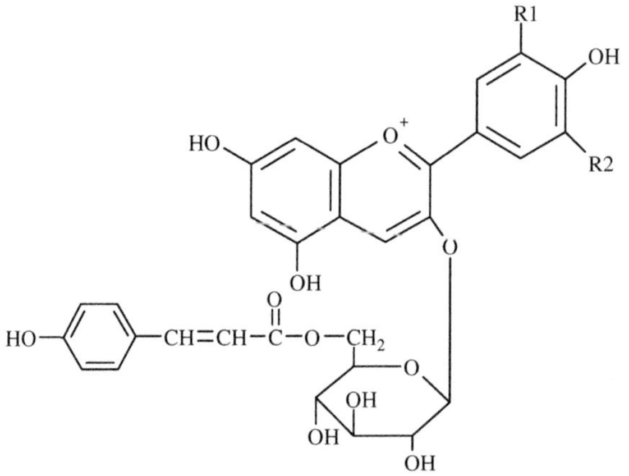

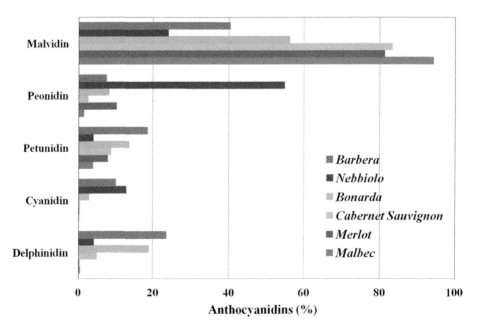

FIGURE 5.4   Monoglycosidic anthocyanin esterified with *p*-coumaric acid.

FIGURE 5.5   Relative Anthocyanidin Composition of Six Grape Varieties.

Other American species (*Vitis riparia* or *Vitis rupestris*) contain diglycosides, and as a result it has been possible to discover fraudulent marketing of Spanish appellation wines that are in fact derived from hybrid varieties. The presence of diglycosides is a genetically dominant character, and consequently, crosses between *V. vinifera* and an American vine species gives rise to plants with this characteristic. Although a second cross between a hybrid producer and a *V. vinifera* variety does not inevitably display the character, most do in fact contain diglycosides, which can be detected even at very low concentrations. This can be used to prevent the fraudulent marketing of wines as being derived from *V. vinifera* when they are in fact produced using other varieties. At one point, this fraudulent activity placed French *appellation d'origine contrôlée* wines at risk, and although nowadays it is of lesser importance, it illustrates the role that can be played by enological research.

4. PROFILE OF TANNINS AND ANTHOCYANINS DURING RIPENING

Phenolic compounds are by-products of sugar catabolism. They are formed from the beginning of development in all of the organs of the vine. From veraison to the ripened grape, the skin becomes progressively enriched in phenolic compounds.

The anthocyanins appear during veraison and are responsible for the characteristic color change observed during this physiological phenomenon; they then accumulate during ripening and are partially degraded at the end of the process. The tannins in the skin exhibit a similar temporal profile although they are already found at notable concentrations during veraison. The concentration of tannins in the seeds reduces from veraison onwards and depends on the ripeness of the grape and the variety. On occasions, the reduction occurs earlier, prior to veraison, and as a result the concentration of tannins remains relatively constant during ripening.

This temporal profile is applicable to most grape varieties and winegrowing regions; nevertheless, the level of accumulation of anthocyanins and the timing of the peak vary according to the environment (winegrowing region), grape variety, and weather conditions.

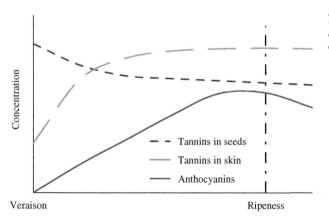

FIGURE 5.6 Temporal profile of tannin and anthocyanin concentrations over the course of grape ripening.

FIGURE 5.7 Relationship between anthocyanin accumulation and ripeness, as defined by the ratio of sugar to acidity.

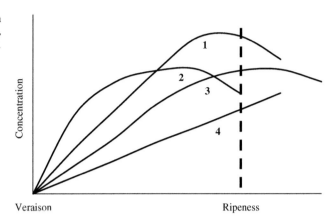

The peak can coincide with ripeness (determined by the ratio of sugar to acidity), indicating optimal ripeness (Figure 5.7, 1). When the peak occurs before ripeness (Figure 5.7, 2), it suggests that the grape variety is not well adapted to the environment in which it is being cultivated, since its phenolic ripeness is premature. When the peak occurs after ripeness (Figure 5.7, 3), the grapes may require a degree of over-ripening or may even not reach phenolic ripeness (Figure 5.7, 4), in which case they would not be suitable for the production of high-quality wines.

The reactivity of the tannins, and therefore their astringency, is usually measured according to the gelatin index, which is based on the capacity of the tannins to react with proteins and form stable complexes. The value for this index is normally between 25 and 80. Values above 60 reveal the presence of highly reactive tannins, which are responsible for hardness and astringency. Values below 40 indicate a lack of balance. Optimum values are therefore between 40 and 60. Table 5.2 shows the temporal profile of this index in the skin and seeds of three grape varieties. All exhibit the same profile of tannins in the skin, becoming progressively less reactive and losing astringency and hardness during ripening. The tannins in the seeds, in contrast, are already highly reactive (>80) at veraison and increase their reactivity slightly during ripening.

4.1. Phenolic Ripeness

The tannin and anthocyan content of grapes can be used to classify varieties and their adaptation to the region in which they are cultivated according to phenolic richness. Theoretically, grapes that are most rich in anthocyans will generate wines with the strongest colors, but phenolic extractability must also be taken into consideration. The diffusion of compounds from the skin into the must during vinification depends, among other factors, on the degradation of cells in the skin; thus, the more extensive the degradation, the easier it is to extract anthocyans. When grapes become slightly over-ripe, the richness of anthocyans in the wine is slightly greater than that obtained with less-ripe grapes, even though the concentration of the pigments in the grapes is lower. Therefore, the anthocyan content will depend on both the concentration in the skin and on the ease of diffusion into the must.

TABLE 5.2 Gelatin Index in the Skin and Seeds of Three Grape Varieties

		Skin	Seeds
Merlot	Veraison	69	80
	Unripe	67	80
	Ripe	54	84
Cabernet Sauvignon	Veraison	69	80
	Unripe	67	80
	Ripe	42	86
Cabernet Franc	Veraison	70	81
	Unripe	69	82
	Ripe	58	86

TABLE 5.3 Concentration of Anthocyanins (mg/L) in Grape Skins and Wine According to the Date of Harvest in Cabernet Sauvignon Grapes
Percentage extraction and color intensity are also shown

	Skin	Wine	% Extraction	Color Intensity
September 13	1500	930	61	0.686
September 20	1743	1046	59	0.812
September 28	1610	1207	75	0.915

TABLE 5.4 Mean Concentration (mg/kg) of Phenolic Compounds in Ripe Grapes

	Pulp	Skin	Seeds
Tannins	Trace	100–500	1000–6000
Anthocyans	0	500–3000	0
Phenolic acids	20–170	50–200	0

5. EXTRACTION OF PHENOLIC COMPOUNDS DURING VINIFICATION

Analysis of the phenolic composition of grapes and wine reveals that not all of these compounds found in wine are derived directly from the berry. Some are specific to wine and are derived not only from the different phases of maceration and fermentation but also from changes to the polyphenols initially present in the grapes.

The diffusion of phenolic compounds from the skin begins with crushing and continues during pressing, which is used to extract the must in the case of white wine vinification, or the wine in the case of red wine vinification. Diffusion depends on a number of factors, such as sulfur content, temperature, and ethanol concentration.

The total concentration of phenolic compounds is between 50 and 350 mg/L in the case of white wines and between 800 and 4000 mg/L in the case of red wines (expressed as equivalent weight of gallic acid). In red wines, around 60% of the total phenolic compounds are extracted, including 38% anthocyans and 20% tannins.

5.1. Benzoic Acids and Flavanols

The concentrations of caftaric acid (caffeic-tartaric acid) and coutaric acid (*p*-coumaric-tartaric acid) in white grape musts correspond to 40% and 20%, respectively, of the concentrations present in the grape. Skin maceration increases these concentrations by 20%, while those of other compounds such as monomeric and dimeric flavanols are increased three-fold. Other processes such as carbonic maceration double the concentrations of cinnamic acids and increase those of dimeric flavanols ten-fold.

5.2. Anthocyans and Tannins

Anthocyans are rapidly extracted in the grape must (in the aqueous phase) and achieve peak concentrations in the first few days. When the ethanol concentration reaches a certain level, the concentration of anthocyans is reduced due to adsorption by yeast and the solid elements of the berry, or as a result of modifications of their structure (formation of tannin-anthocyanin complexes).

The tannins in the skins are extracted along with the anthocyanins, although more slowly. The process thus takes longer. The tannins in the seeds are solubilized when the cuticle has been dissolved by the ethanol produced during fermentation; consequently, the tannin levels of red wine increase with increasing length of maceration. These kinetic differences are linked to both the solubility of the different compounds and to the accessibility or

TABLE 5.5   Mean Concentrations of Phenolic Compounds in White and Red Wines

	White Wine	Red Wine
Benzoic acids	1–5 mg/L	50–100 mg/L
Cinnamic acids	50–200 mg/L	50–200 mg/L
Flavonols	Trace	15 mg/L
Anthocyans	0 mg/L	20–500 mg/L
Monomeric flavanols	Trace	150–200 mg/L
Procyanidols	0–100 mg/L	1500–5000 mg/L

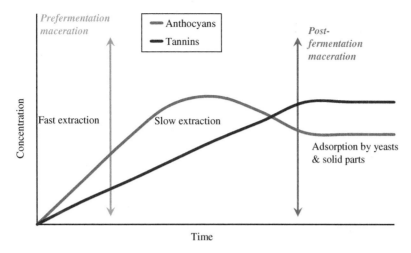

FIGURE 5.8   Relationship between fermentation time and concentration of tannins and anthocyans.

extent of contact between the skin and seeds of the grapes and the must or wine in which they are immersed.

6. VINIFICATION STRATEGIES AND POLYPHENOL CONTENT

Taking into account the differences in extraction kinetics of phenolic compounds, various fermentation strategies can be employed according to the type of wine desired and the health and phenolic ripeness of the grapes.

Type of Wine Desired

The processing of young wines requires a short maceration in order to control the amounts of tannins extracted, since excessive tannin levels can produce a high degree of astringency and vegetal flavors. Careful control of temperature is required to ensure that the wine retains its fruity aromas. The elaboration of wines that are destined for aging requires a longer maceration, designed to extract sufficient tannins to improve color stabilization and obtain an appropriate structure.

Vine Health

If the grapes are healthy, in principle there is no reason that the desired wine cannot be produced; the problem arises when rotted grapes are harvested. In this case, long macerations should be avoided and, if possible, techniques should be employed that allow rapid extraction of color. Aeration should be avoided as far as possible, since the laccase enzyme that is characteristic of rotted grapes rapidly oxidizes a whole host of phenolic compounds; high doses of sulfites should also be added during fermentation, and devatting should be performed as soon as possible. These measures will prevent wine spoilage. Another option is the manual selection of healthy grapes and removal of rotted grapes; this method is expensive but it is the only way to obtain high-quality wines.

Phenolic Ripeness

Grapes that have achieved adequate or optimal phenolic ripeness can be used to produce any type of wine. If the grapes are not sufficiently ripe, however, a short maceration will result in inadequate color extraction, and long maceration will run the risk of extracting excessive levels of tannins from the seeds, with a consequent increase in astringency. In such a case, it is preferable to use techniques that allow rapid extraction of color while avoiding excessive astringency.

7. MODIFICATION OF PHENOLIC COMPOUNDS DURING VINIFICATION

As discussed, phenolic compounds can be modified during the vinification process as a result of endogenous enzyme activity in the grapes, as well as exogenous enzyme activity and chemical phenomena, the latter mainly involving nucleophilic addition reactions.

7.1. Enzymatic Activity

We can distinguish between two main types of enzymatic activity. The first leads to oxidation of polyphenols to form quinones (polyphenol oxidase activity) and the second is caused by hydrolases that cleave ester and glycosidic bonds.

Polyphenol Oxidases

While the must is being obtained, contact with the air leads to spontaneous formation of dull, brownish pigments. These are due to the oxidation of phenolic compounds to quinones. The enzyme responsible is a polyphenol oxidase. Cinnamic acids and their tartaric esters are particularly prone to oxidation, which causes darkening of white grape musts. Two different enzymes with polyphenol oxidase activity have been identified, one endogenous (tyrosinase) and the other exogenous (laccase). Polymerization of quinones leads to the formation of brown pigments that are responsible for darkening of the must.

Hydrolases

A number of enzymes with hydrolytic activity have been found in wine. They are fungal in origin and are derived from the microflora of the grape or from enzyme preparations. Essentially, these enzymes release the aglycones from flavonols and anthocyans (β-glucosidase activity), and this can lead to the precipitation of insoluble flavonol aglycones or a reduction in color caused by instability of anthocyanidols. Free caffeic and p-coumaric acids (not esterified with tartaric acid) can also be obtained as a result of esterase activity. Tannase (tannin acyl hydrolase) activity leads to the cleavage of ester bonds with release of gallic acid from monomeric or polymeric

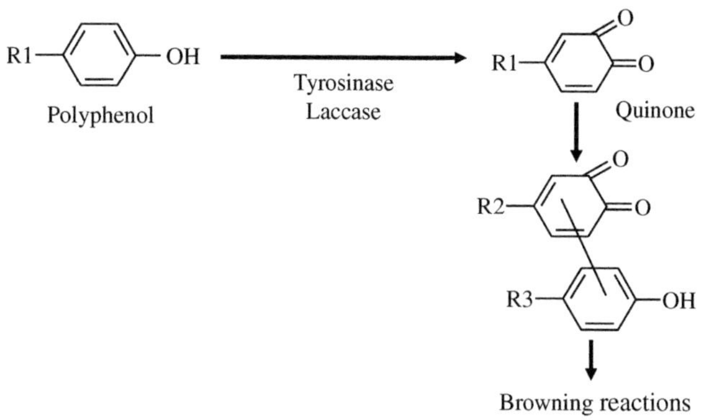

FIGURE 5.9   **Enzymatic oxidation of phenols.**

galloylated flavanols. This enzyme is derived principally from fungi such as *B. cinerea* and *Aspergillus niger,* and the addition of fungal tannase can reduce the excessive astringency associated with young wines.

Hydrolysis can occur spontaneously in acid media, although at a slower rate.

Cinnamate Decarboxylase and Vinylphenol Reductase

The activity of cinnamate decarboxylases generates serious olfactory defects, since they catalyze the transformation of phenolic acids, which do not influence aroma or flavor, into volatile phenols, which can have a negative effect on wine aroma. This enzymatic activity is present in most yeasts and transforms ferulic and *p*-coumaric acids into 4-vinyl guaiacol and 4-vinylphenol. Although red wines contain higher concentrations of the precursors of these volatile phenols, the concentration of the

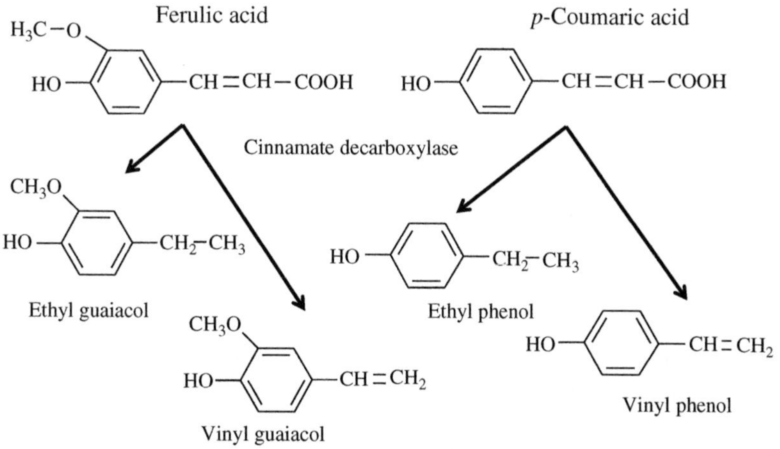

FIGURE 5.10   **Formation of ethyl and vinyl phenols.**

volatile phenols themselves is lower because the activity of cinnamate decarboxylase is inhibited by polyphenols.

Nevertheless, some red wines contain high concentrations of ethyl phenols generated by the activity of a vinylphenol reductase. The origin of these compounds varies, since they can appear during malolactic fermentation due to bacterial metabolism, during aging via effects of *Brettanomyces* sp., or even through extraction from the charred wood on the inside of the barrels.

Metabolic Activity of Yeast

Another of the compounds that is always present in white and red wines is tyrosol (*p*-hydroxyphenyl ethanol), which is found at concentrations of 20 to 30 mg/L. It is formed from tyrosine (*p*-hydroxyphenylalanine) during alcoholic fermentation, and its concentration remains relatively constant during aging.

$$HO-\langle\!\!\!\!\bigcirc\!\!\!\!\rangle-CH_2-CH_2OH$$

Tyrosol

7.2. Chemical Phenomena

Notable among the chemical phenomena is the reaction undergone by the quinones generated during enzymatic oxidation (discussed in Chapter 17) and copigmentation.

Equilibrium of Anthocyanins in Acid Media and Copigmentation

Anthocyans can appear in different forms, depending on the pH of the medium, and this influences their color. At low pH, the predominant form is the flavylium cation (A^+), which is red, but as the pH increases, two reactions can occur:

– A proton-transfer reaction leads to the formation of a quinone base (AO), which has a violet hue:

$$A^+ \longleftrightarrow AO + H^+$$

– Addition of water to the double bond of the heterocycle gives rise to a colorless carbinol base (AOH):

$$A^+ + H_2O \longleftrightarrow AOH + H^+$$

This process can be explained by the formation of conjugate bases, since the chemical species are present in acid-base equilibria governed by the corresponding equilibrium constants.

The carbinol base can be transformed into a yellow chalcone (C), via a keto-enol tautomeric equilibrium. The chalcone exists as *cis* and *trans* isomeric forms in equilibrium.

Both forms have a slight yellow coloration and their formation is favored by high temperatures.

$$AOH \leftrightarrow cis\text{-Chalcone} \leftrightarrow trans\text{-Chalcone}$$

The *trans* isomer can be oxidized to generate phenolic acids (colorless).

$$trans\text{-Chalcone} \rightarrow \text{Phenolic acids}$$

With the exception of the oxidation, all of these reactions are reversible.

The proportions of the different forms that are present depend on the pH, and a curve of their theoretical proportions can be plotted from their equilibrium constants (Figure 5.12). The first observation is that at the pH commonly found in wines there is equilibrium between the red, blue, and colorless forms, and that only 20% to 30% of anthocyans contribute to the color of the wine. At pHs close to 4, the wine should be more blue than red, but this is not the case. Furthermore, wines that have been aged for some time are less sensitive to pH changes. This is because wine is not simply a solution of anthocyans; in practice, more complex equilibria exist. One of these factors is copigmentation, which alters the tonality and intensity of the color in wine compared with pure anthocyan solutions.

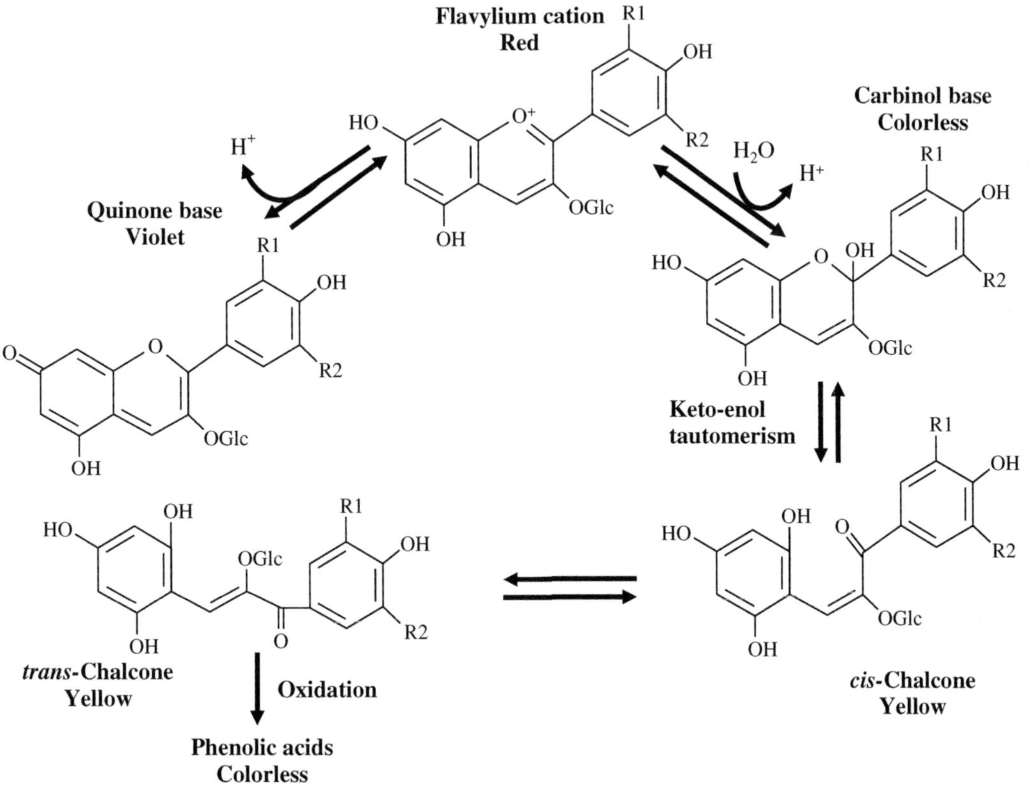

FIGURE 5.11   Anthocyan equilibria as a function of pH.

Copigmentation

The positive charge on the flavylium ion is in practice delocalized over the heterocyclic part of the molecule, and can even be found distributed over both benzene rings. This delocalization is responsible for the red color. The conjugation of a water molecule to the heterocycle and an increase in pH (nucleophilic attack mechanism) leads to neutralization of the positive charge.

Sulfite, specifically the HSO_3^- ion, also acts as a decoloring agent. This ion neutralizes the positive charge on the flavylium cation and leads to loss of color.

$$A^+ + HSO_3^- \longleftrightarrow AHSO_3$$

However, anthocyan molecules can stack up on top of each other and interact weakly (hydrogen bonds and van der Waals interactions) with other colorless, phenolic molecules. This leads to the formation of a hydrophobic macromolecular structure that impedes access by water molecules and, therefore, neutralizes positive charge. As a consequence, a larger proportion of flavylium cations is maintained than would theoretically be expected. This phenomenon is known as copigmentation.

The molecules responsible for copigmentation are generally monomeric phenolic compounds such as hydroxycinnamic acids, catechins, and flavones, and there is no evidence that tannins can function as copigments. These interactions lead to an increase in color intensity known as the hyperchromic effect. The intensification of color can be between two and six times that of anthocyans without copigmentation. This phenomenon accounts for up to 40% of the color of young wines (less than 6 months old). The increase in color caused by copigmentation or the hyperchromic effect is generally accompanied by an increase in the wavelength at which maximum absorbance is observed. This is known as the bathochromic shift and leads to displacement of the absorbance peak by approximately 20 nm. This

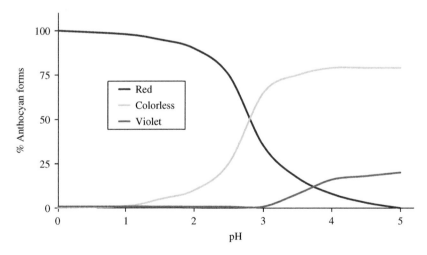

FIGURE 5.12 Theoretical proportions of different anthocyan forms according to pH.

results in the appearance of bluish and purple tones in red wines with a high level of copigmentation.

8. BIOSYNTHESIS OF PHENOLIC COMPOUNDS

The biosynthesis of phenolic compounds is regulated by the enzyme phenylalanine ammonia-lyase, which also regulates the synthesis of proteins. Consequently, the synthesis of compounds that are indispensible for plant growth competes with the formation of polyphenols.

The activity of the enzyme varies across the growth cycle of the plant, with veraison representing the point at which its activity increases. Other factors that influence the activity of the enzyme include the following:

- Hours of sunlight. The activity of the enzyme is greater the longer the exposure to sunlight.
- Temperature. Higher temperatures lead to greater accumulation of colored compounds in the skin.

The two factors that determine whether the activity of phenylalanine ammonia-lyase focuses on the synthesis of polyphenols or proteins are fertility and availability of water. In highly fertile ground with plenty of available water, the vegetative part of the plant grows extensively and protein synthesis is very active. Under these conditions, protein synthesis consumes most of the available phenylalanine, and in these types of soil, abundant harvests are achieved but the berries contain very low levels of phenolic compounds. On the other hand, less fertile ground with limited availability of water leads to reduced protein synthesis, and most of the available phenylalanine can be used in the formation of phenolic compounds. The grape yield will be lower but the quality of these vintages will be improved. The least-fertile ground will have the lowest yield but the highest concentration of sugars, tannins, and anthocyans (Table 5.6).

Reduced availability of water for the vine produces wines with higher alcohol and polyphenol concentrations, although the yield is lower (Table 5.7).

TABLE 5.6 Influence of Soil Fertility on Cabernet Sauvignon Grapes

	Poor Fertility	Fertile
Yield (kg/vine)	7 ± 1	10 ± 1
Sugars (°Brix)	20 ± 1	18 ± 1
Anthocyans (mg/kg)	1167 ± 59	908 ± 109
Tannins (mg/kg)	2054 ± 313	1358 ± 306

TABLE 5.7 Influence of Water Availability on the Composition of
Agiorgitiko Wines

	Poor Availability	Good Availability
Yield (kg/vine)	2.2	3.4
% Ethanol	12.2	11.1
Anthocyans (mg/L)	429	304
Tannins (g/L)	2.69	2.13

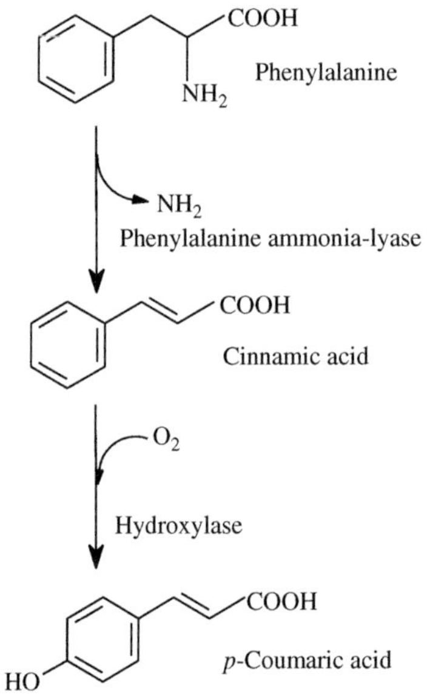

FIGURE 5.13   Biosynthesis of simple phenols.

Although phenolic compounds can be synthesized via a number of different pathways, the most important precursor is shikimic acid, which is produced by the cyclization of glucose.

8.1. Biosynthesis of Simple Phenols

The most simple phenols or non-flavonoid compounds are synthesized from phenylalanine. Phenylalanine ammonia-lyase catalyzes a reaction that generates cinnamic acid. The presence of hydroxylases leads to the formation of other simple phenols such as p-coumaric acid. Cleavage of the side chain, with release of water, generates the corresponding aldehyde, which leads to the production of benzoic acids by oxidation.

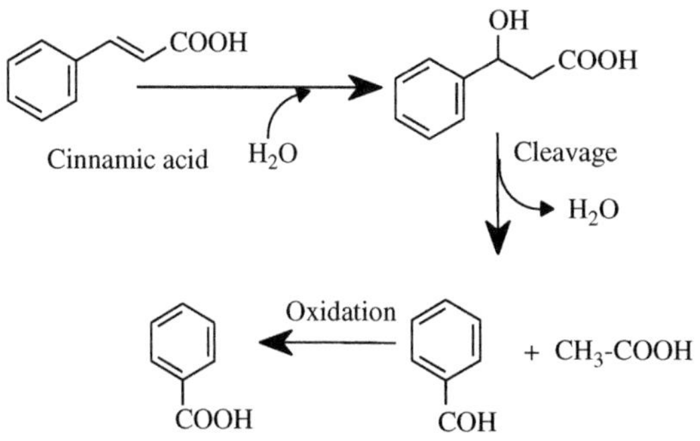

8.2. Biosynthesis of Flavonoid Phenols

As discussed in Section 3, flavonoids have the basic structure C_6-C_3-C_6. The aromatic rings are derived from different sources. Aromatic ring B is derived from p-coumaric acid, whereas aromatic ring A is formed by joining together molecules of acetyl coenzyme A (CoA), which is an intermediate in all biosynthetic reactions. Acetyl CoA is carboxylated to generate malonyl CoA, which is an intermediate in the synthesis of the A ring.

The joining of three molecules of malonyl CoA with p-coumaroyl CoA leads to the formation of a chalcone, which can undergo intramolecular cyclization to generate a flavonoid ring. The flavonoid compounds are formed from flavanone (Figure 5.15).

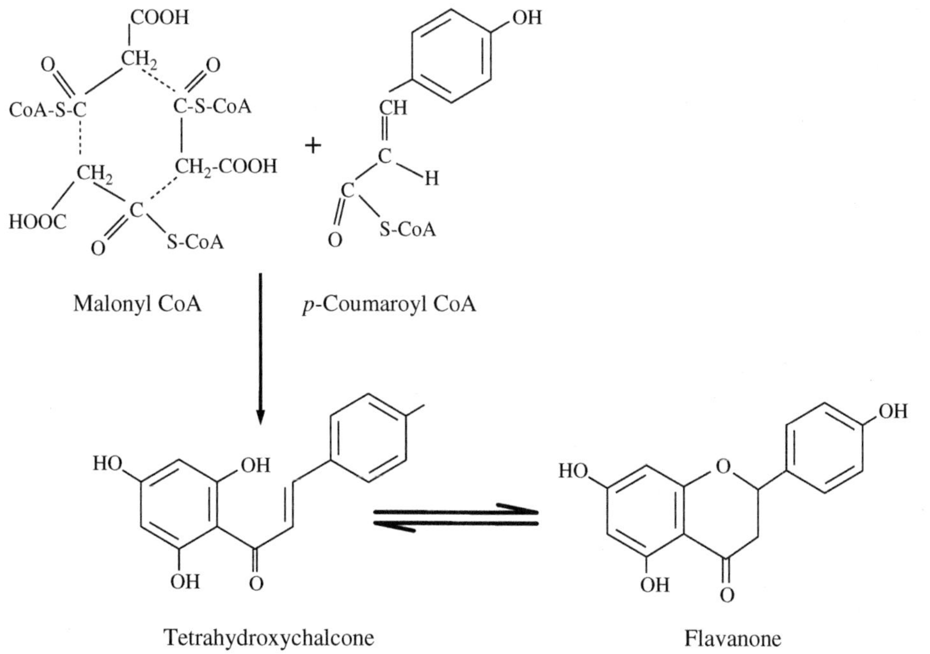

FIGURE 5.14   Formation of flavanone from malonyl coenzyme A (CoA) and p-coumaroyl CoA.

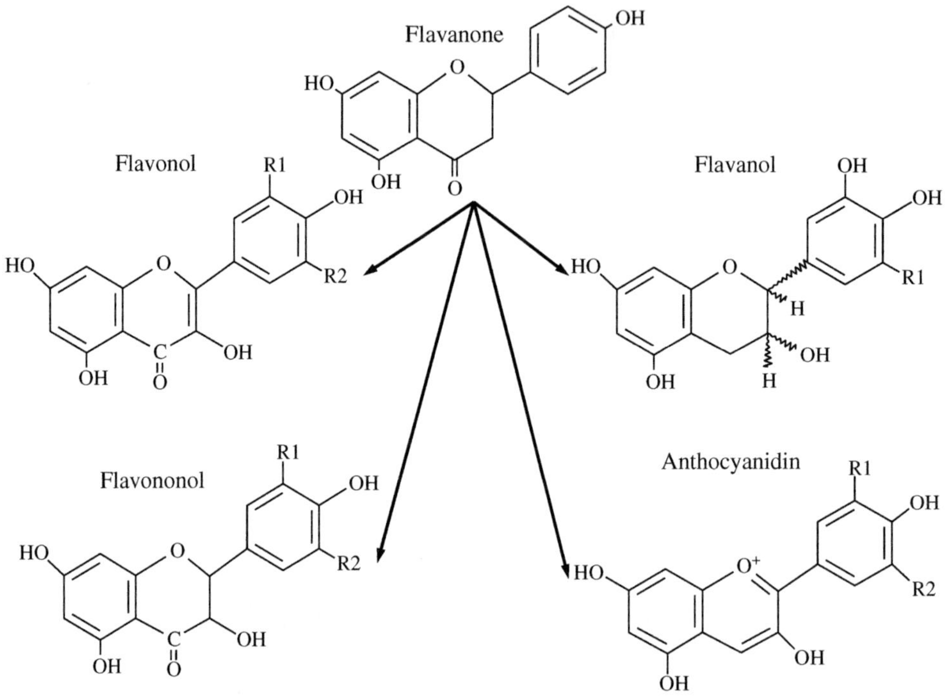

FIGURE 5.15   Formation of phenolic compounds from flavanone.

Sugars: Structure and Classification

1. INTRODUCTION

The generic term *carbohydrates* encompasses a wide variety of very important compounds of highly complex composition which are found in grapes and wine. The carbohydrate content of both grapes and wine is very important in winemaking as these compounds intervene in practically all the molecular processes that occur from the onset of berry ripening to the transformation of must into wine. They are also key to the quality of the final product because they influence alcohol content, flavor and other organoleptic properties, and clarity. They can also affect certain technological processes (the presence of polysaccharides, for example, can interfere with filtration and clarification operations).

TABLE 6.1 Classification of Carbohydrates in Winemaking According
to Origin and Biological Function

Simple sugars	Fermentable
	Non-fermentable
Monosaccharide and oligosaccharide derivatives	Derived from the vine
	Derived from microorganisms
Polysaccharides	Derived from plant cell wall structures
	Derived from microbial cell walls

Among the carbohydrates of interest to winemakers are glucose and fructose, which are the most abundant sugars in grapes. Indeed, it is the alcoholic fermentation of these sugars by yeasts that is responsible for the transformation of must into wine. Wine also contains other sugars and sugar derivatives that are not fermented by yeasts, in addition to several oligomers, formed by numerous monomers joined by glycosidic bonds; these bonds may be between different sugar molecules or between sugar molecules and non-sugar molecules (generally referred to as aglycones). The most common bonds occur between sugars and phenols, forming anthocyans, and between sugars and monoterpene alcohols, giving rise to the precursors of compounds that determine varietal aroma; these precursors are responsible for the so-called hidden aroma of grapes.

When sugar molecules combine with compounds such as amino acids, proteins, or other nitrogen compounds, they form glycoproteins and nucleic acids. Other carbohydrates found in musts and wines are polysaccharides, which are derived from berry or yeast cell walls, or from the microbial flora present on the grape. These materials are formed from an enormous variety and number of monomers, have a high molecular weight, and are extremely complex in terms of composition, structure, and properties.

Carbohydrates can be classified by size, with a distinction being made between monosaccharides, oligosaccharides, and polysaccharides (>10 monomers). Winemakers, however, tend to classify carbohydrates according to their origin and biological function.

2. STRUCTURE OF CARBOHYDRATES

Simple sugars, recognizable by the suffix −ose, are polyalcohols with an aldehyde or ketone group. They are generally referred to as carbohydrates because their empirical formula is $C_n(H_2O)_n \equiv (CH_2O)_n$. They are divided into aldoses and ketoses and are commonly named according to their Fischer projection formula. In the case of aldoses, the carbons are numbered (from 1 upwards) starting from the aldehyde group, while in the case of ketoses, numbering starts at the carbon bearing the ketone group. Most sugars are chiral (meaning that they have asymmetric carbons) and are optically active. This activity is classified as (−) or (+) as follows:

(−) if they rotate plane polarized light to the left (levorotatory molecules)
(+) if they rotate plane polarized light to the right (dextrorotatory molecules)

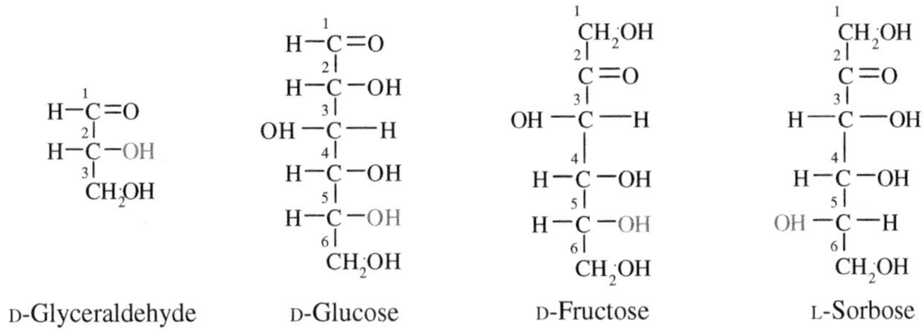

$$
\begin{array}{cccc}
& & \overset{1}{CH_2OH} & \overset{1}{CH_2OH} \\
& & \overset{2}{C}=O & \overset{2}{C}=O \\
\end{array}
$$

Fischer projections:

D-Glyceraldehyde
$$
\begin{array}{c}
H-\overset{1}{C}=O \\
H-\overset{2}{C}-OH \\
\overset{3}{C}H_2OH
\end{array}
$$

D-Glucose
$$
\begin{array}{c}
H-\overset{1}{C}=O \\
H-\overset{2}{C}-OH \\
OH-\overset{3}{C}-H \\
H-\overset{4}{C}-OH \\
H-\overset{5}{C}-OH \\
\overset{6}{C}H_2OH
\end{array}
$$

D-Fructose
$$
\begin{array}{c}
\overset{1}{C}H_2OH \\
\overset{2}{C}=O \\
OH-\overset{3}{C}-H \\
H-\overset{4}{C}-OH \\
H-\overset{5}{C}-OH \\
\overset{6}{C}H_2OH
\end{array}
$$

L-Sorbose
$$
\begin{array}{c}
\overset{1}{C}H_2OH \\
\overset{2}{C}=O \\
H-\overset{3}{C}-OH \\
H-\overset{4}{C}-OH \\
OH-\overset{5}{C}-H \\
\overset{6}{C}H_2OH
\end{array}
$$

FIGURE 6.1   Fischer projection of the main carbohydrates.

The prefixes D and L refer respectively to the (+) and (−) enantiomers of glyceraldehyde. Accordingly, monosaccharides in which the chiral center furthest from the aldehyde or ketone group has the same configuration as D-glyceraldehyde belong to the D series, while those having the opposite configuration belong to the L series. In the Fischer projection of D-glyceraldehyde, the −OH functional group lies to the right of its chiral center. In other words, it belongs to the D series. L-Glyceraldehyde, in contrast, belongs to the L series. Whether a monosaccharide belongs to the D or L series in terms of absolute configuration has no influence on its optical properties. In other words, it can be dextrorotatory or levorotatory.

Simple monosaccharides with increasing numbers of carbon atoms are considered derivatives of glyceraldehydes if they are aldoses or of dihydroxyacetone if they are ketoses. Both types of monosaccharides can have an absolute D or L configuration depending on whether the −OH group of the highest numbered asymmetric carbon is to the right or left of the carbon, respectively. Accordingly, the addition of carbons, one by one, to D-glyceraldehyde will give rise first to two D-tetroses, next to four D-pentoses, and finally to eight hexoses.

In the case of ketoses, the addition of a carbon atom to dihydroxyacetone produces just one tetrose (D series) and the addition of two atoms produces two pentoses. The lengthening of this carbon chain with a third atom gives rise to four ketohexoses. The same procedure applied to L-glyceraldehyde would produce L-aldoses.

2.1. Cyclization of Carbohydrates: Haworth Projection

The Fischer projection is only used to gain a better picture of the configuration of carbohydrates. In solution, the open-chain form of a monosaccharide is in equilibrium with the cyclic hemiacetal form (aldoses) or the hemiketal form (ketoses). This ring can be composed of five or six atoms (furanose and pyranose forms, respectively), and in such cases, the Haworth projection is more useful than the Fischer projection, as it shows the cyclic structure.

The cyclization of carbohydrates results in the appearance of a new chiral center at carbon 1 (anomeric carbon). This anomer is designated α when the −OH group is *trans* with respect to the −CH$_2$OH group which carries the sixth carbon atom (in other words, it is on the opposite side of the plane of the ring). Accordingly, it is designated β if the hemiacetal −OH group is *cis* with respect to the same group (on the same side of the plane).

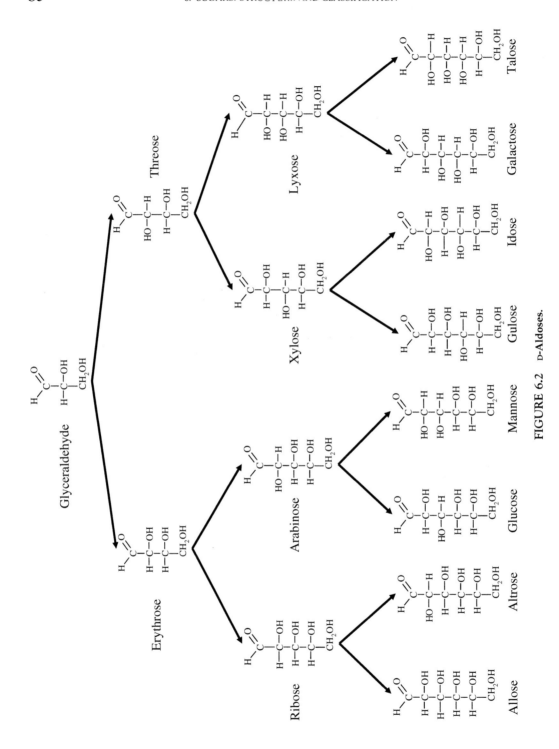

FIGURE 6.2 D-Aldoses.

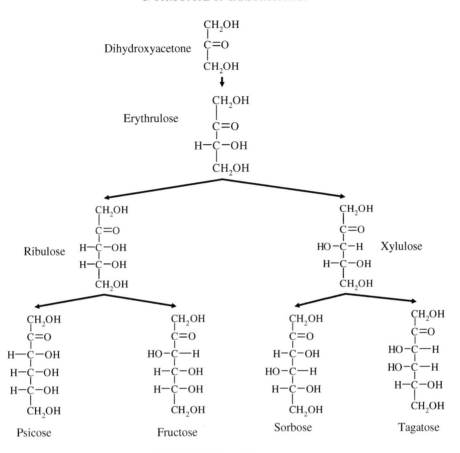

FIGURE 6.3 D-**Ketosugars.**

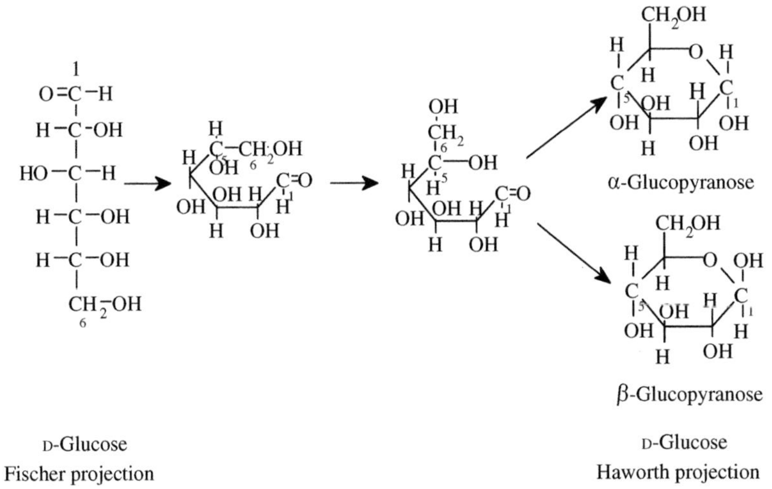

D-Glucose
Fischer projection

D-Glucose
Haworth projection

α-Glucopyranose

β-Glucopyranose

FIGURE 6.4 Haworth projection of glucose.

In solution, the α and β forms of all free monosaccharides are present in an equilibrium involving an open-chain form.

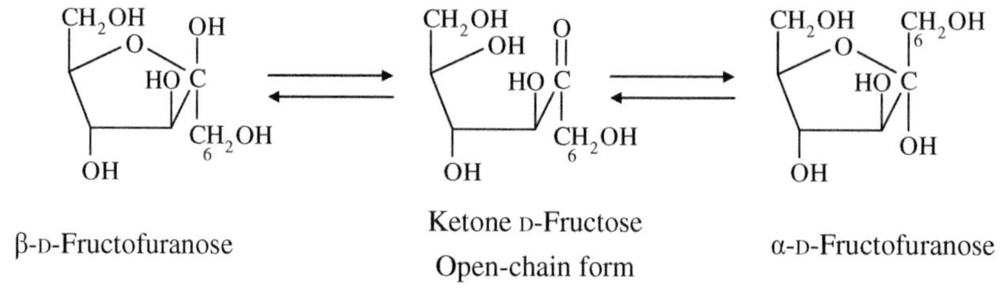

β-D-Fructofuranose   Ketone D-Fructose   α-D-Fructofuranose
                            Open-chain form

This phenomenon is known as mutarotation, as the two anomers do not generally have the same optical rotatory power and therefore an equimolecular solution of both anomers can have optical activity.

Equilibrium between the α and β forms in aqueous solution is reached when the mixture contains 63.6% of the β form. The optical rotatory power for this mixture is $\alpha_D = +52.7°$.

Finally, to complete this brief overview of the structure of monosaccharides, it should be recalled that the actual conformation of a monosaccharide is not planar; accordingly, pyranose (six-membered) rings adopt the most stable carbohydrate form, the chair, while furanose (five-membered) rings adopt the conformation of an open envelope.

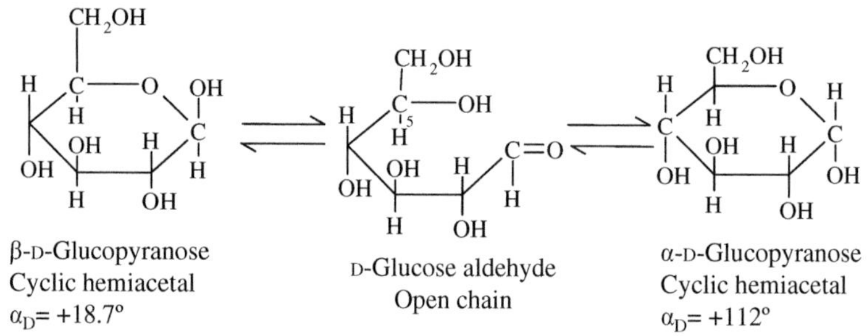

FIGURE 6.5 Mutarotation equilibria for glucose.

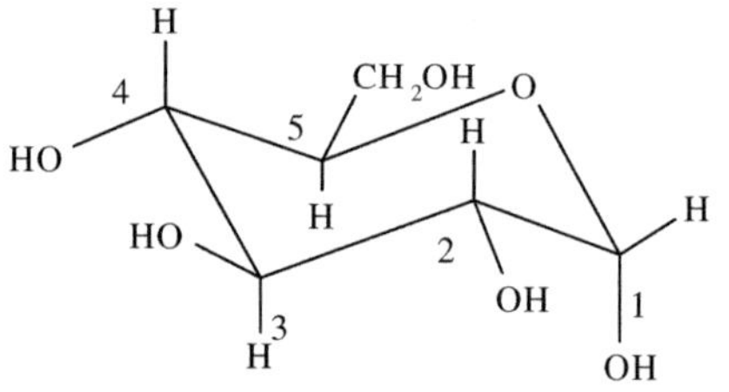

β-D-Glucopyranose
Cyclic hemiacetal
$\alpha_D = +18.7°$

D-Glucose aldehyde
Open chain

α-D-Glucopyranose
Cyclic hemiacetal
$\alpha_D = +112°$

FIGURE 6.6 Chair diagram of α-D-glucose.

Monosaccharide Derivatives

Like simple pentoses and hexoses, oligosaccharides and polysaccharides in grapes and wine are also formed from a wide variety of monosaccharide derivatives such as:

- Deoxy sugars
- Methylated monosaccharides
- Uronic acids
- Aldonic acids
- Aldaric acids
- Amino sugars

A hydrogen atom can also be replaced by an alkyl group ($-R$) to give a branched monosaccharide.

The following symbols, consisting of three-letter abbreviations, tend to be used in complex formulas involving simple monosaccharides.

- Glc → glucose
- Gal → galactose
- Fru → fructose
- Ara → arabinose
- Xyl → xylose
- Man → mannose
- Rha → rhamnose (6-deoxy-l-mannose, which is a methyl-pentose)
- Rib → ribose

The symbols are accompanied by the suffixes p or f (in italics), depending on whether the rings are pyranose or furanose.

2.2. Monosaccharides of Interest in Winemaking

Of the aldoses that participate in the winemaking process, D-glucose (found in concentrations of several grams per liter in musts) and D-galactose (found in concentrations of 100 mg/L in wine) are of particular interest. In the case of C_6-ketoses, fructose is the only sugar that is considered essential. Together with glucose, this is the most important hexose found in grapes. Pentose sugar concentrations in must and wine vary between 0.3 and 2 g/L. The main sugars in this group are D-xylose and L-arabinose, which reach concentrations of several hundreds of milligrams per liter. D-Ribose and L-rhamnose (a methyl pentose), in contrast, do not exceed levels of 100 mg/L.

Glucose, fructose, and mannose, which all belong to the D series, are the most relevant simple hexoses in the study of grapes and wines and they are easily interconvertible thanks to their keto-enol tautomerism.

As mentioned in previous sections, the glucose found in grapes is dextrorotatory and is, accordingly, also referred to as dextrose. Likewise, fructose, which is levorotatory, used to be called levulose.

FIGURE 6.7 Interconversion between monosaccharides via keto-enol tautomerism.

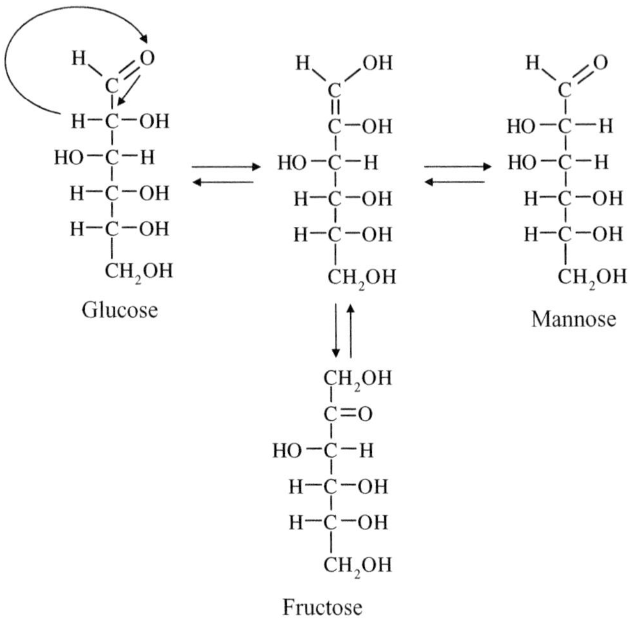

3. THE GLYCOSIDIC BOND: POLYMERIZATION

The hemiacetal hydroxyl group (C1) can form an acetal in the presence of a hydroxyl group from another molecule. Sugar acetals are called glycosides and they are joined by glycosidic bonds. Glucose forms glycosides. The formation of disaccharides, trisaccharides, or polysaccharides simply involves the glycosidic linkage of several monosaccharide molecules.

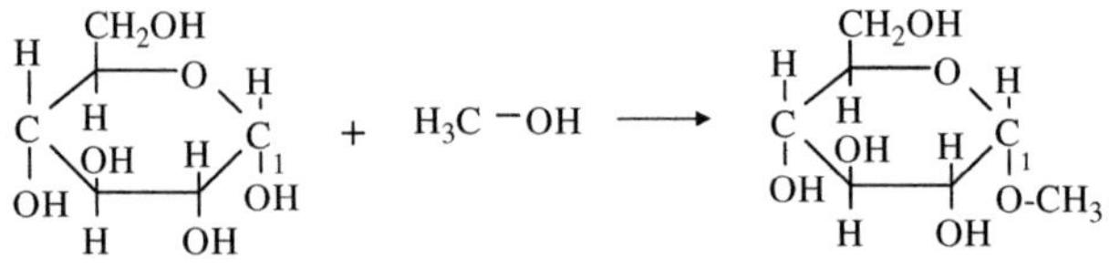

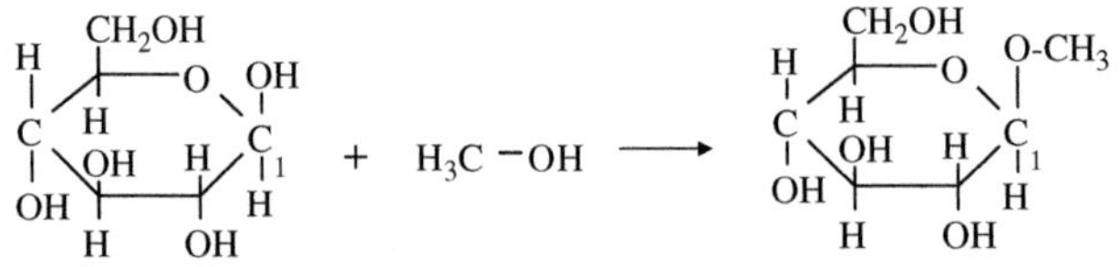

Glycosidic bonds consist of an oxygen bridge between the anomeric carbon of one of the glucose residues and a carbon from the other residue. Glycosides formed in this manner belong to the chemical family of acetals; they are stable to bases and do not reduce Fehling's solution. In acid solutions, however, they hydrolyze easily, giving rise to the original sugars.

Carbohydrates can form glycosidic bonds with any molecule that has a hydroxyl group (−OH) to give rise to compounds generically referred to as glycosides. All glycosides are named according to the following three criteria:

1. The name of the sugar and configuration (pyranose or furanose) of the residues joined by glycosidic bond.
2. The α or β position of the −OH group of the anomeric carbon of the sugar moiety.
3. The position of the glycosidic bond, in terms of the positional number of the carbon in the ring that is taking part in the link.

The position of the carbon involved in the glycosidic bond is important, as it determines whether or not the resulting combinations will have reducing power, and also what other glycosidic combinations can then form. Compounds which have been formed by glycosidic bonding between the anomeric carbons from two glucose residues do not have reducing power because the anomeric carbon, which bears the carbonyl group in each monosaccharide, is blocked by the bond; these compounds will therefore lack the reactivity of the carbonyl group. Furthermore, these compounds do not exhibit mutarotation in neutral or alkaline solutions and are incapable of giving rise to other glycosidic compounds. It should be recalled that the reducing nature of a compound is due to the presence of an aldehyde group in the open chain form, which is measured by its stability towards oxidation in Fehling's solution and the resulting formation of aldonic acids. One example is sucrose (β-D-fructofuranosyl α-D-glucopyranoside), which has no reducing power because both of the anomeric carbons which could potentially form aldehyde groups are bound up in the glycosidic bond.

On the other hand, disaccharides with a glycosidic bond between C1 of one monosaccharide and either C4 or C6 of another have reducing power and are known as reducing disaccharides. They are important in winemaking because they can form glycosidic compounds with other molecules that contain a hydroxyl group, such as terpene alcohols or a range of phenols. They have reducing power because one of the monosaccharides has a free hemiacetal −OH group.

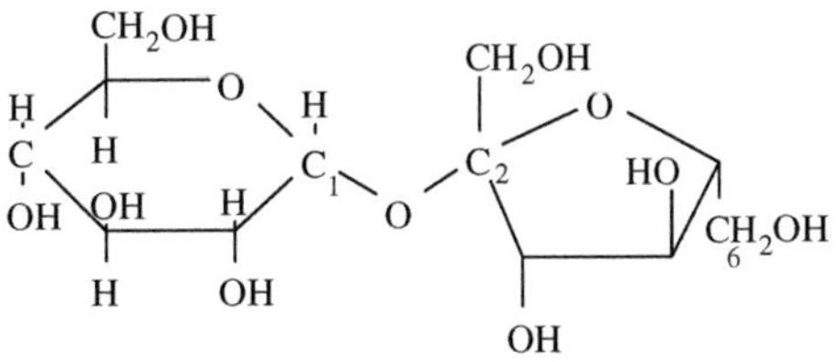

FIGURE 6.8 Haworth projection of sucrose.

3.1. Disaccharides and Trisaccharides

The main disaccharides found in grapes and wine are sucrose and trehalose, although maltose, melibiose, and lactose may also be present. There have also been reports of small quantities of the trisaccharide raffinose.

Sucrose

Sucrose is a non-reducing disaccharide that is generally found in musts and wines at levels of between 2 and 5 g/L, although it can exceptionally reach levels of up to 15 g/L. The levels of this sugar are higher in table grapes.

Sucrose is hydrolyzed from the onset of fermentation to an equimolecular mixture of glucose and fructose known as invert sugar. The disaccharide is hydrolyzed by the enzyme invertase, which is present in considerable quantities in grape berry pulp cells, although it is also produced by yeast.

Sucrose is widely present in the plant kingdom. In the case of grapevines, it is formed in the leaves by photosynthesis. It is dextrorotatory and has an optical rotatory power of 66.5°.

$$\text{Sucrose}: [\alpha]_D{}^{20} = +66.5°$$

The optical power of a solution in which sucrose is hydrolyzed will be inverted due to the formation of an equal number of glucose and fructose molecules. These monosaccharides, which combine to form sucrose, have the following optical rotatory powers:

$$\text{Glucose}: [\alpha]_D{}^{20} = +52.3° \rightarrow \text{dextrorotatory}$$

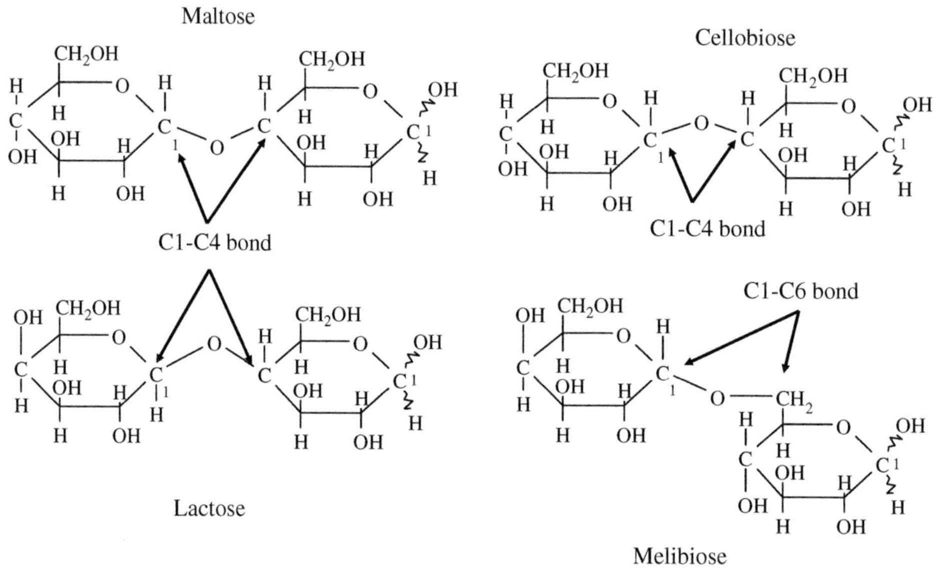

FIGURE 6.9   Disaccharides with C1-C4 and C1-C6 bonds.

Fructose: $[\alpha]_D{}^{20} = -93.0° \rightarrow$ levorotatory

As can be seen in the calculation that follows, the final solution will be levorotatory:

$$[\alpha]_D{}^{20} = -93.0° + 52.3 = -40.7° \rightarrow \text{levorotatory}$$

Accordingly, the quantity of sucrose in a given solution can be determined by the polarimetric analysis of the solution before and after hydrolysis.

Sucrose is highly soluble in water, insoluble in 95% alcohol, and has no reducing power in Fehling's solution. Hydrolysis thus is necessary to test for this sugar.

The practice of enriching the must with sugar, specifically cane sugar, dates back to the 18th century. The process of adding sugar to unfermented wine, which is typical in the north of France, is known as chaptalization, after Count Jean-Antoine-Claude Chaptal, a French industrial scientist and government minister, who recommended this method in 1801 in his book *L'art de faire, de gouverner et de perfectionner les vins* (The Art of Making, Managing, and Perfecting Wines).

Trehalose

Trehalose (α-D-glucopyranosyl α-D-glucopyranoside) is a non-reducing disaccharide found in wine but not must. Yeasts store glucose in the form of trehalose during alcoholic fermentation and it is only released during yeast autolysis, explaining why it is not found in must. Trehalose is a fermentable disaccharide whose levels in wine range from 50 to 600 mg/L. This residual sugar plays as important a role as glucose and fructose in the microbiological stability of wine.

Raffinose

Raffinose is a trisaccharide in which glucose acts as a monosaccharide bridge between galactose and fructose. It has both α and β glycosidic bonds and can therefore be hydrolyzed to D-galactose and sucrose via enzymes with α-glycosidic activity, and to melibiose and D-fructose via enzymes with β-glycosidic activity.

Raffinose

α-D-galactopyranosyl-$(1 \rightarrow 6)$-α-D-glucopyranosyl β-D-fructofuranoside

$C_{1\alpha} C_6$	$C_{1\alpha} C_{2\beta}$
Cleavage A	Cleavage B
(α)	(β)

Cleavage A: Hydrolysis occurs via the action of α-galactosidase to produce D-galactose and sucrose.

Cleavage B: Hydrolysis occurs via β-fructosidase to produce the monosaccharide D-fructose and the disaccharide melibiose.

By analyzing the two enzymes involved, it can be deduced that raffinose is composed of the monosaccharides D-galactopyranose, D-glucopyranose, and D-fructofuranose. This trisaccharide is very common in plant seeds, leaves, stems, and roots. As is evident from its structure (its anomeric carbon atoms are involved in glycosidic bonds), it is a non-reducing sugar.

Raffinose: trisaccharide composed of :

α-D-Galactopyranosyl
α-D-Glucopyranosyl
β-D-Fructofuranoside

Hydrolytic enzyme α gives rise to Galactose
Sucrose

Hydrolytic enzyme β gives rise to Melibiose
Fructose

Oligosaccharides and polysaccharides can also be formed, like trisaccharides, by linking an increasing number of monosaccharide residues by successive glycosidic bonds. In general, linear polysaccharides (or polysaccharide chains) are built by joining disaccharides (e.g., cellobiose in the case of cellulose). Monosaccharides which can form two glycosidic bonds in a linear chain will give rise to branching; in such cases, the longest chain is considered to be the main chain and the others are considered to be side chains.

The complexity of polysaccharides increases with the number of sugar residues that form them (degree of polymerization), the diversity of these residues, the type of bonds involved, and the degree of branching.

4. POLYSACCHARIDES

Polysaccharides are high-molecular-weight carbohydrates that are insoluble in water. They are formed by the polymerization of monosaccharides or monosaccharide derivatives via glycosidic bonds. They have no reducing power, and can be divided into homopolysaccharides (formed by identical monosaccharides) and heteropolysaccharides (formed by two or more types of monosaccharide).

Polysaccharides are still poorly understood because of their enormous configurational and structural complexity. While these compounds do not have a direct influence on the organoleptic properties or visual quality of wine, they are extremely important in winemaking, particularly in processes that use rotten or raisined grapes, as they can adversely affect numerous operations such as filtration, clarification, and stabilization.

Given their structural complexity and enormous molecular mass (they are macromolecules), they cannot be ignored during the production process, as they intervene in colloidal processes such as the filtration of sediment and the formation or prevention of hazes and precipitates. They are also the main reason for filter clogging and play an essential role in foam stability in sparkling and semi-sparkling wines. Several acid polysaccharides can form complexes with toxic heavy metals, such as lead, or with other elements such as calcium, which is the main gelling agent used in wine production.

Polysaccharides are a key component of the cell wall of grapes and are therefore indirectly related to the extraction of the color and odorant compounds from these cells. Their presence in must and wine is due to the degradation and solubilization of a portion of the pectins found in the cells of grape skin and pulp.

The classification of polysaccharides has always been subject to debate. In certain wine fields, they are divided into pectins and gums. Such a system, however, is highly arbitrary and not very accurate as the pectin component classification is based on the precipitation of polysaccharides in 80% (vol/vol) ethanol and the subsequent formation of calcium pectate. The fraction of gums is determined by calculating the difference in weight between the total colloid concentration and the pectin fraction; the presence of gum is attributed to microbial activity.

Pectins contain galacturonic acid almost exclusively, which is partially esterified with methanol. These chains are also known as polygalacturonic acid. The bonds are α-$(1 \rightarrow 4)$. Pectins, which are soluble in must, precipitate rapidly in the presence of calcium chloride, giving rise to calcium pectate.

Gums, for their part, are soluble polysaccharides that occur after the removal of pectins by precipitation. Their main components, in addition to galacturonic acid, are arabinose, rhamnose, and galactose. They are a highly complex mix of heteropolysaccharides with a highly variable molecular weight, ranging from under 1,000,000 to over 2,000,000 Da.

Polysaccharides that are soluble in must can also be classified as acidic or neutral pectins depending on whether or not they contain galacturonic acid.

With the advances that have been made in analytical methods, and particularly with the emergence of the colorimetric method for neutral sugars and uronic acids, acidic and neutral polysaccharides are now classified, respectively, as pectins and gums. In reality, however, polysaccharides are much more diverse and complex than this classification, as numerous polysaccharides contain both acid and neutral residues. Furthermore, the terms *pectin* and *gum* correspond to properties such as gelling and high viscosity, which polysaccharides that are naturally present in wine do not possess.

For winemakers, it is preferable to classify polysaccharides by origin, with a distinction being made between those derived from the grape berry and those from the cell walls of fungi and yeasts.

Finally, the term colloid, which is widely used in winemaking, should not be confused and indiscriminately applied to polysaccharides, as wine contains many types of colloidal substances. The term colloidal is generally used to describe molecules or macromolecules of dimensions of between 1 nm and 1 mm that do not form true solutions in a liquid medium, but rather form colloidal dispersions over time, until the phases separate.

4.1. Polysaccharides in Grape Berry Cell Walls

The composition of the cell walls in the pulp and skin of grapes has not been widely studied, although, in broad terms, their qualitative composition is similar to that of other plants. It can thus be confidently stated that the cells of the pulp and skin are protected by primary walls (which is a characteristic of young growing tissue) and that over 90% of this wall structure is formed by structural polysaccharides, the most important of which are listed below:

1. Cellulose, a β-(1→4)-D-glucose polymer that forms fibrils.
2. Hemicelluloses, xyloglucans, and arabinoxylans (all of which form a coat on the fibrils) and galactomannans.
3. Pectins, the most complex polysaccharides found in the primary walls of all plants. They have three main components, homogalacturonan and rhamnogalacturonans I and II. The less abundant components of particular interest are xylogalacturonans and apiogalacturonans.
4. Structural glycoproteins, which are rich in the amino acid hydroxyproline.
5. Other components specific to each plant.

Lignification is a key process in the transformation of primary walls into secondary walls. When this process starts, tissue growth stops, leading to a thickening of the cells walls (caused by the accumulation of lignin and polysaccharides) and eventually cell death. The pectic polysaccharides in the walls are also esterified by phenolic acids, ferulic acids, or *p*-coumaric acids, adding further strength to the overall structure.

The cell walls of the skin of the grape are much thicker than those of the pulp, making it more difficult for components to be released into the medium. They are, however, a relatively important source of polysaccharides, explaining why wines fermented on grape skins have a richer polysaccharide content.

Surprisingly, the polysaccharide content of the cell walls of red and white grape skins are almost identical, with the only difference being the presence of anthocyanins in the skins of red grapes.

4.2. Polysaccharides from Fungi and Yeast

The cell walls of fermentation yeasts and filamentous fungi (responsible for rot in grapes) are an important source of structural polysaccharides, particularly in harvests containing rotten grapes.

Fungal Cell Walls

Polysaccharides account for 90% of the cell wall structure in filamentous fungi. The most abundant of these polysaccharides are β-D-glucans, which are D-glucans joined together in a linear chain by β-(1→3) glycosidic bonds, and branches joined by β-(1→6) bonds. These

FIGURE 6.10 Cellulose molecule (β-D-glucopyranose polymer with 1→4 linkages).

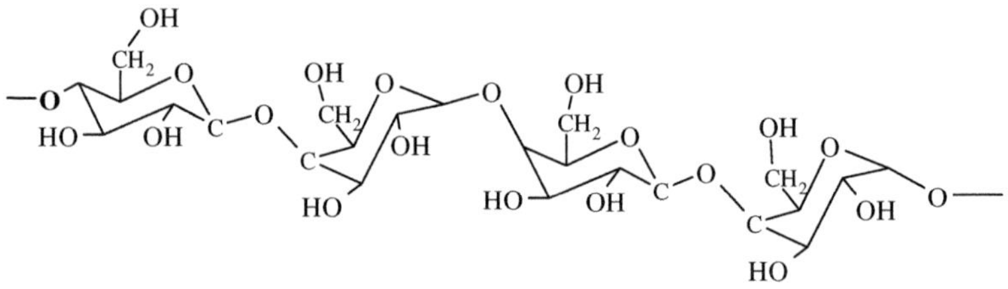

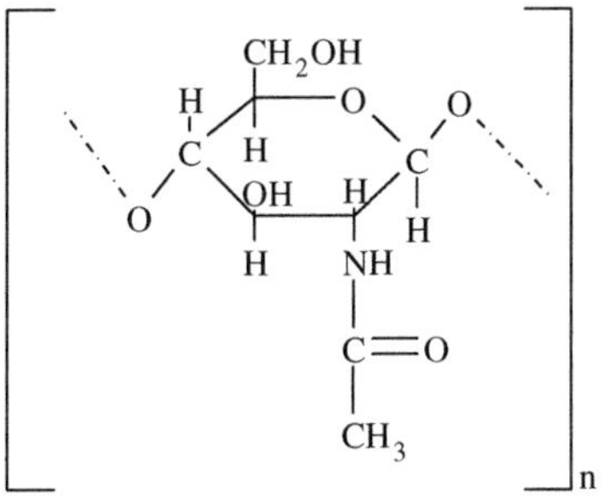

FIGURE 6.11   β-(1→4)-N-acetyl-D-glucosamine monomer.

branches confer a certain degree of solubility, explaining why polysaccharides are soluble in grape juice obtained from rotten grapes. β-D-glucans are found in association with highly insoluble polymers, such as cellulose (β-(1→4)-D-glucose polymer) and chitin (β-(1→4)-N-acetyl-D-glucosamine polymer).

Yeast Cell Walls

Yeast cell walls contain β-D-glucan polymers in association with chitin, as well as a high proportion of mannoproteins, which are proteoglycans (polysaccharides + proteins) formed of approximately 20% protein and 80% D-mannose. The polysaccharide contains several D-glucose and N-acetylglucosamine residues linked by α-(1→6), (1→2), and (1→3) bonds. The resulting chains can attach to the protein component via amide bonds (through asparagine), producing extensively branched N-linked glycans, or via ether bonds with serine or threonine residues, producing linear O-linked glycans. Mannoproteins are highly soluble in aqueous media and are also excreted spontaneously by yeasts during their growth.

5. GLYCOSIDES

The combination of monosaccharides and, particularly, disaccharides with non-sugar molecules via glycosidic bonds gives rise to compounds known as glycosides.

Glycoside = non-sugar compound (aglycone) + sugar residue (carbohydrate)

Glycosylation, which is the process by which glycosides are formed, has two essential functions:

1. It contributes to detoxification by stabilizing radicals and ions, which, due to their high chemical reactivity, can be toxic or harmful to plants or animals.
2. It contributes to transportation by increasing the solubility of compounds that are only slightly soluble in water, favoring their transport in plant or animal fluids from the production to the storage site.

Aglycones (non-sugar compounds) are generally insoluble in water but soluble in ether, chloroform, and other non-polar solvents, while glycosides are soluble in absolute alcohol, low-grade alcohols, and other polar solvents. Glycosylation increases the solubility of agly-cones, with solubility increasing with the number of sugars attached to the aglycone.

Glycosides formed between monosaccharides or disaccharides and phenolic compounds and aromatic compounds such as terpene alcohols and carotenoids have an important role in winemaking. Wine also contains glycoproteins derived from grape and yeast cell walls. Finally, the nucleic acids present at trace levels in wine are those with ribose or deoxyribose in their structure.

6. THE IMPORTANCE OF GLYCOSIDES IN WINEMAKING

The glycosides of interest in winemaking are those formed by certain sugars and phenolic or terpene compounds (particularly monoterpene alcohols). It is important to know whether the compounds have α or β linkages in order to choose the correct enzymes to catalyze hydro-lysis and release the aglycones of interest.

6.1. Phenols

In phenolic compounds, the presence of sugars or unsubstituted −OH groups increases the polarity of the molecule and hence its solubility in water. Methoxylation, in contrast, decreases this polarity.

Anthocyans, which confer color to red and rosé wines, are among the most important of the phenolic compounds responsible for color. Grape anthocyanins are formed by the combi-nation of a sugar with a non-sugar molecule containing a hydroxyl group; the most common phenol found in grapes is anthocyanidin, which links with a β-D-glucopyranosyl moiety at position 3 of the middle heterocycle. This linkage is characteristic of *Vitis vinifera*. The loss of the sugar unit in phenolic glycosides (particularly flavonol 3-glucoside) increases the insta-bility of flavonols.

Phenolic aglycones in grapes preferentially link to glucose and are found in greater concentrations in leaves. While *V. vinifera* contains mostly monoglucoside forms of phenolic compounds, American varieties and direct-producing hybrids also contain diglucoside forms.

6.2. Terpenes

The most common glycosides found in grapes are formed by the combination of a terpene molecule with a glucose molecule or disaccharide. The glycosylation of free forms of terpenes gives rise to compounds that are highly soluble in aqueous media; their main function is transport, hence these aromatic compounds are redistributed around the grape berry. The formation of terpene-sugar combinations leads to a loss of the volatility of these compounds, and they are responsible for the hidden aroma of grapes and wines.

Glycosides containing monoterpenes appear to play an extremely important biochemical role in plants, as they serve as intermediaries in the biosynthesis of terpenes and mainly inter-vene in the stabilization of carbon ions, which could not exist in their free form due to their high reactivity.

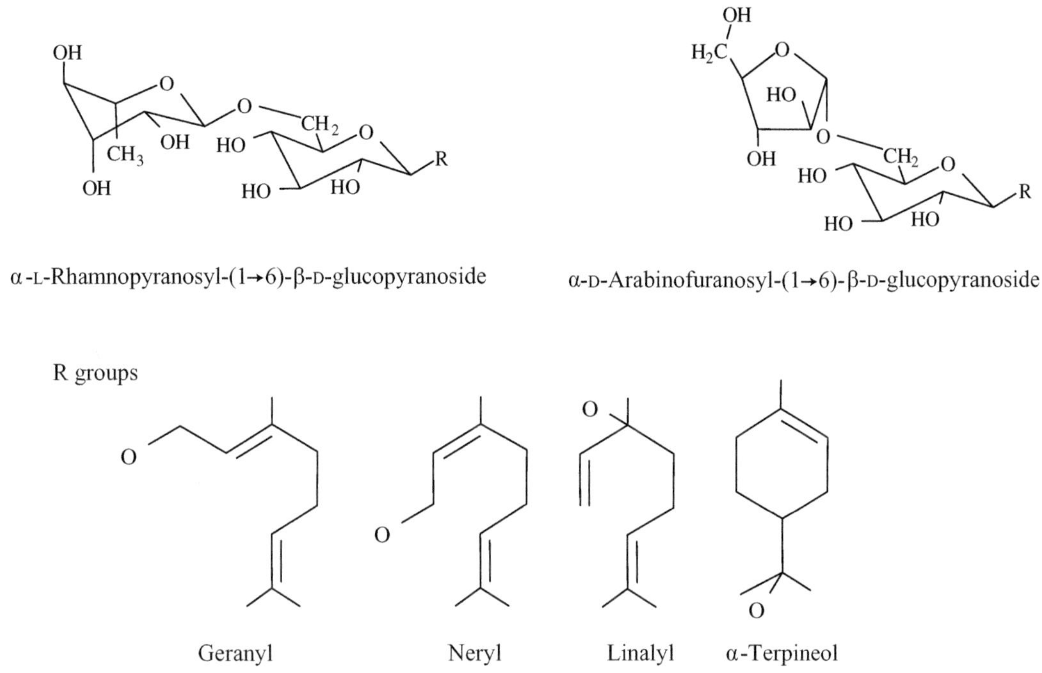

α-L-Rhamnopyranosyl-(1→6)-β-D-glucopyranoside   α-D-Arabinofuranosyl-(1→6)-β-D-glucopyranoside

R groups

Geranyl   Neryl   Linalyl   α-Terpineol

FIGURE 6.12   Terpene glycosides in grapes: most common aglycones and disaccharides.

Plants accumulate terpene glycosides in undifferentiated tissue and use them for transport towards differentiated tissue. Monoterpene glycosides are more soluble and less lipophilic than the free forms and can therefore cross cell barriers more easily and move towards areas of accumulation, usage, or transformation. Glycolated monoterpenes might be involved in the formation of the cell wall, in a similar manner to mannosyl-1-phosphoryl-polyisoprenol in the synthesis of mannans by bacteria.

Aromatic grape varieties such as Muscat are the richest in these glycosidic combinations. The most abundant molecules in these combinations are disaccharides such as:

- Rutinose (α-L-rhamnopyranosyl-(1→6)-D-glucopyranose)
- 6-O-α-L-arabinofuranosyl-β-D-glucopyranose
- 6-O-α-D-apiofuranosyl-β-D-glucopyranose

The above disaccharides produce rutinosides, arabinosyl glucopyranosides, and apiosyl glucopyranosides, respectively, on combining with monoterpene aglycones such as linalool, geraniol, farnesol, and nerol.

Sugars in Must

1. INTRODUCTION

Simple sugars belonging to the hexoses are easily metabolized by yeasts and bacteria, and are also known as fermentable sugars. In contrast, pentose sugars and other polyalcohols with fewer carbon atoms are not readily metabolized by microorganisms and so are known as non-fermentable sugars.

Glucose, fructose, and mannose are well-known fermentable sugars. Galactose can also be fermented by certain strains of yeast, albeit with difficulty and only after a period of adaptation to the medium. Glucose, fructose, and mannose have analogous configurations and are interconvertible due to the equilibrium between their keto and enol forms (keto-enol tautomerism).

The hexoses found in grapes are almost exclusively D-glucose (D-glucopyranose) and D-fructose (D-fructofuranose); the ratio is approximately 1 to 1 in ripe grapes.

2. PROFILE OF FERMENTABLE HEXOSES

Sugars are produced during photosynthesis, a process that occurs in the green, chlorophyll-containing organs of the plant. Most of the sugar produced is transported from the photosynthetic organs to the rest of the plant in the form of sucrose.

Sugar produced by the leaves is first and foremost used to satisfy energy and growth needs; the remaining sugar is then transported to other organs (buds, shoots, trunk, roots, and berries) in an order of priority that varies according to the growth cycle of the plant. In the growth stage, for example, the priority is the tip of the shoots, while in veraison (onset of ripening), it is the berries.

The sugar content of berries never exceeds 20 g/kg of fresh weight before veraison. Berries need sugar to grow, but above all, they need it to develop their seeds. Shortly before veraison, there is an abrupt increase in the sugar content of berries. Veraison is thus characterized by a change in color of the berries (due to the accumulation of anthocyans) and an increase in sugar levels. While the two processes are related, the mechanisms involved are relatively independent.

The berries continue to accumulate sugar as they ripen. The final concentration is related to photosynthetic activity, but it can be adversely affected by factors such as high rainfall or low temperatures.

Sucrose is hydrolyzed in the berries by the enzyme invertase. The glucose to fructose (G:F) ratio in unripe berries is 4:5, as fructose is metabolized more readily than glucose; it also appears that fructose is converted to glucose by an enzyme known as epimerase. The G:F ratio nears 2 coming up to veraison and starts to decrease as soon as the berries begin to ripen. In ripened berries, the ratio tends to be less than 1.

FIGURE 7.1 Sugar content of Pedro Ximénez grapes.

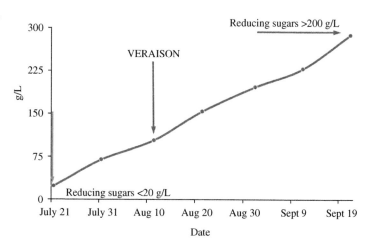

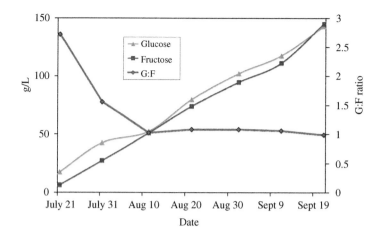

FIGURE 7.2 Glucose and fructose content and glucose to fructose (G:F) ratio during the ripening of Pedro Ximénez grapes.

3. PHYSICAL PROPERTIES OF GLUCOSE AND FRUCTOSE

Both glucose and fructose are whitish in color, have similar densities, and are highly soluble in water and less so in alcohol. One of the most important physical properties of these hexoses is their ability to rotate polarized light. Indeed, this optical activity is responsible for the alternative names for these sugars: dextrose in the case of glucose (from dextrorotatory, which means capable of rotating light to the right) and levulose for fructose (from levorotatory, which means capable of rotating light to the left). The ability of glucose and fructose to rotate the plane of polarized light is important as it means that the concentration or proportion of these sugars in a given substance can be determined by polarimetry.

Main Characteristics of Glucose

Glucose is an aldohexose that crystallizes in an anhydrous form in absolute alcohol and has a density of 1.52 g/mL.
It is highly soluble in water and alcohol.
It hydrates easily and loses this water slowly when heated to 100°C.
Anhydrous glucose melts slowly at a temperature of 146°C.

Main Characteristics of Fructose

Fructose is a ketohexose that exists as a crystalline powder with a density of 1.55 g/mL.
It is highly soluble in water and fairly soluble in hot absolute alcohol.
It has a melting point of 45°C.

3.1. Optical Rotatory Power

Specific optical rotatory power refers to the angle of rotation (which varies from one sugar to the next), measured at 20°C using a polarimeter with a 1-dm path, produced by a solution containing 1 g/mL of sugar at the characteristic spectral line of sodium, known as the D line. It is expressed by $[\alpha]_D^{20}$.

$$[\alpha]_D^{20} = \frac{\alpha \times 100}{l \times C} \quad T = 20°C \quad D \approx 5893 \text{ Å (yellow sodium light)}$$

α: observed rotation (in degrees)
l: length of polarimeter tube (in dm)
C: concentration (in g/100 mL)

Because α is, logically, temperature dependent, these values are tabulated for different compounds at a standard temperature. For example, at 20°C and using a 20-cm polarimeter tube, a 1 g/L solution of glucose and fructose will cause a rotation of +0.1047° and −0.1824°, respectively. Applying the above formula:

$$\text{Glucose } [\alpha]_D^{20} = \frac{0.1047 \times 100}{2 \times 0.1} = 52.4°$$

$$\text{Fructose } [\alpha]_D^{20} = \frac{-0.1824 \times 100}{2 \times 0.1} = -91.2°$$

Typically, the concentration of a solute is calculated by measuring the rotation angle [α] of this solute at 20°C, hence:

$$C \text{ (g/100mL)} = \frac{\alpha \times 100}{[\alpha]_D^{20} \times l}$$

In the case of both glucose and fructose, the rotatory power of the solute changes with time, until equilibrium between the α and β forms is achieved.

The rotatory power of a solution containing the α and β forms of fructose in mutarotation equilibrium is achieved at −91.2° and varies with temperature at a rate of 0.006 divisions per degree celsius.

The hydrolysis of sucrose gives rise to glucose and fructose; an equimolecular solution has a specific rotatory power of $[\alpha]_D^{20} = -91.2 + 52.4 = -38.8°$. As the rotatory power of sucrose before hydrolysis was +66.5°, the net effect of hydrolysis on the rotatory power of the solution is inversion as the sign changes from positive to negative.

3.2. Relationship Between Optical Rotatory Power and Sugar Composition

When measuring sugar content in wine by polarimetry, it is first necessary to remove all other substances that are capable of rotating the plane of polarized light, through the use of

TABLE 7.1   Specific Optical Rotatory Power ($[\alpha]_D^{20}$) of Sucrose, Glucose, and Fructose

Sucrose	Glucose			Fructose
	α	β	α ⇔ β	α ⇔ β
+66.5°	+113.4°	+19.7°	+52.4°	−91.2°

refining agents. The reagent most commonly used for this purpose is a saturated solution of lead acetate (500 g/L), although the Carretz I and Carretz II reagents, potassium ferricyanide (150 g/L) and zinc sulfate (300 g/L), respectively, are also used.

The rotatory power of dry wines is very small. In musts and wines with a high sugar content (such as natural sweet wines), the relative proportions of glucose and fructose can be calculated if the rotatory power of the two sugars and the total sugar content (P) are known. The sugar content of must can be measured using a refractometer or by chemical oxidation, while that of wine has to be measured by chemical oxidation.

The optical rotation, α, observed at the test temperature must be converted to the corresponding rotation at the standard temperature (20°C) using the following equation:

$$\alpha_{20} = \alpha_t + \alpha_t (t - 20)K$$

α_t: rotatory power in degrees at a given temperature
t: temperature in °C
K : correction factor

The correction factor K is obtained from Table 7.2 by calculating the P/α_t ratio (where t is the temperature at which P was measured in the refined wine); this factor is then used in the equation that relates α_t, to α_{20} to calculate α_{20}; once α_{20} is known, it is simple to determine the percentage of fructose from Table 7.2.

Sample problem: Let us take a presumably natural sweet wine with a total sugar content of 72 g/L. The rotatory power measured at 13°C is −6.3. Calculate the percentage of fructose, the G:F ratio, and establish whether the wine is truly a natural sweet wine.

$$P/\alpha_{13} = -11.43$$

TABLE 7.2 Relationship Between Total Sugar Content and Rotatory Power in Degrees (P/α)
Temperature Correction Factor K and Fructose as a Percentage of Total Sugar Content

P/α_{20}	K	% Fructose
−5.5	0.006	100
−6.5	0.007	90
−8.0	0.008	80
−10.4	0.009	70
−12.2	0.010	65
−14.8	0.012	60
−18.7	0.013	55
−25.6	0.016	50
−40.6	0.022	45
−97.0	0.042	40
	0.042	35

The table shows that the corresponding correction factor is K ≈ 0.010.

In the following calculation, we apply the correction for α with the temperature:

$$\alpha_{20} = -6.3 - 6.3(13 - 20) \times 0.01 = -5.86$$
$$P/\alpha_{20} = 72/-5.86 = -12.3 \rightarrow \% \text{ fructose} \approx 65\%.$$

The percentage of glucose is thus 35%, so as we will see in the next section, the wine tested may be a natural sweet wine. The corresponding G:F ratio is 0.54.

The equations below can also be used to calculate glucose and fructose content:

$$P = X + Y$$
$$\alpha_{observed} = \alpha_{glucose} X + \alpha_{fructose} Y$$

P: total sugar content
X: glucose content
Y: fructose content
$\alpha_{glucose}$: specific optical rotatory power of glucose
$\alpha_{fructose}$: specific optical rotatory power of fructose
$\alpha_{observed}$: observed rotatory power in refined wine solution

Sample problem: Let P = 72 g/L and $\alpha_{observed}$ = −6.3 at 13°C. Calculate the proportions of glucose and fructose and the G:F ratio.

The rotatory power of glucose and fructose are 0.1047 and −0.1824, respectively.

Note: If P is expressed in g/L, the rotatory powers must be measured at a concentration of 1 g/L. Applying the above equations, we get:

$$X = 72 - Y$$
$$\alpha_{observed} = -6.3 \text{ at } 13°C; \ \alpha_{20} = -5.86 \text{ (corrected to } 20°C)$$
$$-5.86 = 0.1047(72 - Y) - 0.1824Y \rightarrow Y = \frac{13.398}{0.2871} = 46.7$$

If the fructose content is 46.7 g/L, that of glucose will be 25.3 g/L (72−46.7 g/L) and the G:F ratio will be 0.54.

$$\% \text{ fructose} = 46.7/72 = 64.8\% \rightarrow \% \text{ glucose} = 35.2\%.$$

The results are virtually identical to those obtained using the table.

Using P/α and G:F, it is possible to determine whether a wine is naturally sweet or whether it has been adulterated by the addition of invert sugar syrup or concentrated must.

3.3. Polarimetry in Winemaking

Winemakers are particularly interested in determining glucose and fructose content. These two sugars, which are derived from grapes, are the main sugars found in wine. The G:F ratio of ripe grapes, which is close to 1 (0.95), is of use when calculating the proportion of glucose and fructose in wine.

Natural sweet wines are those wines in which part but not all of the sugars have been fermented by yeast. In such cases, fermentation is interrupted by the addition of ethanol or high doses of sulfites. These wines have a G:F ratio of less than 1, as glucose is more readily metabolized by yeast than fructose. Mistelle, for example, is produced by adding ethanol to non-fermented must. The G:F ratio of this sweet wine is around 1, which is the same as that of the starting must. Other sweet wines are obtained artificially by adding sucrose and ethanol. In such cases, the G:F ratio is 1.

3.4. The Sweetening Power of Sugars

The term *sugar* is automatically associated with sweetness, but not all sugars are sweet, and different sugars have different levels of sweetness. For a substance to provide flavor, it must first be dissolved; the ability of a substance to confer a sweet taste is not necessarily associated with its molecular structure, or with other chemical or physical properties, as substances that are very different to sugars are also sweet.

The sweetness of the sugars found in grapes can be expressed relative to that of sucrose, which has arbitrarily been chosen as the reference sugar.

Sucrose: 1
Glucose: 0.74
Fructose 1.73
Arabinose: 0.40

4. CHEMICAL PROPERTIES OF SUGARS

The chemical properties of glucose and fructose are determined by their aldehyde and ketone groups. Indeed, the main method used to quantify these sugars is based on these properties. Specifically, this method involves analyzing the reaction between cyclic hemiacetals or hemiketals and the Cu^{2+} ion contained in an alkaline solution, such as Fehling's solution, Benedict's reagent, or the Luff-Schoorl reagent.

As mentioned in Chapter 6, glucose and fructose form cyclic hemiacetals and hemiketals, respectively, and are reducing sugars, so they undergo oxidation reactions. The most noteworthy of these reactions are described below.

4.1. Oxidation in the Presence of the Luff-Schoorl Reagent

The reaction that takes place between reducing sugars and the Luff-Schoorl reagent can be expressed as follows:

$$C_6H_{12}O_{6(aq)} + 2\,Cu^{2+}{}_{(aq)} \xrightarrow{\Delta} CH_2OH\text{-}(CHOH)_4\text{-}COOH_{(aq)} + Cu^+ + Cu^{2+}{}_{exc}$$

This reaction is a simplified version of what actually happens, as, in addition to gluconic acid and copper oxide, the redox reaction also gives rise to uronic and aldaric acids.

After the redox reaction has taken place, the excess Cu^{2+} in the solution is reduced by adding a solution containing potassium iodide and sulfuric acid until Cu^+ and I_2

(iodine) form (reaction 1). I_2 is then measured using a standardized solution of thiosulfate (reaction 2).

$$\text{Reaction 1}: \text{formation of } I_2. \quad Cu^{2+}_{exc} + 2\,KI \;\rightarrow\; I_2 + Cu^+$$

$$\text{Reaction 2}: \text{measurement of } I_2. \quad I_2 + S_2O_3^{2+} \;\rightarrow\; 2\,I^- + S_4O_6^{2-}$$

In this sequence of reactions, the iodine formed by the excess Cu^{2+} (i.e., the fraction that did not react with the sugars) is quantified. Because this test measures a non-reactive fraction, a blank in which the sugar solution is replaced with distilled water must be run in parallel. In the control, none of the Cu^{2+} will react with the sugars, so the amount of iodine formed in reaction 1 will be greater than that of the sample and consequently more thiosulfate will be used in reaction 2. This type of testing is known as back titration.

Finally, because the oxidation reaction of sugars is not stoichiometric, it is necessary to consult an equivalence table to relate the amount of thiosulfate used to the milligrams of sugar present in the original sample.

4.2. Enzymatic Oxidation

Both *Botrytis cinerea*, the fungus responsible for gray mold and noble rot, and certain acetic acid bacteria (*Gluconobacter* strains) can oxidize glucose to gluconic acid. This acid is therefore used to measure the level of *Botrytis* infection or bacterial contamination. A level of over 0.5 g/L is considered to indicate insufficient control of rot or the presence of bacterial contamination. Levels of gluconic acid in rotten grapes range between 0.5 and 2.5 g/L, although, in particularly bad years, levels of close to 5 g/L have been reported. Gluconic acid that is not metabolized by yeasts during fermentation remains in the wine and is an important contributor to microbiological instability, particularly in wines that are left to be aged.

Gluconic acid can undergo intramolecular esterification, which can give rise to δ-gluconolactone or γ-gluconolactone, depending on whether it occurs between C1 and C5 or between C1 and C4, respectively.

4.3. Non-Enzymatic Oxidation Reaction

Sugars, either alone or in the presence of amino acids, can undergo non-enzymatic oxidation reactions, which are commonly referred to as Maillard reactions. Maillard reactions are generally favored by temperatures of over 50°C and pH levels of between 4 and 9. They are of particular interest in wines made from raisined grapes or concentrated musts.

The Maillard reaction influences the production of brown pigments, volatile compounds, and antioxidant compounds. The reaction starts with the nucleophilic addition of an amino group, which gives rise to an aldimine (Schiff base) followed by an aminoketone (Figure 7.4).

This reaction is just the first of a series of reactions involving dehydration, fragmentation, and polymerization that ultimately give rise to high-molecular-weight compounds. These polymers, which are brown, are known as melanoidins; they are highly heterogeneous and are poorly characterized due to their complex structure.

Several of the volatile compounds formed from sugars in an acid medium are of interest in winemaking. Examples are furfural, derived from pentose sugars, and 5-methylfurfural and

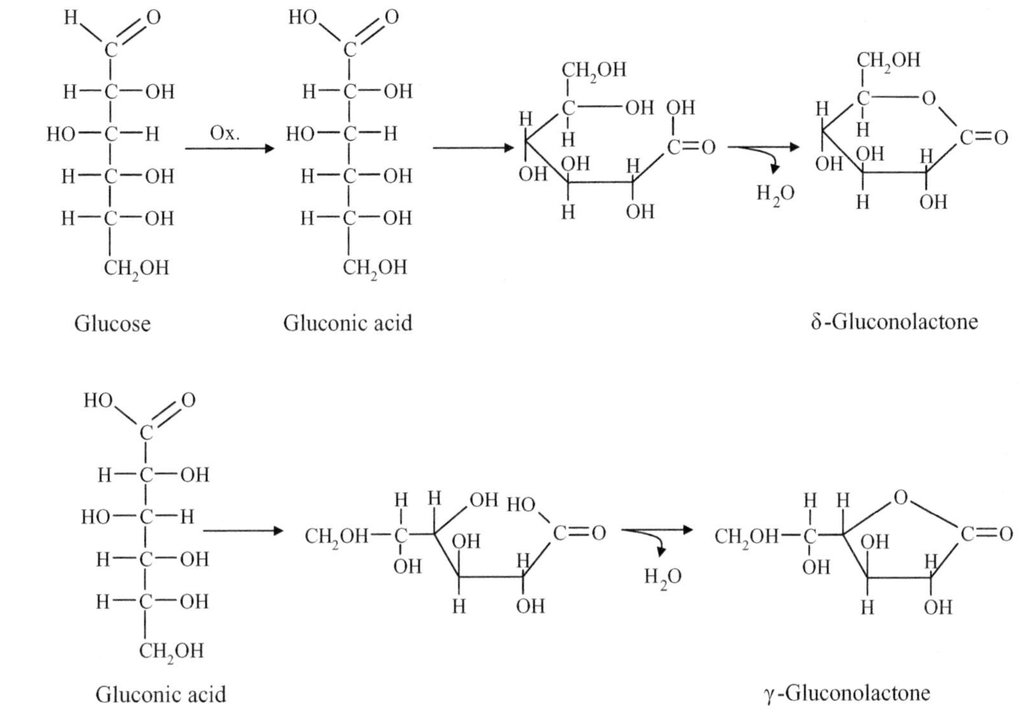

FIGURE 7.3   Formation of gluconic acid and corresponding lactones.

5-hydroxymethylfurfural, derived from hexoses. Furfural and 5-methylfurfural have been detected in musts made from raisined grapes. All three compounds have an almond-like smell but 5-hydroxymethylfurfural has a lower perception threshold.

Furfural and its derivatives are found in wine aged in barrels; they are derived from the wood as a result of Maillard reactions that carbohydrates undergo during the toasting of the barrels. The antioxidant activity that characterizes several of these compounds is possibly due to reductones, which are compounds formed from the Amadori product. A direct

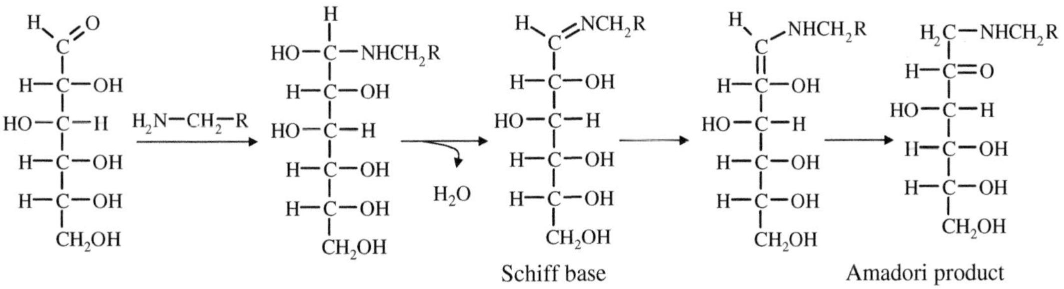

FIGURE 7.4   First steps in the Maillard reaction.

FIGURE 7.5 Simplified diagram of Maillard reaction cascade.

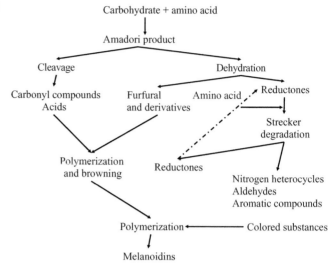

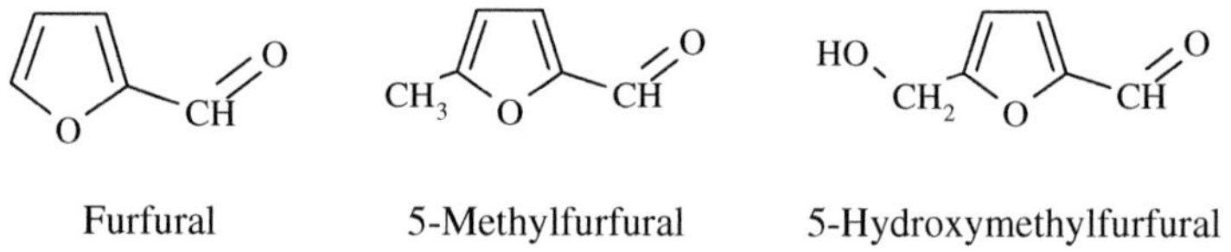

FIGURE 7.6 Furfural and its main derivatives.

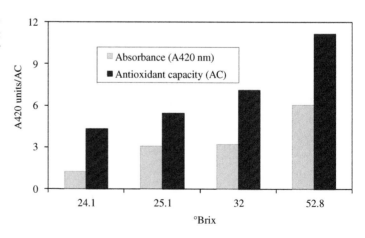

Furfural   5-Methylfurfural   5-Hydroxymethylfurfural

FIGURE 7.7 Keto-enol tautomerism between unsaturated carbonyl derivatives (reductones).

$$R`—C=C—C—R \rightleftharpoons R`—CH-C—C—R$$
$$\quad\;\; | \;\; | \;\; || \qquad\qquad\qquad | \;\; || \;\; ||$$
$$\quad\;\; OH \; OH \; O \qquad\qquad\quad OH \; O \;\; O$$

FIGURE 7.8 Browning (absorbance 420 nm) and antioxidant capacity of musts with varying sugar content made from raisined grapes.

Legend:
- Absorbance (A420 nm)
- Antioxidant capacity (AC)

Y-axis: A420 units/AC (0, 3, 6, 9, 12)
X-axis: °Brix (24.1, 25.1, 32, 52.8)

relationship between the level of raisining of the grapes used to make the must and its anti-oxidant activity has been established recently. Nonetheless, much remains to be done in exploring the antioxidant activity of these compounds.

4.4. Combination with Sulfite

When SO_2 is dissolved in water, the following acid-base equilibria are established:

(1) $SO_2 + H_2O \leftrightarrow HSO_3^- + H^+$; $K_{a1} = 1.7 \times 10^{-2}$; $pK_{a1} = 1.77$

(2) $HSO_3^- \leftrightarrow SO_3^{2-} + H^+$; $K_{a2} = 6.2 \times 10^{-8}$; $pK_{a2} = 7.02$

These equilibria are governed by the value of the corresponding ionization constant.

SO_3^{2-} does not occur in wine due to the pH of the medium, as the pK_a of reaction (2) is 7.02 and acid-neutralization reactions start in a solution with a $pH = pK_a - 2$. As the pH of wine is less than 5, this neutralization reaction does not take place. Molecular SO_2 and the bisulfite ion (HSO_3^-) form what is known as free SO_2. The molecular form of SO_2, known as active SO_2, has the greatest antiseptic properties, antioxidase properties (protection against oxidation enzymes), and antioxidant properties (protection against oxidation). The percentage of active SO_2 in relation to free SO_2 depends on pH and temperature, and is never greater than 10% under winemaking conditions.

The bisulfite ion can combine with molecules containing carbonyl groups, as shown in the following reaction:

This form of SO_2, known as combined SO_2, accounts for a large fraction of the total, but it is not active from a winemaking perspective. The proportion of active SO_2, therefore, decreases as a result of this combination.

It can be seen from the previous reaction that sugars are capable of combining with SO_2. One gram of glucose combines with 0.3 mg of SO_2 when the content of free SO_2 is ≈ 50 mg/L, whereas fructose and SO_2 hardly combine at all. Grapes affected by rot contain gluconic acid and its ketone derivatives (2-ketogluconic acid and 2,5-diketogluconic acid) and 5-keto-fructose (a product of the oxidation of fructose), and have a high fraction of combined SO_2. These grapes need to be treated with a considerably higher dose of SO_2 to prevent, as far as possible, undesirable microbial activity and oxidation.

5. NON-FERMENTABLE MONOSACCHARIDES AND DERIVATIVES

5.1. Non-Fermentable Monosaccharides

Small quantities of pentose sugars such as arabinose, xylose, and ribose are naturally present in must and wine. Arabinose and xylose are dextrorotatory and have a slightly sweet flavor. Because they are aldoses, they combine with SO_2; arabinose, for example, can combine with 10 times more SO_2 than glucose. Xylose, in contrast, seldom combines with SO_2. The concentrations of these sugars range from 0.3 g/L to 2 g/L (exceptional). Galactose, which is the most abundant hexose, reaches concentrations of 100 mg/L.

5.2. Non-Fermentable Derivatives

Polyalcohols

The most abundant polyalcohols in wine are 2,3-butanediol and glycerol. Neither of these alcohols is found in musts extracted from healthy grapes and both are produced by glycero-pyruvic fermentation. In the case of grapes with noble rot, concentrations of glycerol of up to 20 g/L have been reported, as *B. cinerea* can produce glycerol from the sugars in the grapes. Glycerol is also present in musts made from grapes with gray mold, but in lower quantities. Lactic acid bacteria can metabolize glycerol to acrolein, which is a bitter substance that can cause wine spoilage. Butanediol is found at concentrations of under 1 g/L and its concentrations increase with increasing levels of *Botrytis* infection.

Another polyalcohol found is *myo*-inositol, which is derived exclusively from grapes; its concentration in musts ranges from 350 to 700 mg/L.

Sorbitol is a common sweetener used in candy, chewing gum, and ice cream because it produces a sensation of freshness. It is present in numerous fruits (including apples, pears, plums, cherries, and peaches), but is absent in healthy grapes, leaves, and stems. Its presence in all types of wine is a cause of concern, since high levels indicate that apple juice or cider has been fraudulently added to the must or wine, respectively. Sweet wines have the highest levels of sorbitol, followed by red wines, and white wines, explaining the different legal limits imposed by the International Organization of Vine and Wine (OIV) for different types of wines: 600 mg/L for sweet wines, 250 mg/L for red wines, and 150 mg/L for white wines.

Sorbitol formed by yeast during fermentation does not tend to exceed levels of 50 mg/L. Higher levels are thus difficult to attribute to natural processes. The degree of *Botrytis* infection has a considerable influence on the polyalcohol content in wine, with levels being directly related to the proportion of infected grapes used to make the must. Sweet Sauternes wines, for example, which are produced from grapes with noble rot, have the highest concentrations of polyalcohols, which are largely responsible for the sticky, oily texture of these wines.

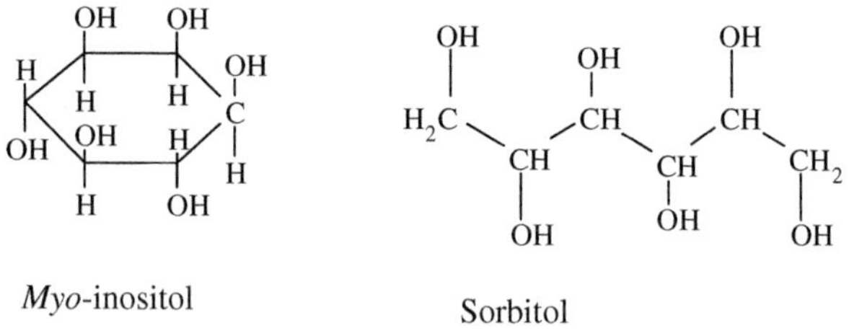

FIGURE 7.9   Structure of *myo*-inositol and sorbitol.

Myo-inositol

Sorbitol

5.3. Uronic Acids

Uronic acids are derived from the oxidation of the hydroxyl group on C6 of aldoses. The main uronic acids found in wines made from healthy grapes are D-galacturonic acid and D-glucuronic acid, whose concentrations range from 40 to 400 mg/L and from 0 to 60 mg/L, respectively. Galacturonic acid is derived from pectins through the action of pectinase enzymes, and glucuronic acid is primarily the result of the enzymatic oxidation of glucose. Wines made from grapes infected by *B. cinerea* have much higher concentrations (several grams per liter) of uronic acids.

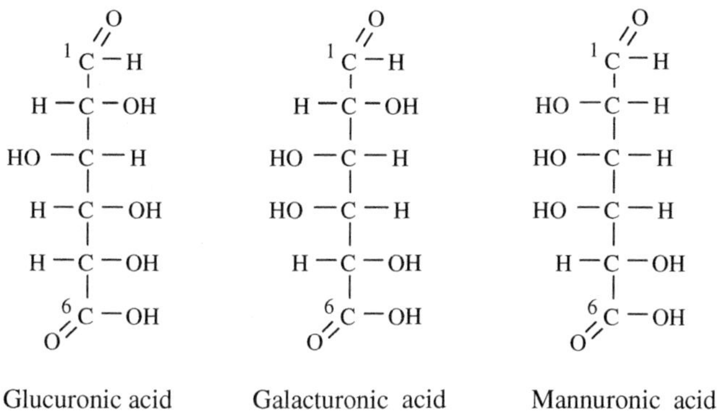

Glucuronic acid   Galacturonic acid   Mannuronic acid

FIGURE 7.10   Structure of several uronic acids.

8

Carboxylic Acids: Structure and Properties

1. INTRODUCTION

Acids have been known to be present in biological material since antiquity. Acetic acid and formic acid, for instance, were isolated in relatively pure forms very early in the history of science. The term *acetic* is derived from the Latin *acetum*, which means vinegar, since the acid was isolated by distillation of vinegar. *Formic*, in turn, is derived from the Latin *formica*, meaning ant, as the acid was obtained via the pyrolysis of red ants.

Both must and wine have a fundamentally acidic flavor. The pH is around 3.5 and they have a high buffering capacity. These properties are explained by the presence of acidic compounds. The acids present in wine are either derived from the grapes or formed during alcoholic and malolactic fermentation. They are weak acids and therefore partially dissociated.

2. THE CARBOXYL GROUP: BASIC CONCEPTS

Traditionally, acids in wine have been classified into two large groups according to their origin: those derived from the metabolism of the vine (of plant origin) and those derived from the metabolism of the microorganisms that control alcoholic fermentation. All common biological acids are organic compounds. The carboxyl group, which is characteristic of this family of compounds, is formed by a carbonyl group (C=O) and a hydroxyl group (−OH). These groups are located at one end of a carbon chain:

Carboxyl or carboxy group	Kekulé structure of carboxylic acids	Condensed formulae
$O=C\overset{OH}{\diagdown}$	$O=C\overset{OH}{\underset{R1}{}}$	$R1\text{-}CO_2H$ $R1\text{-}COOH$

The carboxyl group gives rise to the family of carboxylic acids and is usually represented in the forms shown above, where R1 may be a hydrogen atom, an alkyl radical, or an aryl radical (aromatic).

The carbon atom of the carboxyl group uses three hybrid sp² orbitals. These orbitals are all in the same plane, and the remaining p orbital of the carbon atom forms a Π bond with a p orbital of the oxygen in the carbonyl group, such that the group lies in a single plane and the hydrogen of the hydroxyl group lies outside that plane. It is important to understand the structure of the carboxyl functional group, since this forms the basis for a reasoned understanding of the most important property of these substances, namely acidity.

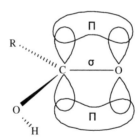

Nomenclature

According to IUPAC naming conventions, linear-chain acids are named by changing the "-e" suffix in the corresponding hydrocarbon to the suffix "-oic" and adding the word

TABLE 8.1 Types of Carboxylic Acid

R1	Acid	Formula
H	Formic or methanoic acid (1st in the series)	H-COOH
R-(alkyl radical)	Aliphatic carboxylic	R-COOH
Ar-(aryl radical)	Aromatic carboxyl	Ar-COOH

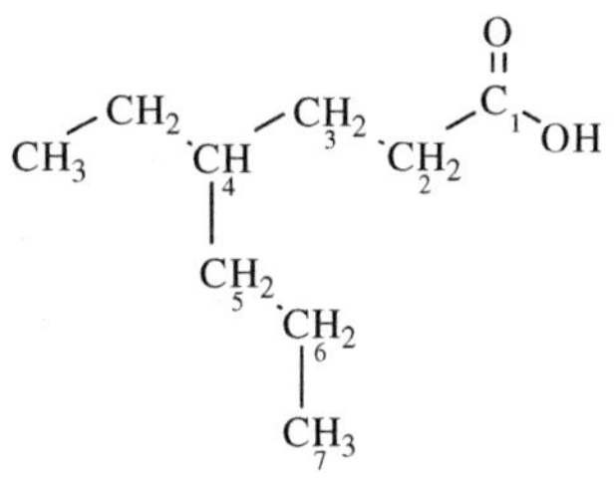

FIGURE 8.1   4-Ethyl heptanoic acid (not 4-propyl hexanoic acid).

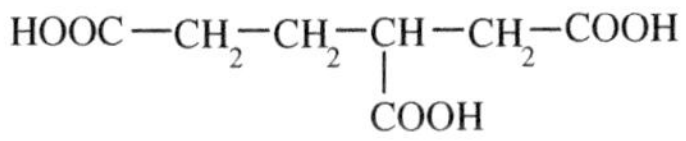

$$HOOC-CH_2-CH_2-CH-CH_2-COOH$$
$$\underset{COOH}{|}$$

FIGURE 8.2   1,2,4-Butanyl tricarboxylic acid.

acid. The carbon atom of the carboxyl group is assigned the number 1 and the subsequent carbons in the longest chain of which it forms a group are numbered accordingly. If common names are applied, Greek letters (α, β, γ, δ, etc.) are used to name the substituents on the main carbon chain.

If there is more than one carboxyl group, the additional groups are treated as substituents of the main carbon chain.

3. MONOCARBOXYLIC ACIDS

3.1. Physical Properties

Both the melting and boiling points (Table 8.2) increase with the number of carbons in the chain and are higher than those of hydrocarbons of a similar molecular weight. This is due to the formation of hydrogen bonds between molecules to generate dimers, trimers, etc. Their solubility in water is explained by the capacity of the carboxyl group to exchange H^+ with water molecules, and the first four carboxylic acids are water soluble in any proportion. However, as the carbon-chain length increases, its properties begin to dominate and the solubility of the compound in polar solvents is reduced.

3.2. Chemical Properties

The most important chemical property of the organic acids is undoubtedly their acidity; in solutions or in aqueous media, carboxylic acids donate a proton to the water and are converted into conjugate bases according to the following general scheme:

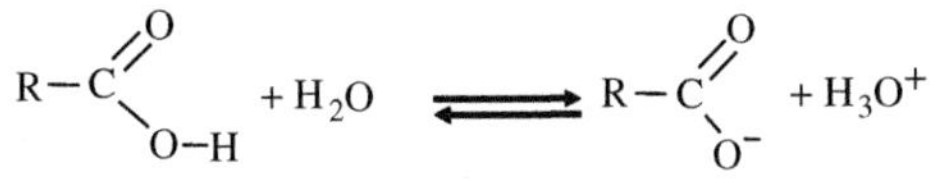

Weak acid   Conjugate base

TABLE 8.2 Structure and Properties of Monocarboxylic Acids

Formula	IUPAC Name	Common Name	Melting Point (°C)	Boiling Point (°C)	Solubility at 20°C (g/100 mL)	Ka (25°C)
$H\text{-}COOH$	Methanoic	Formic	8.4	101	∞	1.77×10^{-4}
$CH_3\text{-}COOH$	Ethanoic	Acetic	16.6	118	∞	1.76×10^{-5}
$CH_3\text{-}CH_2\text{-}COOH$	Propanoic	Propionic	−21	141	∞	1.34×10^{-5}
$CH_3\text{-}(CH_2)_2\text{-}COOH$	Butanoic	Butyric	−5	164	∞	1.54×10^{-5}
$CH_3\text{-}(CH_2)_3\text{-}COOH$	Pentanoic	Valeric	−34	186	4.97	1.52×10^{-5}
$CH_3\text{-}(CH_2)_4\text{-}COOH$	Hexanoic	Caproic	−3	205	0.968	1.31×10^{-5}
$CH_3\text{-}(CH_2)_5\text{-}COOH$	Heptanoic	Enanthic	−8	223	0.244	1.28×10^{-5}
$CH_3\text{-}(CH_2)_6\text{-}COOH$	Octanoic	Caprylic	17	239	0.068	1.28×10^{-5}
$CH_3\text{-}(CH_2)_7\text{-}COOH$	Nonanoic	Pelargonic	15	255	0.026	1.09×10^{-5}
$CH_3\text{-}(CH_2)_8\text{-}COOH$	Decanoic	Capric	32	270	0.015	1.43×10^{-5}
$CH_3\text{-}(CH_2)_{10}\text{-}COOH$	Dodecanoic	Lauric	44	299	0.0055	–
$CH_3\text{-}(CH_2)_{12}\text{-}COOH$	Tetradecanoic	Myristic	54	251 (100 mm)	0.0020	–
$CH_3\text{-}(CH_2)_{14}\text{-}COOH$	Hexadecanoic	Palmitic	63	267 (100 mm)	0.00072	–
$CH_3\text{-}(CH_2)_{16}\text{-}COOH$	Octadecanoic	Stearic	72	–	0.00029	–
(benzene ring)–COOH	Benzenecarboxylic	Benzoic	122	249	0.34 (25°C)	6.46×10^{-5}
(benzene ring with OH)–COOH	o-hydroxybenzoic acid	Salicylic	159	211	0.22 (25°C)	1.1×10^{-3} (19°C)

The values of the ionization constants of the organic acids are all around 10^{-5} ($pK_a \approx 5$); they are therefore weak acids and their conjugate bases, the carboxylate anions, are stronger bases than water. The higher the K_a, the stronger the acid and the lower the pK_a, since $pK_a = -logK_a$.

The acidity of these materials is due to the capacity of the carboxylate ion to distribute its excess negative charge between the two oxygens. In general terms, the more stable the ion produced after donation of a proton, the greater the acidic character of the substance in question. Something similar occurs with aromatic alcohols or phenols, which are more acidic than alcohols because the electron can be delocalized around the aromatic ring.

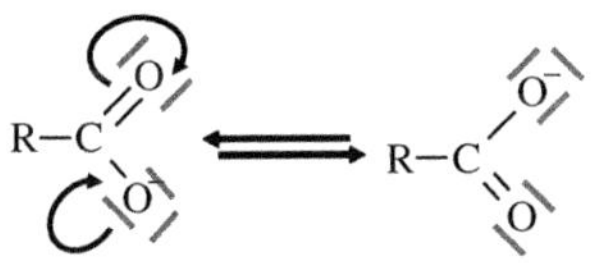

An important reaction of acids is the formation of esters with alcohols:

$$R\text{-}COOH + R'\text{-}CH_2OH \leftrightarrow R\text{-}COO\text{-}CH_2\text{-}R' + H_2O$$

Esterification has an equilibrium constant of approximately 4 for most ethyl esters.

3.3. Factors that Influence Acidity

Aliphatic Carboxylic Acids

The substituents located on the aliphatic chain linked to the carboxyl group have an effect on the stability of the carboxylate ion according to whether they are electron donors or acceptors:

a. Substituents that are electron acceptors attract excess electrons from the carboxylate ion and spread the negative charge more effectively. Consequently, they enhance the stability of the ion and increase acidity.

b. Substituents that are electron donors (alkyl groups) donate electrons to the carboxylate anion and increase the charge density on the ion. Consequently, the attraction towards the H^+ ion is increased and the acidity decreased.

In general terms, the capacity of electron-acceptor substituents to increase acidity is enhanced by the following factors:

1. Greater electronegativity of the substituent
2. Larger numbers of substituents
3. Closer proximity of the substituent to the carboxyl group

The following substituents are placed in order of their capacity to increase acidity:

$$-NO_2 > -CN > -F > -Cl > -Br > -I > -OCH_3 > -C_6H_5$$

TABLE 8.3 Influence of Different Substituents on the Acidity of Aliphatic Acids

Acid	Formula	K_a	pK_a
Acetic	CH_3-COOH	1.8×10^{-5}	4.74
Iodoacetic	$I-CH_2-COOH$	6.7×10^{-4}	3.18
Bromoacetic	$Br-CH_2-COOH$	1.3×10^{-3}	2.90
Chloroacetic	$Cl-CH_2-COOH$	1.5×10^{-3}	2.86
Dichloroacetic	$Cl_2-CH-COOH$	5.5×10^{-3}	1.26
Trichloroacetic	$Cl_3-C-COOH$	0.23	0.64
Fluoroacetic	$F-CH_2-COOH$	2.6×10^{-3}	2.59
Butyric	$CH_3-CH_2-CH_2-COOH$	1.5×10^{-5}	4.82
4-Chlorobutyric	$Cl-CH_2-CH_2-CH_2-COOH$	3×10^{-5}	4.52
3-Chlorobutyric	$CH_3-CHCl-CH_2-COOH$	8.9×10^{-5}	4.05
2-Chlorobutyric	$CH_3-CH_2-CHCl-COOH$	1.39×10^{-3}	2.86

Alkyl substituents are electron donors and reduce acidity to a greater extent in the following situations:

1. Greater chain length or larger number of side chains
2. Closer proximity to the carbon that carries the carboxyl functional group

The following substituents are listed in order of their capacity to reduce acidity:

$$-C(CH_3)_3 > -CH(CH_3)_2 > -CH_2CH_3 > -CH_3$$

Aromatic Carboxylic Acids

In general, substituents linked to an aromatic ring affect the acidity of aromatic acids to the same extent as they do aliphatic acids; in other words, electron-acceptor substituents increase acidity whereas electron donors reduce it. Nevertheless, another effect must be taken into consideration with aromatic acids, namely the resonance effect, which occasionally predominates over the electron donor or acceptor effect.

Either the electron donor/acceptor effect or the resonance effect will predominate, according to the position on the aromatic ring occupied by the substituents in relation to the carboxyl group, since benzene is a typical example of a resonating structure.

- Substituents in a meta position: electron effect predominates
- Substituents in a para position: resonance effect predominates
- Substituents in an ortho position: electron and resonance effects can be equally strong and steric interactions or hydrogen-bond formation resulting from proximity to the carboxyl group can modify the acidity

o-Hydroxybenzoic acid: $pK_a = 2.97$, steric effect
m-Hydroxybenzoic acid: $pK_a = 4.08$, electron-acceptor effect

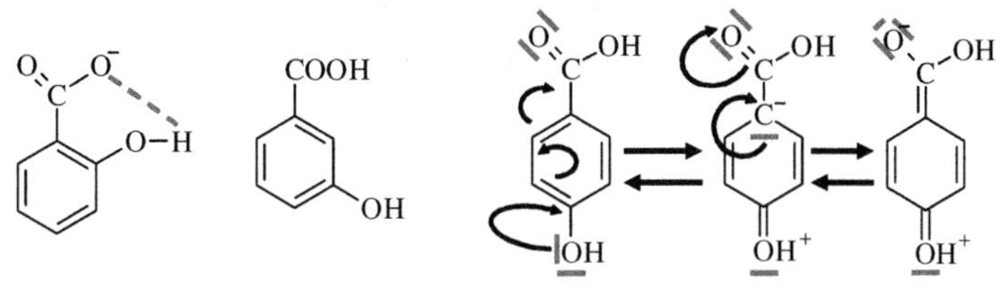

FIGURE 8.3   Steric, electron-acceptor, and resonance effects.

p-Hydroxybenzoic acid: $pK_a = 4.48$, resonance effect
Benzoic acid: $pK_a = 4.20$

The $-OH$ group is an electron-accepting group, and its inductive effect tends to increase acidity compared with benzoic acid: this can be seen in *m*-hydroxybenzoic acid (pK_a of 4.08 versus 4.20). In contrast, if the $-OH$ group is located in a para position, the resonance effect is greater than the inductive effect and is opposite in nature, meaning that the acidity is reduced. This is because the donation of electrons to the aromatic ring by the $-OH$ group leads to a charge redistribution and an excess of electrons in the carboxyl group, which hence has less capacity to donate a proton (Figure 8.3).

The high acidity of the ortho form of hydroxybenzoic acid can only be explained by the existence of a steric interaction, since in this position the inductive and resonance effects of the $-OH$ group should cancel each other out. Given the proximity of the carboxyl and hydroxyl groups in this arrangement, it appears in *o*-hydroxybenzoic acid that an internal hydrogen bond is formed that favors H^+ donation by the acid, and, as a consequence, the acidity increases due to the greater stability of the ion produced.

4. DICARBOXYLIC ACIDS

Aliphatic and aromatic acids containing two carboxyl ($-COOH$) groups are referred to as dicarboxylic acids. Most of them are known by their common names, since they are naturally highly abundant as salts and are easily separated from the other substances with which they are found.

4.1. Physical Properties

Dicarboxylic acids are solids at room temperature and they have melting points that are higher than those of monocarboxylic acids containing the same number of carbon atoms, since stronger associations between molecules exist, mainly as a result of hydrogen bond formation.

4.2. Chemical Properties

Dicarboxylic acids are dibasic or diprotic acids and therefore have two dissociation constants, K_{a1} and K_{a2}:

$$HOOC\text{-}(CH_2)_n\text{-}COOH \leftrightarrow HOOC\text{-}(CH_2)_n\text{-}COO^- + H^+; K_{a1}$$
$$HOOC\text{-}(CH_2)_n\text{-}COO^- \leftrightarrow {}^-OOC\text{-}(CH_2)_n\text{-}COO^- + H^+; K_{a2}$$

Each carboxyl group can ionize independently but, as shown in Table 8.4, the first carboxyl group is generally much more acidic than the second (lower pK_a, dissociates more easily), especially when the two are in close proximity. This is explained by the inductive electron-acceptor effect of the second carboxyl group that increases the stability of the ionized group and therefore also the acidity.

The dipolar structure of the C=O double bond in the carboxylate ion stabilizes the negative charge that is produced by ionization of the carboxyl group; however, K_{a2} is lower than

TABLE 8.4   Structure and Properties of Dicarboxylic Acids

Structure	Common Name	IUPAC Name	Melting Point (°C)	Ka_1	Ka_2
HO-CO-OH	Carbonic			3.02×10^{-7}	6.31×10^{-11}
HOOC-COOH	Oxalic acid	Ethanedioic	189	3.5×10^{-2}	4.0×10^{-5}
HOOC-CH$_2$-COOH	Malonic acid	Propanedioic	136	1.4×10^{-3}	2.2×10^{-6}
HOOC-(CH$_2$)$_2$-COOH	Succinic acid	Butanedioic	185	6.4×10^{-5}	2.5×10^{-6}
HOOC-(CH$_2$)$_3$-COOH	Glutaric acid	Pentanedioic	98	4.5×10^{-5}	3.8×10^{-6}
HOOC-(CH$_2$)$_4$-COOH	Adipic acid	Hexanedioic	151	3.7×10^{-5}	2.4×10^{-6}
	Maleic	cis-2-butenedioic acid	130	1.2×10^{-2}	3×10^{-7}
	Fumaric	trans-2-butenedioic acid	302	9.3×10^{-4}	2.9×10^{-5}
	Phthalic acid	o-benzenedicarboxylic acid	231	1.2×10^{-3}	3×10^{-6}
	Isophthalic acid	m-benzenedicarboxylic acid	348	2.9×10^{-4}	2.7×10^{-5}
	Terephthalic acid	p-benzenedicarboxylic acid	300 (sublimation)	1.5×10^{-4}	—

K_1 because the presence of a carboxylate ion reduces the acidity of the second carboxyl group due to the electrostatic repulsion between the two negative charges on the dicarboxylate ion. This effect will be lessened as the length of the chain separating the carboxyl groups increases. Accordingly, the difference between K_{a1} and K_{a2} will reduce with increasing chain length.

Among other chemical properties of biological interest, all dicarboxylic acids are stable in the presence of oxidizing agents, with the exception of oxalic acid, which can be oxidized to CO_2, and hence functions as a reducing agent.

The effect of heat on these acids depends upon the position of the carboxyl groups in the chain. Thus, acids in which the second carboxyl group is in an α or β position are decarboxylated by heat:

$$HOOC\text{-}COOH \rightarrow CO_2 + HCOOH \text{ (with heating to } 140°C)$$

$$HOOC\text{-}CH_2\text{-}COOH \rightarrow CO_2 + CH_3\text{-}COOH \text{ (with heating to } 150°C)$$

Acids with a second carboxyl group at positions γ or δ generate cyclic anhydrides:

Succinic acid   Succinic anhydride

5. HYDROXY ACIDS

Like dicarboxylic acids, hydroxy acids are naturally abundant and are therefore more generally known by their common names. Examples are lactic acid from sour milk, citric acid from citrus fruits such as oranges and lemons, malic acid from apples, and tartaric acid from grapes.

These acids contain both a hydroxyl and a carboxyl group, forming α-, β-, γ- and δ-hydroxy acids according to the position of the $-OH$ in relation to the carboxyl group. For instance, 3-hydroxybutanoic or β-hydroxybutyric acid:

$$C_\gamma H_3\text{-}C_\beta HOH\text{-}C_\alpha H_2\text{-}C_1 OOH$$

5.1. Physical Properties

With the exception of glycolic and lactic acid, which are liquids, all hydroxy acids are solids. They are more water soluble than comparable alcohols and monocarboxylic acids, and they have higher melting points than monocarboxylic acids. This is because the presence of both hydroxyl and carboxyl groups in the same molecule increases the opportunities to form hydrogen bonds with neighboring molecules.

5.2. Chemical Properties

Acidity is greater than in the corresponding monocarboxylic acids due to the presence of an electron-accepting −OH group within the same molecule.

Hydroxy acids participate in reactions associated with both acids and alcohols, but they also participate in reactions that are specific to hydroxy acids, which depend upon the position of the functional groups within the molecule. These reactions include the formation of lactones and lactides, which progress in ways which depend on the relative position of the hydroxyl and carboxyl groups:

- α-Hydroxy acids do not form stable lactone rings and undergo intermolecular esterification reactions with the same molecules to form dilactones, known generically as lactides.

Lactic acid   Lactic acid lactide

- β-Hydroxy acids cannot undergo intramolecular esterification and therefore under acid conditions they are easily dehydrated to generate α,β-unsaturated acids.

β-Hydroxy acid   α,β-Unsaturated acid

TABLE 8.5 Structure and Acidity of the Most Common Hydroxy Acids

Structure	Common Name	IUPAC Name	pk_a	K_a
CH_2OH-$COOH$	Glycolic	α-hydroxyacetic	3.83	1.48×10^{-4}
CH_3-$CHOH$-$COOH$	Lactic	α-hydroxypropionic	3.86	1.38×10^{-4}
$COOH$-$CHOH$-CH_2-$COOH$	Malic	α-hydroxysuccinic (2-hydroxybutanedioic)	3.46 5.13.	3.47×10^{-4} 7.41×10^{-6}
$COOH$-$CHOH$-$CHOH$-$COOH$	Tartaric	α,β-dihydroxysuccinic (2,3-dihydroxybutanedioic)	3.04 4.37	9.2×10^{-4} 4.26×10^{-5}
$COOH$-CH_2-COH-CH_2-$COOH$ $\|$ $COOH$	Citric	β-carboxy-β-hydroxyglutaric (3-carboxy-3-hydroxypentane-1,5-dioic)	3.08 4.74 5.4	8.4×10^{-4} 1.8×10^{-5} 4×10^{-6}

- γ and δ-hydroxy acids undergo intramolecular esterification (lactonization) very easily, simply through heating in the presence of traces of a mineral acid. The result is an intramolecular ester or lactone:

γ-Hydroxybutyric acid   γ-Butyryl lactone (cyclic ester)

Numerous lactones produced by intramolecular esterification of organic acids have been identified and quantified in wine. The most abundant is γ-butyryl lactone, which is found at concentrations of up to 40 mg/L. These compounds tend to have pleasant fruity aromas:

- Butyryl lactone: caramel, coconut
- δ-octalactone: herbaceous
- γ-nonalactone: coconut
- γ-decalactone: pear
- δ-decalactone: pear
- pantolactone: balsamic aromas, toasted notes, wood smoke
- (Z)-whiskey-lactone: coconut

6. KETO ACIDS

Finally, another group of compounds should be mentioned in this brief review of carboxylic acids. The keto acids contain a carbonyl group in addition to the carboxyl group. Depending on the position of the $-C{=}O$ group in relation to the terminal carboxyl group, α or β-keto acids are formed.

From a biological perspective, only the α-keto acids are of interest, and among these the most relevant is pyruvic or α-ketopropionic acid, which is a key intermediate in numerous biochemical and chemical transformations in both plants and animals. In addition, it is worth mentioning α-ketoglutaric acid, which is an amino acid precursor.

TABLE 8.6   Most Common Keto Acids

Structure	Common Name	IUPAC Name	MW
HOOC-CO-CH$_3$	Pyruvic	2-oxopropanoic	88
HOOC-(CH$_2$)$_2$-CO-COOH	Ketoglutaric	2-oxopentanedioic	146

When α-keto acids are heated, they undergo decarbonylation or loss of carbon monoxide derived from the carboxyl group.

$$R\text{-}CO\text{-}COOH \rightarrow R\text{-}CO\text{-}OH + CO$$

The β-keto acids are easily decarboxylated to form ketones.

$$R\text{-}CO\text{-}CH_2\text{-}COOH \rightarrow R\text{-}CO\text{-}CH_3 + CO_2$$

Grape Acids

1. INTRODUCTION

Grape juice and the must obtained from pressing are acidic, with a pH that generally ranges between 3.2 and 4.0. At harvest time, over 90% of grape acidity is due to malic acid and tartaric acid.

Malic acid is very common in the plant kingdom and is found, for example, in apples, plums, and peaches. The levels of this acid in ripe grapes depend largely on the variety of grape and the climate of the growing region. In cold regions, for example, they can exceed 8 g/L, while in warm regions they lie between 1 and 2 g/L.

While tartaric acid occurs in a range of plants, it is typically associated with grapes, which are the only fruit that contain it in large quantities. Its concentration in musts and wines ranges between 3 and 9 g/L, depending on grape variety and environmental conditions, particularly the availability of water. Because tartaric acid is stronger than malic acid and is found at higher concentrations, it has greater influence on the pH of the resulting musts and wines.

Citric acid also occurs naturally in grapes, albeit in smaller quantities. In normal, healthy grapes, levels do not exceed 300 mg/L but in grapes infected with *Botrytis cinerea*, they can reach 1 g/L.

Other acids are also commonly found in must, but at much lower levels. Examples are α-ketoglutaric acid, fumaric acid, oxalacetic acid, pyruvic acid, and gluconic acid. Because these acids occur at lower concentrations than tartaric acid and are also weaker, they do not make a significant contribution to acidity, but they do participate in several biological processes, such as the Krebs cycle. Ascorbic acid (commonly known as vitamin C) is of particular interest. It has antioxidant properties but its levels (approximately 50 mg/L) are insufficient to prevent oxidation by polyphenol oxidases such as tyrosinase and laccase.

2. TARTARIC ACID

Tartaric acid is the most abundant acid found in grapes and wine. It can be produced by treating the pomace of the wine with strong mineral acids. It is a typical example of an organic compound that contains several asymmetric carbon atoms, with four different substituents. These carbons are called chiral centers and they determine the number of possible enantiomers, or optical isomers. Thus, if there are n chiral centers, the number of enantiomers will be 2^n. In the case of tartaric acid, for example, there are four optical isomers ($2^2 = 4$). Each enantiomer corresponds to a specific configuration of the substituents around each chiral center.

2.1. Chemical and Physical Properties of Tartaric Acid

The physical properties of the four tartaric acids are summarized in Table 9.1. The (+) and (−) enantiomers are identical except for the fact that they have positive and negative specific rotation, respectively. A racemic mixture (or racemate) is an equimolecular mixture of (+) and (−) enantiomers and is denoted by the symbol (±). The racemate of tartaric acid is optically inactive due to external compensation. By contrast, mesotartaric acid, which is a diastereomer of (+) and (−) tartaric acids is optically inactive due to internal compensation. Diastereomers are optical isomers that are not enantiomers of each other.

(+)-Tartaric acid　　　(−)-Tartaric acid　　　　　Mesotartaric acid

TABLE 9.1 Physical and Chemical Properties of Tartaric Acid Isomers

	(+)-Tartaric Acid	(−)-Tartaric Acid	(±) Tartaric Acid (racemate)	Mesotartaric Acid
Melting point (°C)	170	170	206	140
α_D (20°C, water)	+12.0	−12.0	0.0	0.0
Solubility (g/100 mL water, 20°C)	139	139	20.6	125
Density (g/cm³, 20°C)	1.760	1.760	1.679	1.666
pK_{a1}	2.93	2.93	2.96	3.11
pK_{a2}	4.23	4.23	4.24	4.80

Although mesotartaric acid has two chiral centers, it is optically inactive because the overall molecule is not superimposable on its mirror image. On analyzing the mirror image of mesotartaric acid, it can be seen that it can be made to overlay the original image by simply rotating it 180° in the same plane. Checking for the presence of a plane of symmetry in the molecule is another rapid method for determining this.

The configuration of tartaric acid in grapes is L-(+), although there are also small proportions of the D form present.

Tartaric acid has a gram formula weight (molecular weight) of 150.09 g/mol and is a dihydroxy dicarboxylic acid. It therefore has two dissociation constants determined by two K_a values:

$$H_2T \leftrightarrow H^+ + HT^-; \quad K_{a1} = \frac{(H^+)(HT^-)}{(H_2T)} = 1.16 \times 10^{-3}; \quad pk_{a1} = 2.93$$

$$HT^- \leftrightarrow H^+ + T^{2-}; \quad K_{a2} = \frac{(H^+)(T^{2-})}{(HT^-)} = 5.83 \times 10^{-5}; \quad pk_{a2} = 4.23$$

Thanks to its two hydroxyl groups, tartaric acid has very important physical and chemical properties that affect the acidity of must (and wine) and can contribute to wine haze via the formation of precipitated salts. Of particular note are calcium tartrate tetrahydrate and potassium bitartrate (potassium tartrate), which is the main component of cream of tartar.

TABLE 9.2 Properties of Two Tartaric Acid Salts

Salt	Molecular Weight	Solubility (g/L) 20°C, in Water	Solubility (g/L) 20°C, in 12% (vol/vol) Ethanol
$(C_4O_6H_5)^-K^+$	188	4.91	2.76
$(C_4O_6H_4)^{2-}Ca^{2+} \cdot 4 H_2O$	220	0.206	0.103

3. MALIC ACID

Malic acid can be synthesized in three ways:

1. Fixation of CO_2 by pyruvic acid in a reaction catalyzed by malic enzyme (NAD = nicotinamide adenine dinucleotide).

Pyruvic acid   Malic enzyme   Malic acid

2. Fixation of CO_2 by ribulose 1,5-diphosphate. A series of reactions yields phosphoenolpyruvate, and the enzyme phosphoenolpyruvate carboxylase, in turn, catalyzes a reaction that generates oxalacetic acid. Finally, the action of malate dehydrogenase gives rise to malic acid.

Phosphoenolpyruvate   Oxalacetic acid   Malic acid

3. Transformation of citric acid. Citric acid is synthesized in the roots and transported to the upper parts of the vine (e.g., leaves and grapes), where it is converted to malic acid within the Krebs cycle.

3.1. Chemical and Physical Properties of Malic Acid

Malic acid is present in grapes in the L(−) form. This acid has two carboxyl groups and just one hydroxyl group, explaining why it is much less acidic than tartaric acid. Accordingly, the first dissociation constant (K_{a1}) for malic acid is three times lower than that of tartaric acid, while the second dissociation constant is six times lower.

The salts formed by malic acid with the cations typically present in musts and wines are readily soluble. Nonetheless, unlike tartaric acid, malic acid is easily degraded by microorganisms and also easily metabolized within the Krebs cycle as part of respiration.

4. CITRIC ACID

Citric acid, which is present at low concentrations in grapes and leaves, is mainly synthesized by the roots, from where it is transported to the upper parts of the plant. It is formed in

TABLE 9.3   Physical and Chemical Properties of
                    L-Malic Acid

	L-Malic Acid
Molecular weight	134.09
Melting point (°C)	100
α_D (20°C, water)	−2.3°
Solubility (g/100mL water, 20°C)	55.8
Density (g/cm^3, 20°C)	1.609
pK$_{a1}$ and K$_{a1}$	3.46; 3.47 × 10^{-4}
pK$_{a2}$ and K$_{a2}$	5.13; 7.41 × 10^{-6}

the Krebs cycle. While it can be derived from any compound that participates in this cycle, its main precursors are pyruvic acid and its derivatives (phosphoenolpyruvate, acetyl coenzyme A, etc).

4.1. Chemical and Physical Properties of Citric Acid

Citric acid has three carboxyl and one hydroxyl group. It thus has three possible acid dissociation constants and, consequently, three K$_a$ and pK$_a$ values.

The first dissociation constant of citric acid is similar to that of tartaric acid; the second constant is half that of tartaric acid but three times greater than that of malic acid. Considering that the pH of must and wine generally lies between 2.8 and 3.8, it can be assumed that the third dissociation does not occur. Therefore, for all practical purposes, citric acid acts as a diprotic acid, with an acidity between that of tartaric acid and malic acid (medium acidity).

TABLE 9.4   Physical and Chemical Properties of
                    Citric Acid

	Citric Acid
Molecular weight	192.13
Melting point (°C)	156
Density (g/cm^3, 20°C)	1.665
Solubility (g/100 mL water, 20°C)	56.7
pK$_{a1}$ and K$_{a1}$	3.08; 8.4 × 10^{-4}
pK$_{a2}$ and K$_{a2}$	4.74; 1.8 × 10^{-5}
pK$_{a3}$ and K$_{a3}$	5.4; 4 × 10^{-6}

Citric acid is the third most abundant acid in must made from healthy grapes, exceeded only by tartaric acid and malic acid, in that order. In the past it was used as an acidifying agent, and although it is still used in certain food products, its use is severely limited by the fact that it is easily degraded by many microorganisms, including lactic acid bacteria.

5. CHANGES IN ACID CONTENT AND ACIDITY DURING GRAPE RIPENING

It is essential to monitor changes in acid content and acidity during the ripening of grapes as this provides key data that will guide decisions on the time and order of grape harvesting.

Monitoring activities, however, should not be limited to measuring acidity and pH close to harvesting. As we have already seen, the acidity of grapes is primarily determined by their tartaric and malic acid content, and each of these acids behaves differently throughout the ripening process. This behavior determines not only the acidity of the grapes at the time of harvest (and therefore the acidity of the resulting musts and wines), but also the final composition of the wine and the organoleptic properties of the final product.

5.1. Ways of Expressing the Acid Content of Musts

The acid content of grapes and must can be expressed in different ways, each of which is useful at different stages of the process. It is thus simply a question of choosing the form that provides the information that is required at any given time. The main forms of expression are:

- Grams of acid per 1000 grape berries or meq of acid per 1000 grape berries
- Grams of acid per liter of juice or meq of acid per liter of juice

The choice of expression (per liter or per 1000 grape berries) depends on the purpose of the measurement. As grapes ripen, their volume increases, and may even decrease slightly towards the end of ripening. Therefore, the number of berries required to produce a liter of must varies with the degree of ripeness. The levels of the different compounds present in grapes also vary due to translocation, combustion, and even dilution by water transported to the berry. The calculation of acid content per 1000 berries primarily serves to determine what is occurring in the berries, regardless of their volume. For winemakers, it is more relevant to express this content per liter of grape juice, as this provides information about the raw material that will ultimately be converted into wine. Furthermore, the winemaking process principally deals with volumes of liquid.

The reduction in free acid content is mainly due to the effects of dilution and respiratory combustion. Another possible cause that has been proposed is the increased formation of acid salts, but this is not very likely as the alkalinity of ash (which is related to the quantity of acid salts) does not increase significantly in such cases, and in others, it even decreases.

The tartaric to malic acid ratio at veraison (the onset of ripening) remains relatively constant from year to year but it varies considerably during ripening. Thus, depending on the climate in a particular year, the vine preferentially synthesizes malic acid or tartaric

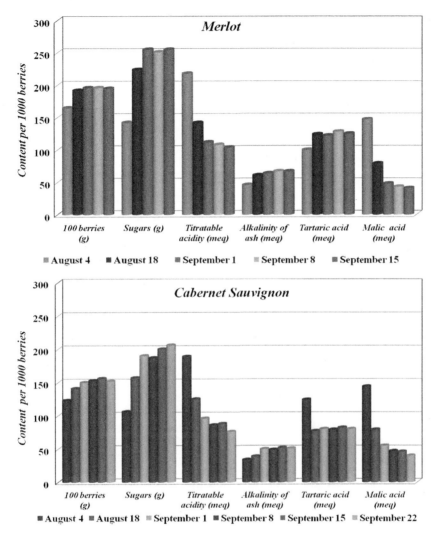

FIGURE 9.1 Evolution of 100 berries-weight, sugars, titratable acidity, ash-alkalinity and tartaric and malic acids contents in Merlot and Cabernet-Sauvignon grapes during the ripening period. Contents are expressed by 1000 grape berries.

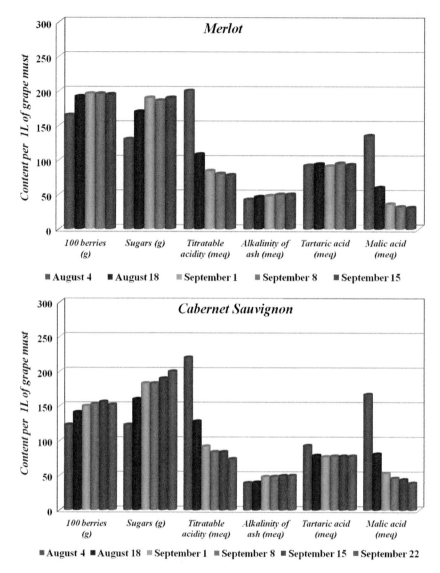

FIGURE 9.2   Evolution of 100 berries-weight, sugars, titratable acidity, ash-alkalinity and tartaric and malic acids contents in Merlot and Cabernet-Sauvignon grapes during the ripening period. Contents are expressed by 1 liter of grape-must.

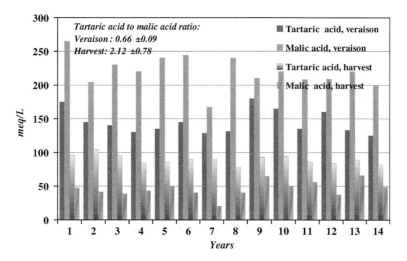

FIGURE 9.3   Tartaric and malic acids in Cabernet-Sauvignon grapes in the veraison and harvesting dates during 14 years.

acid. Another possible explanation for why there are *malic acid years* and *tartaric acid years* is that the tartaric acid content varies less than that of malic acid, which is more dependent on climate. What is clear, however, is that variations in the tartaric to malic acid ratio are largely due to the metabolism of malic acid within the grape.

As can be seen in Figure 9.3, these variations are greater and very similar for both acids at veraison, whereas at harvesting, malic acid levels vary more than those of tartaric acid from year to year.

5.2. Changes in Malic Acid Content During Grape Ripening

Malic acid levels reach a peak at veraison and decrease as the grapes ripen. This decrease is greatly influenced by temperature, because as the weather becomes warmer, grape cells use stored malic acid increasingly to satisfy their growing energy requirements. Hence, grapes grown in cold regions have higher levels of malic acid than those grown in warm regions. The same is true of grapes that ripen in the shade of the vine's leaves compared to those exposed to direct sunlight.

5.3. Changes in Tartaric Acid Content During Grape Ripening

Tartaric acid levels can reach 20 g/L of grape juice at veraison. Levels remain relatively constant from the end of veraison through to harvesting time, although they may vary somewhat during ripening. Tartaric acid levels in ripe grapes tend to lie between 3.5 and 11 g/L and they vary according to the variety of grape and above all the availability of water for the vine.

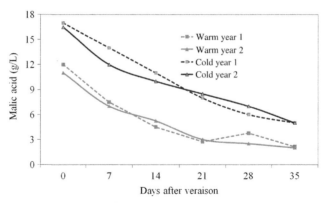

FIGURE 9.4   Changes in malic acid levels in two cold years and two warm years.

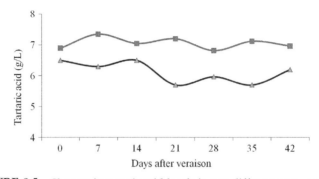

FIGURE 9.5   Changes in tartaric acid levels in two different grape varieties.

6. OTHER GRAPE ACIDS

Grapes also contain small quantities of acids other than those described so far. Their contribution to acidity is comparatively small but they are important as they are involved in numerous biological processes. Grapes also contain phenolic acids, which are discussed in Chapter 5.

6.1. Gluconic Acid: A Special Case

Grapes infected by the fungus *B. cinerea* (which causes noble rot or gray mold) contain elevated concentrations of certain acids such as gluconic acid. This acid is formed by the oxidation of sugars and can exceed levels of 5 g/L in these cases. Its concentration is used as an indicator of the degree of *Botrytis* infection and only musts with levels of under 1 g/L are used to make quality wines. Those with levels of over 1.5 g/L are generally used for distilled beverages.

The oxidation of glucose to gluconic acid occurs via the metabolism of *B. cinerea* or the action of bacteria of the genera *Acetobacter* and, more commonly, *Gluconacetobacter*. This oxidation is carried out by two enzymes: glucose oxidase and glucose-6-phosphate dehydrogenase.

TABLE 9.5 Other Acids in Healthy Grapes

Structure	Common Name	IUPAC Name	MW	pK$_{a1}$	pK$_{a2}$
HOOC-COOH	Oxalic acid	Ethanedioic acid	90	1.27	4.28
HOOC-CH$_2$-OH	Glycolic acid	2-Hydroxyethanoic acid	76	3.83	
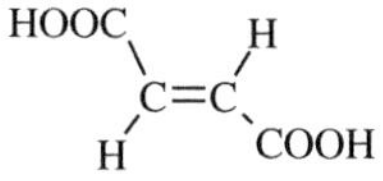	Fumaric acid	(E)-Butenedioic acid	116	3.03	4.44
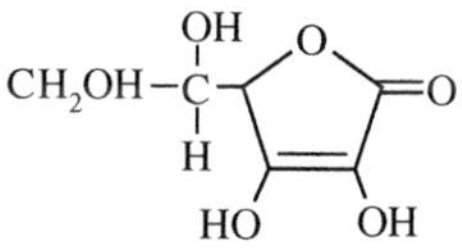	L-(+)-Ascorbic acid	(5R)-5-[(1S)-1,2-dihydroxyethyl]-3,4-dihydroxyfuran-2(5H)-one	176	4.10	

The mechanism of attack used by *B. cinerea* in grapes is highly specific and involves physical and chemical penetration phases. The fungus causes an increase in sugar levels but it also affects acid, terpene, and polyphenol levels. *B. cinerea* has both a desirable and undesirable form. The former is known as noble rot and the latter as gray mold, and whether one or the other develops depends on climate. In Spain, all *Botrytis* infections are gray mold and the external damage caused provides a route of entry for other harmful microorganisms.

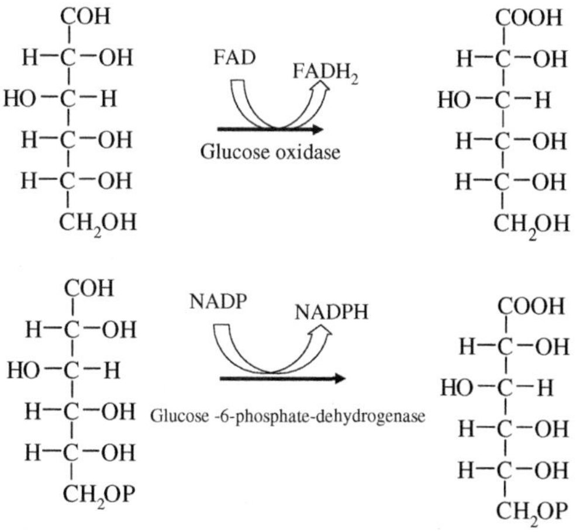

FIGURE 9.6 Formation of gluconic acid from glucose. FAD = flavin adenine dinucleotide; NADP = nicotinamide adenine dinucleotide phosphate.

TABLE 9.6 Other Acids in *Botrytis*-Infected Grapes

Structure	Common Name	IUPAC Name	MW
HOOC-(CHOH)$_4$-CHO	Glucuronic acid	(2S,3S,4S,5R)-2,3,4,5-tetrahydroxy-6-oxohexanoic acid	194.1
HOOC-(CHOH)$_4$-CHO	Galacturonic acid	2,3,4,5-tetrahydroxy-6-oxohexanoic acid	194.1
HOOC-(CHOH)$_4$-CH$_2$OH	Gluconic acid	2,3,4,5,6-pentahydroxyhexanoic acid	196
HOOC-(CHOH)$_4$-COOH	Galactaric acid	2,3,4,5-tetrahydroxyhexanedioic acid	210
HOOC-CO-(CHOH)$_3$-CH$_2$OH	2-ketogluconic acid	(3S,4R,5R)-3,4,5,6-tetrahydroxy-2-oxohexanoic acid	194
HOOC-(CHOH)$_3$-CO-CH$_2$OH	5-ketogluconic acid	(2R,3S,4S)-2,3,4,6-tetrahydroxy-5-oxohexanoic acid	194
HOOC-CO-(CHOH)$_2$-CO-CH$_2$OH	2,5-diketogluconic acid	(3S,4S)-3,4,6-trihydroxy-2,5-dioxohexanoic acid	192

The characteristic metabolism of *Acetobacter* bacteria is the oxidation of ethanol to acetic acid. *Gluconacetobacter* bacteria, in contrast, tend to oxidize sugars. In brief, *Acetobacter* are more common in alcoholic media while *Gluconacetobacter* are more common in sugar-rich media (musts and grapes).

Bacteria of the genus *Gluconobacter* oxidize gluconic acid, giving rise to 2-ketogluconic acid, 5-ketogluconic acid, and 2,5-diketogluconic acid. These molecules bind quite readily to SO$_2$, thus reducing free SO$_2$ levels. Although ketoses are less easily oxidized by bacteria, the oxidation of fructose gives rise to gluconic acid and 5-ketofructose. The carbon backbone of these sugars can also be cleaved, yielding succinic acid, glyceric acid, and glycolic acid.

Yeasts commonly found in grapes and musts do not metabolize gluconic acid during fermentation, explaining why this acid is found in wine. Gluconic acid, however, can be metabolized by other microorganisms such as lactic acid bacteria, resulting in excessive levels of acetic acid and considerable numbers of bacteria which can attack other substrates in the wine and produce undesirable changes.

7. ANALYSIS OF ACIDS

7.1. Measurement of Total Acid Content

Several variables can help to determine the total acid content of must or wine.

pH levels, for example, provide a measure of the concentration of protons released into the liquid medium by acids. As we have already mentioned, tartaric acid is the strongest and most abundant acid in musts and wines. Accordingly, the pH is largely determined by the concentration of this acid and its pK_a value. pH is determined in the laboratory using a potentiometer and a combined glass electrode.

Titratable acidity is a measure of the total acid content of must and wine. Specifically, it measures the level of free, non-neutralized acids.

It is measured directly in must or wine following the removal of CO_2 by shaking and the application of a partial vacuum. The resulting solution is titrated against standardized 0.1 N NaOH to a pH endpoint of either 7 (according to the method of the International Organization of Vine and Wine (OIV) or 8.2 (according to the method advocated by the American Society for Enology and Viticulture (ASEV)).

Titratable acidity is normally expressed as grams of tartaric or sulfuric acid per liter. It is calculated by simply multiplying the number of equivalents per liter after determining the equivalent weight of the acid to which it is to be converted. Generally:

Acid	Molecular Weight	Equivalent Weight
Tartaric acid	150 g/mol	75 g/eq
Sulfuric acid	98 g/mol	49 g/eq

Titratable acidity, which measures the overall acid content of wine, tends to distinguish between two types of acidity:

- Volatile acidity
- Fixed acidity

Volatile acidity refers to the total content of short-chain volatile acids removed from wine by steam distillation. Acetic acid (boiling point of 118°C at 1 atm) accounts for 95 to 99% of this acidity and the rest is due to small quantities of lactic acid, propionic acid, butyric acid, and formic acid. To prevent the distillation of these acids as much as is possible, the rectifying column is placed between the wine to be distilled and the condenser. This is particularly important for lactic acid (boiling point of 121°C at 1 atm), which is extremely abundant in wines that have undergone malolactic fermentation.

Fixed acidity is calculated by simply subtracting the volatile acidity from the titratable acidity. Fixed acids are not removed from wine by steam distillation. The most important of these acids are tartaric acid, malic acid, and other acids derived from grapes and fermentation, such as succinic, lactic, and pyruvic acid.

The levels of neutralized acids are measured by determining the alkalinity of ash. The ash contained in musts or wines is obtained by evaporating a known volume of wine and then incinerating the residue after evaporation. This process generally yields carbonates of the cations that partially neutralize the acids in the wine. The resulting ash is dissolved in a known volume of hydrochloric acid and the excess acid is then measured against a known volume of sodium hydroxide.

Total acidity is a measure of the total content of free and neutralized acids found in grapes, musts, or wines.

Total acidity (meq/L = titratable acidity (meq/L) + alkalinity of ash (meq/L).
Total acidity = total acid content
Titratable acidity = free acid content
Ash alkalinity = neutralized acid content

The pH and titratable acidity in musts and wines depend on several factors, but particularly on grape ripeness, the growing region, and type of wine. Musts made from grapes grown in cold regions will generally have a lower pH than those made from grapes grown in hot regions. Acidity can then be adjusted, depending on the type of wine being made. An important factor to bear in mind during this operation is that the tartrate precipitation that occurs during fermentation alters both pH and titratable acidity. These two variables are also affected by malolactic fermentation. Broadly speaking, titratable acidity ranges between 4.5 and 7 g of tartaric acid per liter. Wine generally has a pH of between 2.8 and 3.8.

Low levels of volatile acidity are desirable. Levels of over 1 g/L of acetic acid indicate the possible presence of bacterial contaminants, with higher levels indicating greater contamination.

Although the addition of small quantities of wines with high volatile acidity to must during fermentation has been proposed as a method of correcting spoiled wines, this practice is not recommended due to the risk of contaminating otherwise healthy wine. This recommendation is based on the scientifically proven fact that yeasts consume considerable quantities of acetic acid produced during the early stages of fermentation, or derived from other sources such as the addition of these spoiled wines.

7.2. Measurement of Specific Acids

The most common method used to measure levels of individual acids is high-performance liquid chromatography (HPLC), which can be used to analyze all the acids present in a sample of wine in a single operation. The technique works by driving a small volume of the test wine through a suitable column.

Another common technique, which is widely used because of its selectivity, is enzymatic analysis, which can be used to determine the concentrations of lactic acid, malic acid, and succinic acid. Using this method, acid concentration is measured indirectly by spectrophotometry at a wavelength of 340 nm to assess changes in absorbance by the NADH formed or consumed in the corresponding enzymatic reaction.

In the enzymatic analysis of lactic acid, for example, lactic acid is converted to pyruvic acid in a reaction catalyzed by lactate dehydrogenase (LDH). This reaction generates NADH, and the absorbance at 340 nm is proportional to the concentration of the initial lactic acid.

$$CH_3\text{-}CHOH\text{-}COOH + NAD^+ \xrightarrow{LDH} CH_3\text{-}CO\text{-}COOH + NADH$$
$$\text{L-}(+)\text{-lactic acid} \qquad\qquad\qquad \text{Pyruvic acid}$$

$$[\text{Concentration}] = \varepsilon \times F \times Abs.$$

F is the dilution factor of the sample
ε is the molar extinction coefficient at 340 nm
Abs is the absorbance of NADH measured at 340 nm

Tartaric acid can also be quantified using the Rebelein colorimetric method, which is based on the formation of an orange complex with metavanadate and the Ag^+ ion. Another method of measuring tartaric acid is to precipitate the acid by adding its enantiomer, L-tartaric acid, which is not found in wine.

These specific techniques do not distinguish between the molecular and dissociated forms of the acids; when considered together, the result will be very similar to total acidity.

The Relationship Between Must Composition and Quality

1. THE HARVEST

Given the human and physical resources that are required for a short period of time during the harvest, it is clearly valuable to the wine industry to be able to determine the time of harvest in advance. However, it is not easy to do this, since a number of different factors must coincide to ensure that the grapes are harvested in an appropriate state of ripeness and health.

The following are the main factors that should be taken into consideration when establishing the date of the harvest:

1. Type of wine (young wines, dessert wines, etc.)
2. Health of the grape (can lead to early harvesting)
3. Weather conditions at harvest time

The choice of a date for harvesting should not be empirical, i.e., based solely upon the appearance of the grape, its firmness, its acidity in the mouth, or the color of the woody parts. Instead, it should be based on monitoring the ripening process through accurate measurement of certain components of the grape.

There are two main ways in which to determine the date of harvest. The first is a long-term method and is based on the length of the growth cycle, while the second involves monitoring the ripening process at defined time intervals.

1.1. Long-term Prediction

Long-term prediction of harvest times is based on the consistency of the growth cycle of the vine and takes into account the periods between flowering or veraison and ripening. Since the length of the growth cycle varies according to the vine and the region in which it is cultivated, predictions should be based on experience obtained in a specific vineyard, with a comparison of observations from the current year with those from previous years.

Some grape varieties ripen sooner than others under the same conditions. Thus, based on comparison with the Chasselas variety, grapes can be classified as first-, second-, or third-phase varieties, according to whether they ripen at the same time, 12 to 15 days later, or 24 to 30 days later than Chasselas grapes grown under the same conditions. Table 10.1 shows the ripening phase of some red and white grape varieties.

Taking into account the length of the growth cycle from mid-flowering to mid-ripeness, an approximate harvesting date can be determined up to 3 months ahead, since approximately 100 days separate the two phenomena, depending, of course, on the specific variety and the region in which it is grown. If the prediction is made from the midpoint of veraison, it will be sufficient to add a mean of 40 to 50 days according to the characteristics of the year and the duration of the ripening period observed in previous years with the same variety in the same growing region.

These predictions are most accurate in warm regions, where the grapes ripen well year after year, and where there is little likelihood of sudden meteorological changes that threaten the quality of the grapes. However, in most cases the harvest begins before the theoretical date. This is particularly true in years in which ripening is delayed or in wet years with a risk of rotting, although the harvest may also be brought forward in dry or very hot years due to fears of excessive over-ripening. In years with a high yield, it may also be brought forward as more time will be required to collect the grapes. Nevertheless, the general recommendation is to delay the harvest as long as possible (the closer to optimal ripeness the better) and to collect the grapes as quickly as possible.

TABLE 10.1   Classification of Grape Varieties According to Ripening Phase

Ripening Phase	Red Varieties	White Varieties
First phase (at the same time as the Chasselas variety)	Gamay Pinot Noir	Chardonnay Melon Traminer
Second phase (12–15 days after Chasselas)	Cabernet Franc Cabernet Sauvignon Cinsault Malbec Merlot Syrah	Altesse Chenin Muscadelle Riesling Roussanne Sauvignon Sémillon Sylvaner
Third phase (24–30 days after Chasselas)	Aramon Carignan Grenache	Clairette Folle Blanche Macabeo Ugni Blanc

1.2. Predictions Based on Monitoring of the Ripening Process

Periodic follow-up of grape ripening is the most logical and reliable method with which to rationally predict harvesting date. It also allows the composition of the harvested grapes to be determined in advance and therefore helps to choose the most appropriate vinification technique.

Monitoring of the ripening process is based on periodic collection of samples and subsequent analysis of the results. The process used for sampling is critically important and must fulfill the following criteria:

1. The sample must be representative. In other words, it should generate the same analytical results as would the entire crop if this were harvested at that moment in time.
2. A sufficient, yet manageable, volume should be obtained to ensure that all necessary analyses can be carried out.
3. The sampling procedure should be sufficiently clear, simple, and detailed so that anybody can collect the sample.
4. The cost of sampling in terms of the time required and the quantity of grapes used should be reasonable.

Due to differences in the level of variation among different factors, the number of points at which samples of grapes should be taken varies according to the analyses required. To accurately determine the values of more heterogeneous variables, it is necessary to collect a larger number of samples to ensure representative data.

To determine the anthocyan content with an error of less than 5%, 221 sampling points are required, whereas to determine the sugar content to the same level of error, only 19 samples are required. In practice, the number of sampling points needs to be relatively large (between 100 and 200 points per plot).

TABLE 10.2   Components of the Grape and Number of Samples

	Mean	Coefficient of Variation (%)	Number of Samples $\alpha < 0.05$
Sugars (g/L)	174.9	11	19
Acidity (g/kg)	4.7	22.6	82
Weight of the berry (g)	1.3	26.5	112
Anthocyans (mg/kg)	6.9	37.2	221

Adapted from Blouin and Guimberteau, 2004.

The concentration of anthocyans and tannins depends on the position of the grape cluster on the vine and of the grape within the cluster. For instance, grapes that are exposed to the sun have higher concentrations of both types of compound than those in the shade.

Sampling to determine the harvesting date can be carried out according to the following protocol:

1. Select a number of vines distributed throughout the plot, taking care to ensure that they are as representative as possible.
2. Begin sampling the chosen vines when the external characteristics of the grape clusters are appropriate (approximately 20 days after the midpoint of veraison). The frequency of sampling should initially be low (7 days) and then move progressively towards shorter time intervals (daily) as ripening proceeds.
3. Grape clusters should be selected from the upper, middle, and lower parts of the vine, from areas exposed to the sun and areas in the shade, and alternating between grapes from the outer and inner parts of the cluster.
4. Samples should be taken after the dew has disappeared but before the sun has become strong, ideally between 11 am and 1 pm.
5. Neither the strongest nor the weakest vines should be chosen, nor those at the highest or lowest topographical sites.
6. Clusters containing rot should be sampled if affected clusters are to be used alongside healthy grapes.

Must should always be collected from the berries in the same way. A number of devices have been developed for this purpose, each with its advantages and disadvantages. Generally speaking, the results obtained for a given variable depend on the type of device, with the exception of sugar concentration, which is relatively independent of the device used. Methods that involve homogenization of the grape (Table 10.3, methods 6 and 7) reproducibly release tannins and anthocyans (so long as the devices work similarly) but cause a considerable reduction in the titratable acidity and an increase in pH, both linked to the release of potassium from the cell walls of the berries broken during the treatment.

The method used to obtain the must should be chosen according to the variable to be monitored. For traditional monitoring of ripening (sugars and/or acidity), traditional presses yield good results. If the aim is to analyze phenolic compounds, centrifugation of the fruit followed by collection of the skins, cleaning, and drying is a good option. The reproducibility

TABLE 10.3   Different Methods for Obtaining the Must from Grape Samples

Method	Advantages	Disadvantages
1. Manual pressing	Extremely simple	Very slow, poorly reproducible, and too gentle
2. Mouli legume-type press	Simple, economical	Crushing of the seeds
3. Vertical, manual press with a central screw	Traditional, similar to a winery press	Slow, poor reproducibility
4. Horizontal, automated laboratory press	Similar to a winery press, rapid, reproducible	Harsh, expensive
5. Pneumatic press	Gentle, similar to a winery press Easy to use and standardize	Slow, can be too gentle
6. Centrifuge for juices	Rapid, reproducible, releases color	Very harsh, powerful extraction from the skin
7. Hand blender	Rapid, easy to clean, releases a large amount of color	Very harsh, powerful extraction from the skin

of this extraction method is good and the error is less than 5%, although the use of a traditional press is required to obtain representative values for acidity and pH.

Information collected in different years represents a very valuable source of data for the optimization of harvest time. The results obtained for the different variables can be used to generate indicators of ripeness for a given vineyard plot or grape variety. When the ripeness index remains constant over a period of time, or else reaches a desirable level, harvesting can begin. As before, data obtained from a number of years will help to improve optimization in subsequent vintages.

If there is no clear correlation between the assessment of ripening and the results obtained in the wine, new vines that are more representative should be selected for sampling and the process restarted.

2. PHENOLIC COMPOUNDS AND SAMPLING

The concentrations of phenolic compounds vary substantially between different samples, and a large number of samples is required to obtain an error of less than 5%. Based on the results of numerous studies designed to address this problem, it has been concluded that the margin for total error is very wide for anthocyans (42–84 mg/L) and total polyphenols (0.6–1.3 mg/L). These margins must be taken into account when interpreting the results.

To be valid, monitoring of phenolic ripeness should be based on the following sampling protocol:

1. Rigorous, random sampling must include at least 200 points in the vineyard plot.
2. Sampling should begin 30 days after the midpoint of veraison. Closer to veraison, the error is larger because not all grapes enter this phase at the same time.

3. Samples must always be obtained according to the same protocol: sampling at the same time of day, always collecting whole berries, and using the same mode of transport and delay between collection and analysis.
4. The extraction method for phenolic compounds should be predefined: total phenolic compounds extracted in acidic media, only phenolic compounds that are easily extractable at the pH of wine, or mechanically extracted phenolic compounds (blender, etc). The last option produces a strong correlation with the extraction observed during vinification.
5. The results from different plots sampled on the same date should be compared to establish an order of harvesting, which is more difficult to establish with other variables such as sugar content or acidity.

3. RIPENESS

Everyone has an idea of what a ripe grape is, but ripeness is not a concrete, well-defined state, and various types of ripeness have therefore been established:

1. **Physiological ripeness:** This is the moment at which the seeds are capable of reproducing to form a plant. It is achieved soon after veraison, but the grape is still bitter and cannot be eaten or used to make wine.
2. **Industrial maturity:** This usually refers to the point at which the grape has the highest sugar content and lowest acidity. The definition is mainly of interest to growers who sell their crop based on the sugar content of the grapes. Nevertheless, with few exceptions, sugar content and/or acidity should not be the only criteria used to determine the date of harvest.
3. **Aromatic ripeness:** This is established according to the level and quality of aromas in a given variety. To establish the optimal moment for harvesting, tasting should be performed by chewing the skins well to assess the quantity and quality of varietal and vegetal aromas. This procedure requires extensive training, but the tasting process is simple and a vegetal character can be avoided in the wine as a result.
4. **Phenolic ripeness:** This is linked to the concentration of anthocyans and tannins. These compounds are more easily extracted with increasing phenolic maturity and, as a consequence, the color of the wine is enhanced.
5. **Enological maturity:** This corresponds to the optimum moment for harvesting that will allow the best wine to be obtained in a given year and under given conditions. Establishing the harvest time according to enological maturity involves selecting options that may appear contradictory, taking into consideration a range of variables such as sugar content, acidity, primary aromas, and polyphenol concentrations. Identifying the point of enological maturity requires the collection of large amounts of information in order to adequately control the process.

3.1. Indicators of Ripeness

Numerous indicators have emerged to monitor ripeness and predict harvest time. The simplest, and most widely used, are based on the ratio of sugar to acidity. This ratio is known as the ripeness index and is based on the observation that sugar concentrations are inversely correlated with acidity as ripening proceeds. Since this has no physiological basis, however, many exceptions have been observed in which a given increase in sugar levels does not correspond to the same reduction in acidity.

The ripeness index is not appropriate for comparisons between different grape varieties, since some varieties have high sugar content and high acidity. Nevertheless, it is a straightforward, practical measure that has some degree of comparative value.

The ratio of sugars to acidity is calculated based on the sugar content of the must expressed in grams per liter, measured directly or calculated indirectly from the density of the must, and the titratable acidity expressed as grams of sulfuric acid per liter:

$$\text{Ripeness index} = \frac{\text{Sugars (g/L)}}{\text{Acidity (g/L)}}$$

Based on the relationship between the density of the must and the acidity, the ripeness factor, R, has also been developed and is calculated using the following formula:

$$R = \frac{\text{Degrees Oechsle}}{\text{g(tartaric acid)/L}} \times 10$$

The Oechsle scale is calculated by subtracting 1000 from the density of the must. According to this formula, values can be obtained for R:

- $R > 100$ allows wines of exceptional quality to be produced.
- $80 < R < 100$ allows the production of very high-quality wines.
- $70 < R < 80$ can generate high-quality wines.
- $R < 70$ is sufficient only for the production of table wine.

These values are merely a guide and should not be taken as a reference except in growing areas that have similar conditions to those found in the German regions in which this indicator of ripeness is most widely used.

Based on the observation that the concentration of malic acid diminishes constantly throughout grape ripening, whereas that of tartaric acid essentially remains constant, the Baragiola coefficient was defined as the percentage of tartaric acid in relation to the titratable acid:

$$\text{Baragiola coefficient} = \frac{\text{g tartaric acid}}{\text{titratable acidity (g } H_2T/L)} \times 100$$

The Baragiola coefficient was later modified by Ferré, who added the alkalinity of ash to the denominator. This correction allows the percentage of tartaric acid to be considered in relation to the total anion content of the must, and circumvents the requirement to measure the malic acid concentration, which is a rather laborious process. Consequently, the Ferré coefficient produces very similar results to those that would be obtained with the ratio of tartaric acid to tartaric + malic acid.

$$\text{Ferré coefficient} = \frac{\text{g tartaric acid}}{\text{titratable acidity } + \text{ alkalinity of ash}} \times 100$$

Ferré also proposed the coefficient of base saturation, which relies on the phenomenon of continuous accumulation of cations during ripening alongside a progressive decline in the concentration of organic acids. This accumulation, as we know, is essentially independent of the combustion of acids and the accumulation of sugars, and as a result, the coefficient is not much more illustrative than a simple measurement of pH.

$$\text{Coefficient of base saturation} = 2 \times \frac{AA}{TA + AA} \times 100$$

AA = alkalinity of ash, TA = titratable acidity

Other coefficients are based exclusively on acid content:

$$R_1 = \frac{\text{g tartaric acid}}{\text{g malic acid}} \qquad R_2 = \frac{\text{acidity}}{\text{tartaric acid} + \text{malic acid}} \ (mEq/L)$$

Index based on the ratio of glucose to fructose and on the pH:

$$R_3 = \frac{\text{sugars}}{(10^{-pH}) \times 10^4} \times \frac{\text{fructose}}{\text{glucose}}$$

Although there are many indexes based on the ratio of sugars to acids, they must always be interpreted in the context of data obtained on ripening in the specific growing region and under similar weather conditions. This is important since the sugar content is strongly dependent upon the hours of effective sun exposure and the mobilization of carbohydrate reserves. Tartaric acid concentration, in contrast, depends fundamentally upon rainfall, while malic acid concentration depends on intracellular combustion, which is linked to temperature. The concentration of mineral bases varies according to the circulation of water in the plant.

TABLE 10.4 Measures of Ripeness in Different Grape Varieties and Growing Regions

Variety	Growth Period	R	R_1 x 10	Baragiola	R_2
Pedro Ximénez	Veraison	17.9	12.2	76.4	67.2
Montilla	Ripeness	128.1	64.9	218.8	37.9
Pedro Ximénez	Veraison	13.4	18.1	87.6	69.8
Jerez Frontera	Ripeness	59.3	51.4	166.4	49.1
Palomino Fino	Veraison	19.2	29.5	104.6	68.8
Jerez Frontera	Ripeness	82.9	171.4	212.2	44.1
Merlot	Veraison	5.6	10.6	54.6	89.3
Bordeaux	Ripeness	40.6	27.3	104.0	68.2
Cabernet Sauvignon	Veraison	8.4	12.2	57.6	90.4
Pauillac, Medoc	Ripeness	34.8	26.4	98.2	71.5

3.2. Other Variables Used to Determine Harvesting Date

Monoterpene Alcohols

The concentration of monoterpene alcohols, in their free forms or bound to sugars, is an important factor to take into consideration in the production of wines with a marked varietal aroma. Consequently, their levels must be taken into consideration when determining the date of harvest.

When the grapes are still green, all of the bound terpene alcohols are present at high concentrations (250–500 µg/kg), whereas the free forms are either completely absent or present at very low concentrations (30–90 µg/kg). Beginning at veraison, the concentration of free terpenes begins to increase, and that of some, such as linalool and α-terpineol, diminishes during over-ripening. Geraniol and nerol concentrations follow the same temporal profile, possibly as a result of the similarity of their molecular structure.

The concentration of bound terpenes is important from a technological perspective, since it can be up to 15 times that of the free form, accounting for concentrations of up to 6 or 7 mg/L. This fraction changes little during the vinification process and is essentially unaltered in young wines. Consequently, it represents a natural resource that can potentiate the varietal aroma of the wine. To this end, enzymes with glycosidic activity have been used to hydrolyze bound terpenes.

In contrast to the terpenes, which have pleasant aromas and are therefore desirable, other compounds have undesirable properties and must therefore also be taken into account. For instance, some C_6-alcohols and aldehydes have vegetal aromas and flavors, and are derived from the oxidation of fatty acids that can occur during treatment of the grapes prior to fermentation. Other compounds, such as pyrazines, which have the aroma of green peppers, are highly abundant in musts from under-ripe grapes belonging to some Sauvignon varieties.

Phenolic Compounds

Phenolic compounds are key to the flavor quality of red wines and consequently a better understanding of these compounds and their organoleptic properties can help to identify indicators of ripeness that, when applied to the harvesting date, help to improve the quality of the wines obtained. Polyphenols are linked to numerous organoleptic and enological characteristics, including harshness, softness, balance, and aging, but these can be difficult to define based only on the composition of the grape at the time of harvesting. Currently, quantitative methods are used that take into account all polyphenols together, which is clearly inadequate, since if this criterion alone were applied, the best wines would be those that had the strongest color and the highest tannin levels (e.g., press wine), and this is clearly not the case. Nevertheless, this is an essential criterion in varieties such as Grenache, Pinot, and Cabernet Franc, which do not achieve sufficient levels of phenolic compounds except when grown in regions with ideal climatic and soil conditions.

In cold regions, the ratio of phenolic compounds to malic acid is used. This is an improvement on the ratio of sugar to acidity based on the observation of an inverse relationship between malic acid content and anthocyan concentrations.

It has recently been proposed that a range of variables relating to the skin of the grapes should be monitored, as these are ultimately responsible for the polyphenol content of must and wine. In this regard, the richness of the skin in easily extractable anthocyans and tannins and the limited amounts of tannins in the seeds appear to be appropriate indicators with which to determine the harvesting date for red grape varieties. Studies have linked the concepts of phenolic, cellular, and seed ripeness with the results obtained following extractions of both the seeds and skin of the grapes at pH 3.2 and pH 1.

Other Compounds

There are other compounds and phenomena that depend on the dynamics of the ripening process and that can be considered indicators of setbacks which occur during ripening. Examples are compounds that act as markers of stress or those that are produced by the plant as a response to diseases or adverse conditions.

Stress indicators include proline content. This amino acid functions as a universal marker of stress in plants and reflects conditions of drought or salinity. Its presence in the cytoplasm appears to be linked to the maintenance of osmotic pressure against the vacuoles that accumulate most of the solutes in the cells (sugars, acids, and ions). Its concentration in the cells of the grapes can double and even quadruple during ripening and it can therefore be used as a marker for the development of the grape at the same time as being an indicator of quality. High concentrations of proline in the grape indicate that something has gone wrong during ripening. Unfortunately, accumulation of proline in the grapes is highly dependent upon the level of nitrogenous fertilizer; when this increases from 0.25 to 8 mM NO_3^-, the content of proline in the must shifts from 100 to 600 mg/L, and as a consequence proline is only a good indicator of the quality of the grapes when the provision of nitrogen to the plant is well defined or remains essentially constant.

Recent studies have analyzed substances produced by the plant as a defensive response to extreme conditions or infection by harmful microorganisms and assessed their usefulness as indicators of quality. The following are examples:

- Resveratrol, pterostilbenes, and flavonoids are phytoalexins that have a substantial antifungal activity.
- Glyoxylic acid is produced during respiration of the berries and damages the cell membrane of *Botrytis cinerea.*
- Defense proteins (or pathogenesis-related proteins) and also chitinases and glucanases have distributions that vary according to the extent of development of the grapes.
- Inhibitors of *B. cinerea* polygalacturonases are present in the berries of some varieties of wine grape that are relatively resistant to attack by this microorganism.

Physical Characteristics of Grape Clusters

The physical characteristics of the fruit, such as the volume of the berry and the fragility of the cell walls in the skin, pulp, and outer layer of the seeds, are of great interest at the time of the harvest. The volume of the grapes can be a criterion when related to sugar content and malic acid concentration, since alone it has no value. In fact, the most interesting physical characteristics are those related to the resistance of the skin and seeds to crushing and pressing, since this can influence the extraction of polyphenols during fermentation.

4. FACTORS AFFECTING THE QUALITY AND RIPENING OF THE GRAPE

4.1. Invariant Factors

The effect of invariant factors is more or less constant and does not vary from one year to the next. Notable among these factors are the following:

Variety

Different varieties of grape behave differently during ripening and the musts that are obtained can have marked differences in composition. Thus, there are varieties that accumulate aromatic substances in the skins, endowing the grapes and the wine with a particular aroma. Nevertheless, from an enological perspective, it is the acidity, in particular the malic acid concentration, that varies most from one variety to another.

Rootstocks

The rootstock plays an important role in the ripening process and the final degree of ripeness. It transmits its vigor to the plant and it is known that the most vigorous plants are also the slowest in ripening, producing fruit that contains less sugar and higher acidity. The choice of rootstock must take into consideration the winegrowing region and the climate. Vigorous rootstocks, for instance, are inappropriate in cold climates as they are slow ripening and generate very acidic fruits.

Age of the Plant

The age of a plant affects its vigor and influences the timing of ripening, in that older vines are less productive and ripen earlier. These vines are better adapted to the terrain, since they have a well-developed root system and are also more resistant to drought and sudden changes in temperature. Their fruits are the first to ripen, producing grapes with a higher sugar content, lower acidity, deeper color, and stronger aroma. On the other hand, very young vines are often harvested first since their fruit is less rich in all components and has a tendency to rot, leading to the production of light wines lacking in body.

Climate (Latitude)

In cold climates, varieties are sought that ripen rapidly before the cold of autumn arrives and that need little hot weather from the beginning of plant activity until ripeness, producing fruit with low acidity. In these climates it is more difficult to obtain grapes with balanced acidity and sufficient color than it is to achieve sufficient sugar content to produce an acceptable wine. Consequently, white grape varieties are generally grown in these climates, since red wines require sunny conditions for the synthesis of anthocyans and tannins.

In regions with hot climates, later-ripening (third- or fourth-phase) vines are used, and these can produce high yields in terms of sugar content or crop weight. In temperate climates, first- or second-phase varieties are used.

4.2. Climatic Factors

Temperature, sunlight, and humidity are the three main external factors that influence the ripening of grapes, which is therefore related to the weather conditions in a given year. These factors are highly variable and it is almost impossible to obtain the same weather conditions in any two years, and therefore vintages are similarly variable.

In order to determine the conditions required to achieve adequate ripeness in a given region, data are required on temperature, sunlight, and humidity over a number of years. Comparison of such data with the quality of the vintages obtained in traditional winegrowing regions has allowed the basic conditions necessary for good ripening of the grapes to be determined. In the Bordeaux region, for example, the following conditions are thought to be necessary for an acceptable vintage:

- A cumulative temperature from April to September, inclusive, of at least 3100°C, since good vintages have not been achieved below this level.
- A minimum of 15 days during which the temperature reaches at least 30°C.
- Rainfall of between 250 and 300 mm.
- At least 1250 hours of sunlight.

Nevertheless, it should be remembered that these are minimum requirements and must also occur at the appropriate time. The character of a vintage will be determined by the microclimate of a given region. This is linked to the French concept of the *cru*, which embraces geography, soil, and climate in a given region.

4.3. Modifiable Factors

Modifiable factors include those that are linked to the maintenance of the vine and to the provision of fertilizers.

Pruning influences the shape and yield of the vine. It should be carried out after the leaves have fallen. The way in which it is done influences the productivity of the plant in the next growth cycle.

Shoot topping involves removal of the ends of the growing branches and diverts more resources to the ripening fruit, although studies have shown that the greatest differences between vines in which shoot topping has been carried out and those in which it has not relate to the malic acid content. The lower concentrations of malic acid found in grapes from vines that have not been topped is due to the greater respiratory combustion of this acid in the leaves.

Leaves are sometimes removed during ripening to help accelerate the process, and to prevent fungal infections and rot. However, the procedure reduces yield and quality due to the reduction in the area available for photosynthesis. Leaf removal is only recommendable a few days before harvesting, and should only involve the leaves at the base of the shoots, which become less and less active at this time. It is performed almost exclusively to prevent rot by allowing air to circulate more easily around the grape clusters, to facilitate ripening by increasing the exposure of the grapes to the sun, and even to facilitate harvesting.

Fertilization has an important influence on the quantity and quality of the grapes that are harvested. The grape crop, the leaves, and the wood that makes up the base of the vine require fertilizers that must be replenished to maintain adequate production by the vine. Excessive or insufficient feeding of the plant cause defects that can affect the grape clusters and therefore the wine. The most striking result of intensive fertilization is the reduction in color of red wines; wines produced from fertilized land are also lower in tannin content and have less body.

4.4. Accidental Factors

Grape quality and ripening can also be affected by disease or weather events.

Mildew can affect the grape clusters at various points during their development. If it attacks the cluster during flowering and fruit setting, it causes complete loss of the fruit; if infection occurs after fruit setting, ripening is delayed and there is a reduction in sugar content and an increase in the acidity of the grapes; if the pedicel is reached by the infection, the grapes may even fail to ripen.

Powdery mildew usually attacks the skin of the grape, causing it to turn brownish-gray, halting its growth, and causing a bulging of the berry, which becomes deformed, bursts, and splits open, on occasions leaving the seeds visible. Nevertheless, ripening still occurs and grapes attacked by powdery mildew are less juicy and may have higher sugar content, but they often also have a higher-than-normal acidity.

Rot is caused by the growth of fungi and molds on the grapes during or just prior to ripening; the most common cause is *B. cinerea*. The disease spreads easily when the berries split, and rupture of the skin can also lead to the growth of other microorganisms such as acetic acid bacteria. The most serious cases occur when *B. cinerea* infection occurs simultaneously with infection by other fungi such as *Penicillium*, *Aspergillus*, or *Mucor*. Rot always leads to loss of color in the grapes and the wines produced always have oxidative precipitation.

5. CORRECTION OF THE HARVEST

Corrections of the harvest are intended to modify the sugar content or acidity according to the type of wine to be produced. These variables are the most affected by weather conditions during ripening of the grapes. Two types of correction exist: traditional corrections, based on mixing of different grape varieties or grapes with different degrees of ripeness to compensate for deficits in some components of the crop, and chemical corrections, which are designed to correct only a single component of the harvest. Here we will focus on chemical corrections, which can be classified essentially as corrections of the sugar or acid content.

5.1. Correction of the Sugar Content

Increasing the sugar content of the must is known as enrichment. Reducing the sugar content via the addition of water to the must is prohibited. There are various ways in which enrichment can be achieved:

- Addition of concentrated must
- Addition of rectified concentrated must
- Addition of beet sugar (chaptalization)
- Elimination of water (subtractive enrichment)

The use of more than one procedure in the same must is prohibited. The degree of enrichment that is permitted is determined by legislation in each growing region.

Addition of Concentrated Must

Enrichment by addition of concentrated must involves the use of natural musts that have been enriched by vacuum distillation until a sugar content of at least 762 g/L has been achieved. Given that the process used to concentrate these musts does not eliminate acids, low acid content, and particularly low malic acid content, is used as an indicator of the quality of the must to avoid obtaining highly acidic enriched wines.

Addition of Rectified Concentrated Must

Rectified concentrated musts are obtained by passing the must through anion exchange resins to remove acids. Cation exchange resins can also be used to remove minerals and other compounds that could later be found in the dry extract of the wine.

Rectified concentrated must is the best option to artificially increase the alcohol content of the wine, since it is a pure solution of grape sugar that contains glucose and fructose at a 1:1 ratio and does not alter the acidity of the wine. Its purity is controlled by European Union regulations and the purchase of a product with a satisfactory quality can be guaranteed. In contrast, no such guarantees are available for concentrated must, the quality of which is essentially dependent upon the starting material.

The rectified concentrate, whatever its origin, can be used to enrich wines recognized as *Denominación de Origen*, since it only contributes grape-derived sugar.

Using the following formulae, the volume of concentrated must required to obtain a given increase in the alcohol concentration of the wine can be calculated:

Law of dilutions: The volume and the concentration must be expressed using the same units.

$$\text{Vol. Mix} \times [CC] = \text{Vol. Must 1} \times [CC] + \text{Vol. Must 2} \times [CC]$$

Rule of mixtures:

$$\begin{array}{ccc} A \searrow & & \nearrow (C{-}B) \\ & C & \\ B \nearrow & & \searrow (A{-}C) \end{array}$$

$(C{-}B)$ are the components with characteristics A that must be used to obtain the mixture.

$(A{-}C)$ are the components with characteristics B that must be used to obtain the mixture. C is the desired characteristic.

Example: A 100-hL volume of must with a predicted ethanol content of 12% is to be increased to 14% ethanol using a concentrated must of 67°Brix. What quantity of concentrated must should be mixed with the original?

Based on conversion tables, 67°Brix ≡ 54.4% ethanol. According to the law of dilutions:

$$100 \times 12 + 54.4 \times X = (100 + X) \times 14; \ X = 4.95 \text{ hL}$$

Percentage of concentrated must in the mixture:

$$V_{concentrated\ must} = \frac{4.95}{104.95} \times 100 = 4.7\%$$

Using the rule of mixtures:

$$54.4 \searrow \qquad \nearrow (14-12) = 2 \text{ parts concentrate}$$
$$14$$
$$12 \nearrow \qquad \searrow (54.4-14) = 40.4 \text{ parts original must}$$

The percentage of concentrated must in the mixture is:

$$V_{concentrated\ must} = \frac{2}{40.4 + 2} \times 100 = 4.7\%$$

The volume to be added would be:

$$\frac{2\ \text{Concentrated must}}{40.4\ \text{Initial must}} \times 100\ \text{hL} = 4.95\ \text{hL}$$

Addition of Beet Sugar (Chaptalization)

The practice of adding beet sugar is prohibited in a number of countries, including Italy, Portugal, Spain, and Greece, as well as in the southern winegrowing regions of France, and California.

White sugar should be used, since brown sugar contributes a flavor of molasses to the wine. Cane sugar was traditionally preferred to beet sugar due to its greater purity, but nowadays the purity of beet sugar does not justify such a position. Theoretically, to increase the alcohol concentration of the wine by 1% (vol/vol), 17 g of sucrose is required per liter of must (1.7 kg/hL). Nevertheless, 2 kg/hL is usually added during the production of red wines to compensate for the loss of alcohol due to evaporation, particularly in barrels or small open-top fermentation tanks.

Artificially increasing the alcohol content by enrichment of the must with sugar is regulated in various winegrowing regions within the European Union, and increases of more than 2% (vol/vol) are generally not permitted. Only in exceptional cases are increases of 4.5% (vol/vol) allowed.

Analytical techniques such as proton nuclear magnetic resonance spectroscopy and mass spectrometry are used to test for the fraudulent addition of beet sugar. Both methods are based on the relative abundance of hydrogen isotopes in the alcohol obtained from cane or beet sugar and that obtained from grape-derived sugars.

Chaptalization has the following effects on wine:

- An increase in the products of glyceropyruvic fermentation (glycerol, succinic acid, and 2,3-butanediol).
- A slight reduction in total acidity and tartaric acid content compared with wine that has not undergone chaptalization.

- An increase in the ratio of alcohol to dry extract is usually observed since the dry extract does not increase in the same proportions as the alcohol concentration. This can therefore be employed to detect the use of chaptalization.

Elimination of Water (Subtractive Enrichment)

Subtractive enrichment involves the elimination of part of the water contained in the must destined for fermentation. Usually, the must can be concentrated by up to 20%, but the best results have been obtained with concentrations of 10%. Various techniques can be used to concentrate the must in this way:

- Direct osmosis
- Evaporation at atmospheric pressure
- Evaporation under partial vacuum, also known as reduced-pressure evaporation
- Reverse osmosis

The advantage of this approach is that it ensures an effective reduction in the volume of wine produced, in contrast to the addition of sugar or concentrated must; however, the process is costly, requiring substantial initial investment and high running costs.

5.2. Correction of Acidity

Acidity can be corrected to maintain the pH within the desired range. Acidity affects, among other characteristics, the clarity and shine, and the concentration of active sulfite. It inhibits the proliferation of certain microorganisms, such that wines with a higher acidity tend to have fewer problems in relation to storage and stability.

Deacidification of Musts

A high level of acidity does not favor the growth and activity of yeast, and fermentation does not proceed normally below a pH of 2.5–2.6. High concentrations of tartaric acid and, especially, malic acid are common in grapes grown in cold climates. In such cases, the use of deacidifying agents is authorized. The products used are potassium or calcium salts of acids that are weaker than the tartaric acid present in musts and wines.

These deacidifying agents work through a mechanism involving the replacement of the cation of an acid salt with another stronger acid. This equilibrium is enhanced if the new salt that is formed is insoluble and precipitates, displacing the equilibrium in the direction of interest, namely the elimination of the acid that is present in the must.

1. Deacidification with Potassium Bicarbonate

$$COOH\text{-}CHOH\text{-}CHOH\text{-}COOH + KHCO_3 \rightarrow COOH\text{-}CHOH\text{-}CHOH\text{-}COOK\downarrow + H_2CO_3$$
$$\quad\quad 150 \text{ g/mol} \quad\quad\quad\quad 100 \text{ g/mol}$$

On the other hand, in the acid medium of the wine the following occurs:

$$H_2CO_3 \rightarrow CO_2\uparrow + H_2O$$

To remove 1 g of tartaric acid (H_2T) $\Rightarrow \dfrac{100}{150} = 0.67$ g of $KHCO_3$ are required.

2. Deacidification with Calcium Carbonate

$$COOH\text{-}CHOH\text{-}CHOH\text{-}COOH + CaCO_3 \quad \rightarrow \quad CHOH\text{-}COO^- + H_2CO_3$$

$$150 \text{ g/mol} \qquad 100 \text{ g/mol} \quad | \quad Ca^{2+}\downarrow \qquad \downarrow\uparrow$$

$$CHOH\text{-}COO^- \qquad CO_2\uparrow + H_2O$$

To remove 1 g of H_2T requires $\Rightarrow \dfrac{100}{150} = 0.67$ g $CaCO_3$

In both cases the reaction is equimolar and produces a poorly soluble salt that precipitates and an acid (H_2CO_3) that breaks down into CO_2 and H_2O in the acid medium. The result is the elimination of excess acid.

In reality, the effect of precipitation of the salts is neither very rapid nor very effective in isolation. In order for precipitation to be effective, a procedure known as cold tartaric stabilization must be used. Calcium carbonate is less effective than potassium bicarbonate; nevertheless, the effect of calcium carbonate on pH is more rapid due to the insolubility of calcium tartrate. In addition, the use of calcium carbonate or potassium bicarbonate at the pH usually found in must has the drawback of only acting on tartaric acid, which is a minor component compared with malic acid in regions where deacidification is required. As a result, the wines are left with very low levels of tartaric acid.

3. Deacidification by Precipitation of Calcium Tartromalate

Precipitation of the double salt calcium tartromalate allows part of the malic and tartaric acid content to be eliminated. There are two ways of carrying out this type of deacidification, both based on the precipitation of the acids as a calcium tartromalate double salt at a pH greater than 4.5. To calculate the required amount of deacidifying agent, the concentrations of tartaric and malic acid initially present in the must need to be determined. The treatment is initially carried out in part of the total volume of must to be deacidified and the rest is added later.

3a) Precipitation using Calcium Carbonate.

This system is limited by the quantity of tartaric acid present in the must to be deacidified, since this usually contains more malic than tartaric acid.

The stoichiometry (Figure 10.1) indicates that for each gram of tartaric acid precipitated, 0.9067 grams of malic acid will also precipitate. Deacidification calculations need to take into consideration the grams of tartaric acid that we want to remain in the must or wine that has been treated.

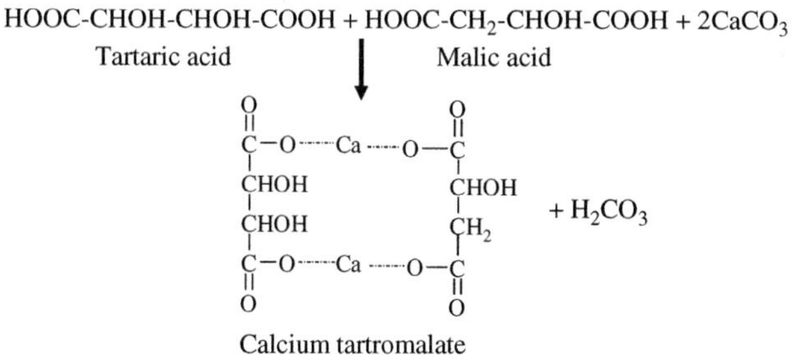

FIGURE 10.1 Formation of calcium tartromalate from tartaric and malic acid.

The grams per liter to be eliminated (D) as a function of tartaric acid would be as follows:

$$X = g/L \text{ tartaric acid} \quad Y = g/L \text{ malic acid}$$
$$D = X + Y \qquad\qquad Y = 0.8933X$$
$$D = X + 0.8933X$$
$$D = 1.8933X$$

The quantity of $CaCO_3$ required for the treatment is based on the requirement for 0.67 and 0.75 g of $CaCO_3$ to precipitate 1g of tartaric and malic acid, respectively.

$$\text{Grams of } CaCO_3 = (0.67X + 0.75Y) \times V_T$$

V_T = total volume of must or wine

Given that tartaric acid is limiting for deacidification, the volume in which the treatment is carried out should be calculated as follows:

$$V_D = \frac{X}{\text{Initial } H_2T \ (g/L)} V_T$$

V_D = volume used for treatment

3b) Precipitation with Calcium Carbonate and Calcium Tartrate.

$CaCO_3$ deacidification has the drawback of being limited by the quantity of tartaric acid and the failure to eliminate greater quantities of malic acid despite it being present in a higher proportion than tartaric acid. Precipitation with calcium carbonate and calcium tartrate is also based on the formation of calcium tartromalate. Deacidification is carried out using a mixture of $CaCO_3$ and calcium tartrate as a reagent; at pHs above 4.5, malic acid forms a double salt with calcium tartrate. Tartaric acid in the must also precipitates as calcium tartrate due to the addition of carbonate.

The grams of carbonate can be determined according to the degree of deacidification required (D):

$$CaCO_3(g) = \left(\frac{D \times (H_2T)_{initial}}{(H_2T)_{initial} + (H_2M)_{initial}} \times 0.67 + \frac{D \times (H_2M)_{initial}}{(H_2T)_{initial} + (H_2M)_{initial}} \times 0.75 \right) \times V_T$$

Calcium tartrate is added at a 2:1 ratio to calcium carbonate. The volume to be treated is determined by the desired level of deacidification. The content of tartaric and malic acid in the volume that has been treated (V_D) would be zero; this does not occur in treatment with carbonate, as malic acid is always left over when that method is used. Consequently, a residual acidity of around 1.5 g/L is required. The volume of must necessary to carry out the deacidification is determined as follows:

$$V_D = \frac{D}{g/L \ H_2T \ initial \ + H_2M \ initial \ - 1.5} \times V_T$$

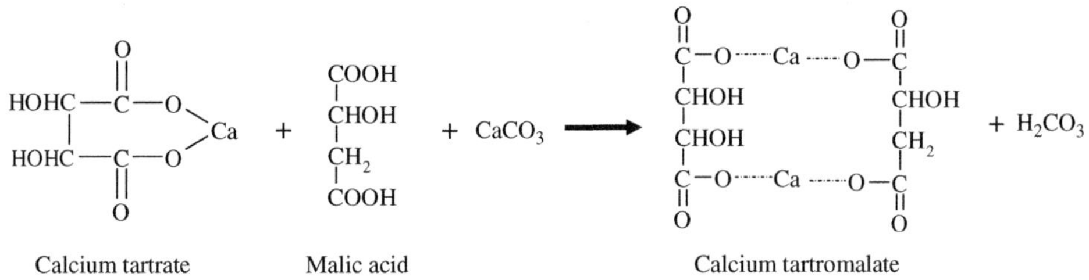

$$HOHC-\overset{\overset{\displaystyle O}{\|}}{C}-O \diagdown \underset{\diagup}{Ca} \quad + \quad \begin{matrix} COOH \\ | \\ CHOH \\ | \\ CH_2 \\ | \\ COOH \end{matrix} \quad + \quad CaCO_3 \quad \longrightarrow \quad \begin{matrix} \overset{O}{\|} \\ C-O\cdots Ca\cdots O-\overset{O}{\underset{\|}{C}} \\ | \qquad\qquad\qquad | \\ CHOH \qquad\qquad CHOH \\ | \qquad\qquad\qquad | \\ CHOH \qquad\qquad CH_2 \\ | \qquad\qquad\qquad | \\ C-O\cdots Ca\cdots O-C \\ \overset{\|}{O} \qquad\qquad\qquad \overset{\|}{O} \end{matrix} \quad + \quad H_2CO_3$$

$$HOHC-\underset{\underset{\displaystyle O}{\|}}{C}-O$$

Calcium tartrate   Malic acid   Calcium tartromalate

FIGURE 10.2 Formation of calcium tartromalate from calcium tartrate and malic acid.

Special care should be taken with pH, since below 4.5 only precipitation of calcium tartrate from the must would occur as a result of calcium tartrate addition. Furthermore, precipitation of potassium bitartrate can be induced.

Treatment of musts with one of these methods involves taking 10% of the volume of must to be deacidified (V_D) and adding the deacidifying salts. The remaining must is added little by little with shaking to promote the release of carbon dioxide formed during the treatment.

In all of the treatments, it should be remembered that the appearance of ethanol during fermentation reduces the solubility of tartaric acid salts and that precipitation of those salts will affect the acidity of the wine.

Acidification of Musts

Increasing the acidity of musts is a common practice in hot regions; indeed this is necessary in order to obtain high-quality wines that are clear, with a strong color and good flavor, and that age well. European Union regulations allow for the acidification of new musts and wines during fermentation in certain winegrowing regions.

The use of citric and tartaric acids to acidify musts is permitted. Citric acid has the disadvantage of being easily degraded by microorganisms and it is easily broken down by the bacteria responsible for malolactic fermentation. As this leads to an increase in volatile acidity, citric acid is of little interest as an acidifying agent in musts. Furthermore, the legal limit for this acid is 1 g/L, which is insufficient to generate a notable increase in acidity. It is preferable to use tartaric acid in must, since it increases hardness in the mouth when added to wine.

It is difficult to define a general rule for acidification, although acidification of musts is recommended when their total acidity is below 4 g/L tartaric acid (53.3 meq/L). In practice, however, it is worth establishing pH values as a reference, with a pH of around 3.5 being a reasonable cutoff. The choice of final pH depends on the type of wine to be produced and the preferences of the winemaker. Thus, wines with a pH close to 3.2 are more microbiologically stable and stand long periods of aging, but they are also harder on the palate. In contrast, wines with a pH above 3.7 are less microbiologically stable and have a shorter shelf life, but their flavor is more pleasant.

Traditionally, correction has been carried out to obtain a titratable acidity of between 4.5 and 5 g/L. The addition of tartaric acid leads to only a slight reduction in pH, since the must is strongly buffered, and part of the added acid precipitates due to formation of potassium bitartrate. It is estimated that around 30% of the added acid is precipitated as a salt. Generally speaking, addition of 1 g/L tartaric acid leads to a pH reduction of 0.2.

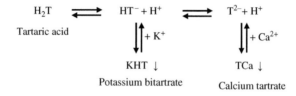

FIGURE 10.3   Equilibrium between tartaric acid and its salts.

Increasing acidity by passing part of the must through ion exchange resins is prohibited in most regions of the European Union; nevertheless, it is one of the methods that least affects the concentration of organic acids in wine, since it only modifies the salts by exchanging K^+ or Na^+ for H^+ ions.

The use of mineral or inorganic acids is strictly prohibited. However, the addition of plaster of Paris (calcium sulfate) continues to be used in some regions to reduce the pH. This practice, which is employed prior to clarification of the must, allows cleaner, clearer wines to be obtained by promoting sedimentation of residues. The legal limit for $CaSO_4$ in the finished wine is 2 g/L.

Ca^{2+} precipitates as calcium tartrate, as indicated in the equilibria shown earlier. The loss of tartrate is compensated by the dissociation of molecular tartrate with the release of protons. Thus, the reaction would be as follows:

$$CaSO_4 + 2\,HKT \; \rightarrow \; K_2SO_4 + TCa\!\downarrow + H_2T$$
$$\updownarrow$$
$$HT^- + H^+$$

The Transformation of Must into Wine

Enological Chemistry. DOI: 10.1016/B978-0-12-388438-1.00011-X

1. INTRODUCTION

Enology, which is the scientific study of wine and winemaking, took a giant step forward with the work of Louis Pasteur, who converted an almost magical phenomenon — the transformation of grape must to wine — into a rigorous science. Pasteur demonstrated that all fermentation processes were due to the action of live organisms and were not the product of spontaneous generation. He used the term *fermentation* to refer to all microbial processes that take place under anaerobic conditions.

Nowadays, fermentation could be defined as the enzyme-catalyzed oxidation of organic compounds in the absence of an external electron acceptor. The role of the electron acceptor in such cases is fulfilled by a compound generated during fermentation; it is therefore an endogenous oxidant.

Based on the above definition, several differences can be established between fermentation and respiration. Respiration, for example, is the enzyme-catalyzed oxidation of organic compounds in which external molecular oxygen acts as the final electron acceptor, but always in aerobic conditions. In anaerobic conditions, in contrast, respiration (or fermentation) is the enzyme-catalyzed oxidation of organic compounds in which NADH (nicotinamide adenine dinucleotide) is not oxidized by oxygen in electron transport systems but by external compounds. These compounds can be inorganic (NO_3^-, Fe^{3+}, SO_4^{2-}, CO_3^{2-}, CO_2) or organic (fumarate $\rightarrow$ succinate).

The biochemical reactions by which microorganisms (yeasts and bacteria) convert the components of grape juice or wine can be induced (by alcoholic or malolactic fermentation) or they can occur accidentally (as the result of bacterial contamination). The type of conversion that takes place determines the chemical composition and organoleptic characteristics of the final product.

The microorganisms that participate in the conversion of grape must to wine use components present in the must for cell growth and multiplication. Also, yeasts that grow in grapes must remain in the converted wine after all the fermentable sugars have been consumed. They form a large biomass that settles at the bottom of the fermentation tank, where diverse substances can be transferred to the wine via excretion or cell lysis. Much remains to be discovered about the role played by microorganisms in these aspects of winemaking; yet they are very important and must be taken into account when describing the true balances associated with the transformations that occur.

Yeasts can grow in aerobic conditions, producing CO_2 and H_2O, or in anaerobic conditions, producing ethanol and CO_2. The difference between the two pathways is shown in the equations below, which show that, compared to fermentation, respiration provides the cell with 20 times more energy.

The processes of fermentation and respiration are shown in the following equations:

General fermentation equation

$$C_6H_{12}O_6 \; \rightarrow \; 2\,CH_3CH_2OH + 2\,CO_2 + 40 \text{ Kcal/mol (2 ATP} + 25.4 \text{ Kcal)}$$

where ATP indicates adenosine triphosphate.

General respiration equation

$$C_6H_{12}O_6 + 6\,O_2 \; \rightarrow \; 6\,CO_2 + 6\,H_2O + 686 \text{ Kcal/mol (38 ATP} + 408.6 \text{ Kcal)}$$

The following relationships can be derived for the mass balance for alcoholic fermentation:

Gay-Lussac equation (1)

$$C_6H_{12}O_6 \rightarrow 2\,CH_3\text{-}CH_2OH + 2\,CO_2$$
$$180\,g \qquad 2 \times 46\,g \qquad + 2 \times 44\,g$$

Production of ethanol (conversion of sugars to alcohol) (2)

15.5 g hexose $\rightarrow$ 7.9 g ethanol + 7.6 g CO_2 (7.9 g ethanol/L $\equiv$ 1%(vol/vol))

Equations (1) and (2) show the theoretical mass balance for alcoholic fermentation. In fact, experiments have shown that the actual mass balance is:

(3) 17 g hexose $\rightarrow$ 7.9 g ethanol + 7.6 g CO_2 + 1.5 g other substances.

The empirical rule of thumb frequently used by winemakers is that 17 g of sugar will produce 1 unit of alcohol. Therefore, a must containing 220 g/L of sugar will produce a wine with 12.9% (vol/vol) alcohol.

2. THE PASTEUR AND CRABTREE EFFECTS

Yeasts are facultative microorganisms. In other words, they can survive in aerobic or anaerobic conditions (with or without oxygen). In the presence of oxygen, yeasts use the respiratory pathway, which is more beneficial to them than the fermentation pathway in terms of energy. The inhibition of fermentation in the presence of oxygen was first described by Pasteur and is known as the Pasteur effect. It is easily explained considering that cells obtain 38 ATP molecules when using aerobic respiration compared to just 2 using alcoholic fermentation.

In a medium with a high glucose concentration, such as grape must, yeast will only use the fermentation pathway, since under these conditions respiration is impossible, even in the presence of oxygen. This phenomenon is known as the Crabtree effect, the counter-Pasteur effect, or catabolic repression by glucose. In the case of *Saccharomyces cerevisiae,* this effect is observed in musts with a glucose concentration of 9 g/L and above. Hence, in grape must, yeasts are obliged to use the fermentation pathway, regardless of the oxygen level, because the Crabtree effect is very strong.

However, the presence of oxygen in must stimulates alcoholic fermentation as it favors the synthesis of fatty acids and sterols by increasing the permeability of cell membranes and thus the transport of sugar. The same effect would be obtained by adding sterols to the must.

From a technological perspective, the use of sugar by the respiratory pathway in yeast is applicable to the industrial production of dry yeast but not to winemaking.

3. GLYCOLYSIS

Glycolysis refers to a series of reactions by which living cells convert C_6 sugars (glucose and fructose) into pyruvic acid. The most common form of glycolysis is the

Embden-Meyerhof-Parnas pathway. The reactions can occur in the presence or absence of oxygen, and there are at least two clearly distinguishable stages.

Stage 1. This starts with the phosphorylation of glucose via an ATP molecule, thus providing the energy needed to activate glycolysis. The phosphate groups of the phosphorylated metabolites are ionized, preventing their passage through biological membranes once they are in the yeast cytosol. The stage ends with the cleavage of fructose 1,6-diphosphate, which has a rather unstable furan-type ring, into two triose-phosphate isomers, which exist in equilibrium: dihydroxyacetone-phosphate (96.5%) and glyceraldehyde-3-phosphate (3.5%). Only glyceraldehyde-3-phosphate participates in the subsequent reactions, meaning that the balance shifts towards this product. Indeed the overall result is as if a C_6 sugar molecule had yielded two glyceraldehyde-3-phosphate molecules (Figure 11.1). The energy balance at this stage is negative as no ATP molecules have been produced.

Stage 2. The second stage involves the formation of pyruvic acid from glyceraldehyde-3-phosphate (Figure 11.2). For each molecule of glyceraldehyde-3-phosphate that is converted to ethanol, two molecules of ATP are obtained. Since each C_6 sugar molecule is converted into two glyceraldehyde-3-phosphate molecules, a total of four ATP molecules are obtained for each sugar molecule.

At the end of glycolysis there is a net balance of two ATP molecules as two molecules were used for the initial phosphorylation of the C_6 sugar molecule.

$$C_6H_{12}O_6 \rightarrow 2\,CH_3\text{-}CO\text{-}COOH + 2\,ATP$$
$$\text{Glucose} \qquad \text{Pyruvic acid}$$

4. ALCOHOLIC FERMENTATION

In aerobic respiration, pyruvic acid is oxidized by Krebs-cycle reactions according to the following equation:

$$CH_3\text{-}CO\text{-}COOH + \frac{5}{2}\,O_2 \rightarrow 3\,CO_2 + 2\,H_2O$$

The NADH coenzymes formed are reoxidized to NAD^+ in the mitochondria, with oxygen acting as the final proton acceptor. In the absence of oxygen, however, NADH needs to find another proton acceptor: pyruvic acid. Because this acid acts as a proton acceptor, it cannot be oxidized as in the equation above. It is therefore directly reduced to lactic acid (homolactic fermentation) or, following decarboxylation, to ethanol (alcoholic fermentation).

Glycolysis is the first step in alcoholic fermentation. The pyruvic acid formed is decarboxylated to acetaldehyde (ethanal) and subsequently reduced to ethanol (ethyl alcohol). This reaction occurs via the reduced form of NAD^+ produced during the oxidation of glyceraldehyde-3-phosphate to 3-phosphoglycerate. These two reactions are linked and represent an oxidation-reduction (redox) system. It is therefore clear that NADH needs to be reoxidized in order to prevent glycolysis from coming to a halt once all the NAD^+ in the cell has been reduced.

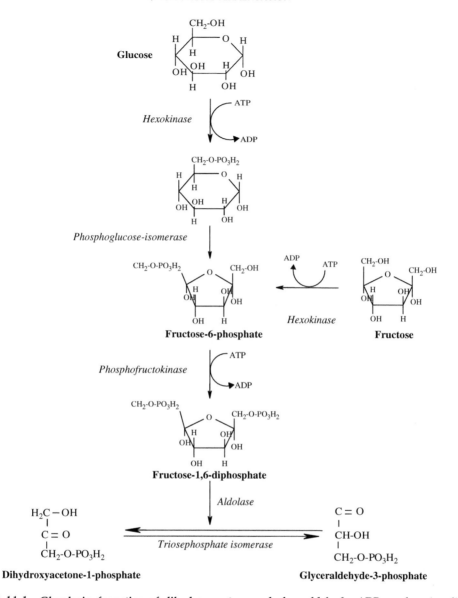

FIGURE 11.1   Glycolysis: formation of dihydroxyacetone and glyceraldehyde. ADP = adenosine diphosphate; ATP = adenosine triphosphate.

The energy balance of alcoholic fermentation is identical to that of glycolysis as no new ATP molecules are formed. The chemical balance of fermentation can be expressed as follows:

$$C_6H_{12}O_6 + 2\,ADP + 2\,H_3PO_4 \rightarrow 2\,CH_3\text{-}CH_2OH + 2\,CO_2 + 2\,ATP + 2\,H_2O$$

The change in free energy for the chemical transformation of one molecule of glucose to CO_2 and ethanol is −40 Kcal, and the energy required to create an ATP molecule is

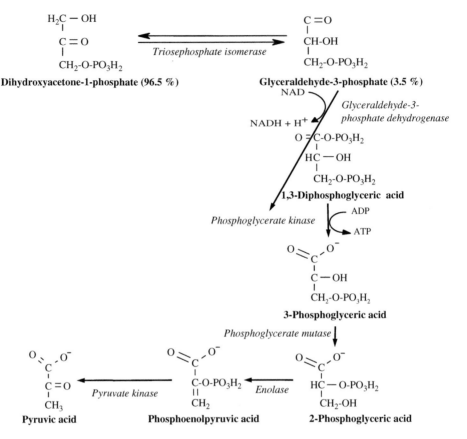

FIGURE 11.2   Glycolysis: formation of pyruvic acid from glyceraldehyde. ADP = adenosine diphosphate; ATP = adenosine triphosphate; NAD = nicotinamide adenine dinucleotide.

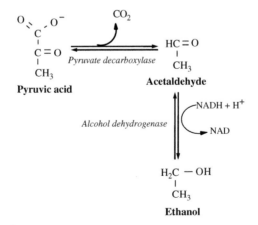

FIGURE 11.3   Formation of ethanol from pyruvic acid. ADP = adenosine diphosphate; ATP = adenosine triphosphate; NAD = nicotinamide adenine dinucleotide.

7.3 Kcal. This means that of the 40 Kcal released, 14.6 Kcal are used by the yeast cells to maintain their vital functions, and in particular, to ensure reproduction. The remainder (25.4 Kcal) is released into the fermentation medium in the form of heat, causing an increase in temperature in the fermentation tank.

5. GLYCEROPYRUVIC FERMENTATION

In the above sections, it was seen that the NADH formed during glycolysis needs to be oxidized back to NAD^+. This occurs when acetaldehyde is reduced to ethanol. At the beginning of fermentation, however, the enzyme pyruvate decarboxylase is only weakly expressed, meaning that there is no acetaldehyde in the medium. Furthermore, when acetaldehyde is eventually formed, it initially combines with SO_2 and cannot be reduced to ethanol. The NADH coenzymes therefore need to be reoxidized in another manner, which is via glyceropyruvic fermentation.

In this process, NADH is reoxidized by the reduction of a dihydroxyacetone-3-phosphate molecule — which is derived from the cleavage of the sugar molecule in glycolysis — to glycerol. In glyceropyruvic fermentation, for every glycerol molecule formed, one molecule of pyruvic acid cannot be reduced to ethanol because the coenzymes needed for this reaction are used to form glycerol. We will see later that this pyruvic acid is the source of several secondary products.

The amount of glycerol present at the end of alcoholic fermentation varies but generally ranges between 4 and 8 g/L. Most of it is formed from the first 100 g of converted sugar. Glyceropyruvic fermentation is a continuous process that is closely linked to alcoholic fermentation. The glycerol produced is used by yeasts to maintain their redox potential, that is, their NAD^+/NADH balance. This serves to prevent the accumulation of NADH from, for example, the synthesis of amino acids or proteins or from oxidation reactions that yield other secondary products.

Accordingly, the general equation for glyceropyruvic fermentation can be expressed as:

$$C_6H_{12}O_6 \rightarrow CH_2OH\text{-}CHOH\text{-}CH_2OH + CH_3\text{-}CO\text{-}COOH$$

Hexose   Glycerol   Pyruvic acid

5.1. Secondary Products of Glyceropyruvic Fermentation

As previously indicated, glyceropyruvic fermentation leads to the formation of glycerol and pyruvic acid. This acid is converted into other products including succinic acid, butanedione, acetoin, 2,3-butanediol, lactic acid, and acetic acid.

Formation of Acetaldehyde

Part of the pyruvic acid formed in glyceropyruvic fermentation is converted to acetaldehyde by pyruvate decarboxylase. The final concentration of acetaldehyde largely depends on fermentation conditions, and on the amount of SO_2 added, but it generally does not exceed 100 mg/L.

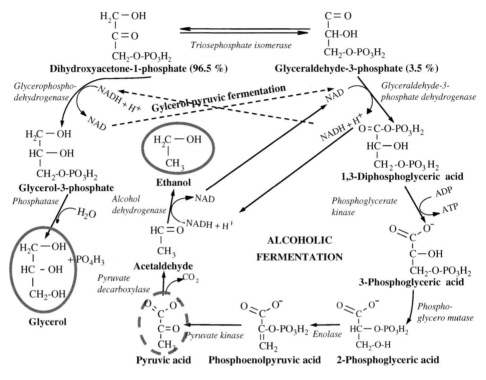

FIGURE 11.4   Alcoholic and glyceropyruvic fermentation: relationship with glycolysis. ADP = adenosine diphosphate; ATP = adenosine triphosphate; NAD = nicotinamide adenine dinucleotide.

Formation of Succinic Acid

Several pathways have been proposed to explain the formation of succinic acid during alcoholic fermentation.

The first of these pathways is the Krebs cycle, which would be considerably slowed as it takes place in the mitochondria, whose functionality is very limited during fermentation. The cycle, however, would stop at the succinic acid step, as the oxidation of succinic to fumaric acid needs the FAD (flavin adenine dinucleotide) coenzyme, which is strictly respiratory. The cycle thus would come to a halt once succinic acid had been formed.

Another pathway by which succinic acid is possibly produced is the formation of oxalacetic acid from pyruvic acid by the enzyme complex pyruvate carboxylase.

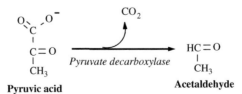

FIGURE 11.5   Formation of acetaldehyde (enzyme involved: pyruvate decarboxylase).

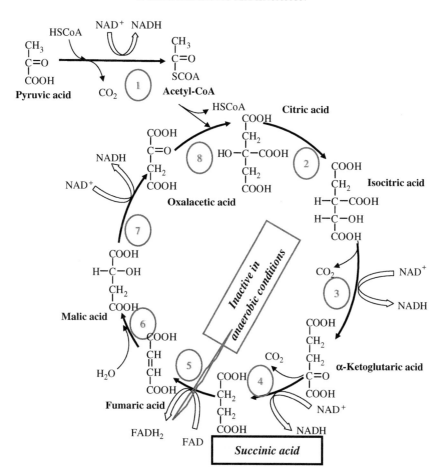

FIGURE 11.6 **Formation of succinic acid through the Krebs cycle.** (1) = pyruvate dehydrogenase; (2) = aconitase; (3) = isocitrate dehydrogenase; (4) = α-ketoglutarate dehydrogenase complex and succinate dehydrogenase; (5) = succinate dehydrogenase; (6) = fumarase; (7) = malate dehydrogenase; (8) = citrate synthase. CoA = coenzyme A; FAD = flavin adenine dinucleotide; NAD$^+$ = nicotinamide adenine dinucleotide.

Oxalacetic acid would be reduced to succinic acid via malic and fumaric acid (the reverse of what occurs in the Krebs cycle). Unlike the Krebs cycle, which is oxidative, this is a reductive pathway in which fumaric acid is converted to succinic acid by NADH and not FADH$_2$. While this pathway exists in bacteria, it is probably limited in yeast during fermentation.

A third possible pathway, proposed by Thunberg, is the formation of succinic acid through the condensation of two acetyl coenzyme A (CoA) molecules.

A fourth possibility is that succinic acid is synthesized from glutamic acid, although this pathway does not involve pyruvic acid. Whatever the case, the concentration of succinic acid at the end of fermentation ranges between 0.5 and 1.5 g/L.

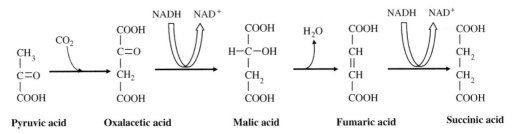

FIGURE 11.7 Formation of succinic acid via the reductive pathway. NAD = nicotinamide adenine dinucleotide.

Formation of Acetic Acid

Acetic acid is the main volatile acid in wine, and, while it is associated with wine spoilage (*piqûre lactique* and acetic spoilage), it is also formed by yeast during fermentation (100–300 mg/L). The final concentration of acetic acid in wine depends on fermentation conditions and the yeast strains that participate in fermentation. High levels can seriously compromise the quality of the final product.

The mechanisms by which acetic acid is formed have not been fully elucidated but there are two possible pathways. The first involves the formation of acetyl-CoA from pyruvic acid and the subsequent hydrolysis of this molecule. This pathway is limited in anaerobic conditions because acetyl-CoA is synthesized in the mitochondria, which, as mentioned previously, have limited functionality in the absence of oxygen. The second possibility is the oxidation of the acetaldehyde previously converted from pyruvic acid by the $NADP^+$ coenzyme; the resulting NADPH can then be used in the synthesis of fatty acids. This pathway also serves to form acetyl-CoA from acetic acid when this is not possible via pyruvic acid. In anaerobic conditions, using standard solutions, the concentration of acetic acid decreases with increasing activity of acetyl-CoA synthetase. In view of the limited mitochondrial functionality in the absence of oxygen, the second pathway is probably the most likely explanation for the formation of acetyl-CoA under anaerobic conditions.

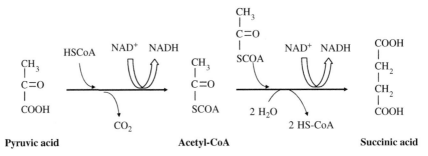

FIGURE 11.8 Formation of succinic acid according to Thunberg. CoA = coenzyme A; NAD = nicotinamide adenine dinucleotide.

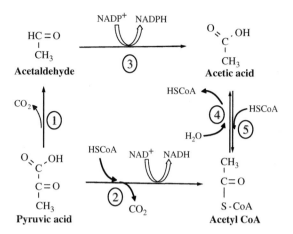

FIGURE 11.9 **Mechanisms of acetic acid formation.** Enzymes: (1) = pyruvate decarboxylase; (2) = pyruvate dehydrogenase; (3) = aldehyde dehydrogenase; (4) = acetyl-CoA hydrolase; (5) = acetyl-CoA synthetase.

The fermentation of musts with a high sugar content results in wines with abnormally high levels of acetic acid (and glycerol) due to the osmotic stress to which the yeasts are subjected; this explains why dessert wines have higher levels of these compounds than other wines. The production of acetic acid is also affected by abnormally low (<3.1) or high (>4) pH levels, by the absence of certain vitamins or amino acids, and by high temperatures (25–30°C) during cell multiplication.

Acetic acid is not continuously produced by yeast. Rather, that part of the acid produced during the fermentation of the first 50 to 100 g of sugar is subsequently metabolized by these microorganisms. The rest is not metabolized and therefore remains in the must until alcoholic fermentation is complete. It has been shown that the addition of acetic acid to musts reduces the production of glycerol, as the acid is reduced to acetaldehyde and can therefore be converted to ethanol, thus preventing the reoxidation of NADH by dihydroxyacetone. In such cases, however, acetoin and 2,3-butanediol levels increase. The refermentation of spoiled wines is one example of how winemakers take advantage of the ability of yeasts to consume acetic acid. In this process, a wine with a high concentration of acetic acid is mixed with a must such that the total acetic acid content does not exceed 0.75 g/L. The mixture is left to ferment and the levels of this acid tend to drop to under 0.4 g/L.

Formation of Lactic Acid

Lactic acid is another product of pyruvic acid (Figure 11.10). It is formed during the reduction of this acid by lactate dehydrogenase. This enzyme basically produces D-lactic acid (200–300 g/L). L-Lactic acid, in contrast, is produced by the metabolism of malic acid by malolactic bacteria. The same bacteria, however, produce D-lactic acid when they metabolize sugar. By analyzing D-lactic acid levels, it is possible to determine whether this acid is a product of yeast metabolism (<300 mg/L) or bacterial metabolism (several grams per liter).

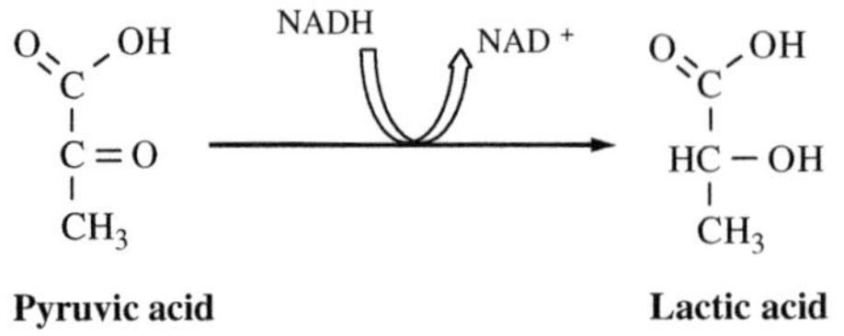

FIGURE 11.10 **Formation of lactic acid (enzyme involved: lactic dehydrogenase).** NAD = nicotinamide adenine dinucleotide.

Formation of Other Acids from Pyruvic Acid

Other products of pyruvic acid are citramalic acid (0−300 mg/L) and dimethylglyceric acid (0−600 mg/L), both of which are formed by the condensation of pyruvic acid and acetyl-CoA. Both citramalic and dimethylglyceric acid have little impact on the organoleptic characteristics of wine.

FIGURE 11.11 Formation of citramalic acid and dimethylglyceric acid from pyruvic acid. CoA = coenzyme A.

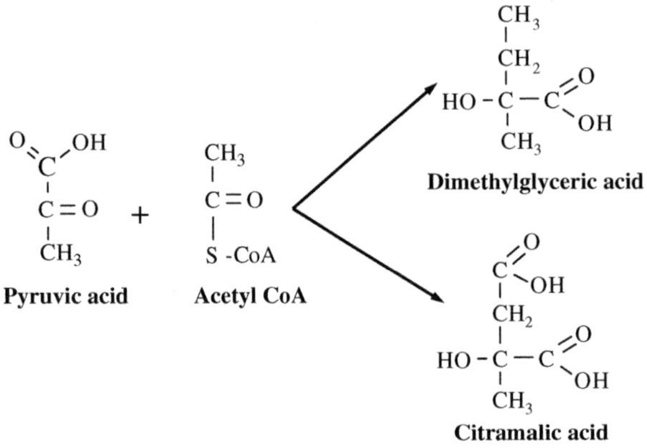

Formation of Butanedione, Acetoin, and 2,3-butanediol

Yeasts form butanedione, acetoin, and 2,3-butanediol from pyruvic acid (Figure 11.12). 2,3-Butanediol binds to a molecule of thiamine pyrophosphate (TPP) and undergoes decarboxylation, giving rise to TPP-C$_2$, also known as active acetaldehyde. The condensation of a molecule of acetaldehyde yields acetoin through pathway (1). The TPP-C$_2$ complex can also bind to a molecule of pyruvic acid, giving rise to α-acetolactic acid. The non-oxidative decarboxylation of this acid also leads to the formation of acetoin through pathway (2). Butanedione, in turn, is derived from the oxidative decarboxylation of α-acetolactic acid. The reduction of butanedione gives rise to acetoin through pathway (3). Finally, 2,3-butanediol is derived from the reduction of acetoin, although this reaction is reversible.

Butanedione is produced at the beginning of fermentation, but is rapidly reduced to acetoin and 2,3 butanediol; this reduction can even continue for several days after fermentation in wines aged on lees. Butanedione and acetoin impart a buttery aroma to wine. An excessive

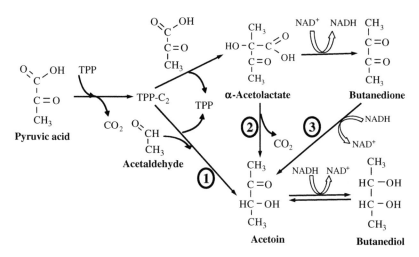

FIGURE 11.12 Synthesis of butanedione, acetoin, and 2,3-butanediol. NAD = nicotinamide adenine dinucleotide; TPP = thiamine pyrophosphate; TPP-C_2 = Active acetaldehyde. (1) = condensation; (2) = decarboxylation; (3) = reduction.

production of these compounds can have a negative effect on aroma, but this does not occur under normal fermentation conditions. The degradation of citric acid by lactic acid bacteria, in contrast, can give rise to very high levels of butanedione and acetoin.

Finally, wine may contain variable quantities of oxalacetic acid, malic acid, and fumaric acid, as these are intermediates in the formation of succinic acid. Small quantities of short-chain fatty acids incorporating acetyl-CoA as a building block are also formed. Finally, small quantities of formic acid (mostly esterified) and propanoic acid have also been detected.

5.2. Mass Balance in Glyceropyruvic Fermentation

The relationship between the main secondary products of glyceropyruvic fermentation can be shown experimentally by progressively adding small quantities of acetaldehyde to a fermentation medium. The result is an increase in secondary products and a decrease in glycerol. The experiment can also be conducted inversely by blocking acetaldehyde as it is produced; this stimulates glyceropyruvic fermentation and the formation of glycerol but results in a decrease in secondary products as there is no acetaldehyde.

Considering the mechanisms by which the different secondary products are formed, as well as the redox equilibrium, it is possible to establish a mass balance based on the reactions in which these products are formed. Acetaldehyde is considered reactive in all cases. To simplify matters, it is necessary to presume that all pyruvic acid (acetaldehyde) not converted to ethanol (due to glyceropyruvic fermentation) has already been formed and that the redox reaction is balanced at this moment. Let us look at the stoichiometry of the formation of the main secondary products of glyceropyruvic fermentation.

The hypothetical chemical reactions in which these substances are formed from acetaldehyde are as follows:

1. Formation of acetic acid

$$2\, CH_3\text{-}CHO + H_2O \;\rightarrow\; CH_3\text{-}COOH + CH_3\text{-}CH_2OH$$

One NADH coenzyme (which subsequently needs to be converted back to NAD^+) is formed for every molecule of acetic acid. The conversion (reoxidation) of NADH involves the formation of a molecule of ethanol from a molecule of acetaldehyde formed during glyceropyruvic fermentation.

2. Formation of succinic acid

$$5\, CH_3\text{-}CHO + 2\, H_2O \;\rightarrow\; HOOC\text{-}CH_2\text{-}CH_2\text{-}COOH + 3\, CH_3\text{-}CH_2OH$$

In this case, it is assumed that succinic acid is formed through the Thunberg pathway. Two NADH coenzymes are formed in the synthesis of the two acetyl-CoA molecules from pyruvic acid. The condensation of the acetyl-CoA molecules leads to the formation of another NADH coenzyme. Accordingly, to balance the redox potential (i.e., to reoxidize the three NADH coenzymes formed), in addition to the two acetaldehyde molecules needed to form succinic acid, another three molecules need to be reduced to ethanol.

3. Formation of acetoin (acetyl methyl carbinol)

$$2\, CH_3\text{-}CHO \;\rightarrow\; CH_3\text{-}CHOH\text{-}CO\text{-}CH_3$$

There is no net gain of NADH in the formation of acetoin.

4. Formation of 2,3-butanediol

$$CH_3\text{-}CHO + CH_3\text{-}CH_2OH \;\rightarrow\; CH_3\text{-}CHOH\text{-}CHOH\text{-}CH_3 \;(1)$$

$$2\, CH_3\text{-}CHO + NADH \;\rightarrow\; CH_3\text{-}CHOH\text{-}CHOH\text{-}CH_3 + NAD \;(2)$$

$$CH_3\text{-}CH_2OH + NAD \;\rightarrow\; CH_3\text{-}CHO + NADH \;(3)$$

One NADH molecule is used in the synthesis of 2,3-butanediol (reaction 2). To maintain the redox balance, one molecule of acetaldehyde will not be converted to ethanol, and its presence in wine will not be due to glyceropyruvic fermentation. In other words, it is as if an already formed molecule of ethanol was converted to acetaldehyde to maintain the redox balance (reaction 3). The net change in acetaldehyde can be calculated by adding reactions 2 and 3 (reaction 1).

As occurs in the case of 2,3-butanediol, the formation of other secondary products such as lactic acid involves the use of NADH. Once again, acetaldehyde will be found in the medium, but not as a result of glyceropyruvic fermentation. The net change for this reaction will therefore be zero.

These equations show the formation of secondary products of acetaldehyde at a general level, without taking into account intermediate stages, or possible changes in the degree of oxidation/reduction of compounds. Their purpose is simply to assess the mass balance of the secondary products of acetaldehyde.

In a fermentation medium, the above compounds are measured as shown below:

- **a** — the molecular concentration of acetic acid
- **s** — the molecular concentration of succinic acid
- **m** — the molecular concentration of acetoin
- **b** — the molecular concentration of 2,3-butanediol
- **h** — the concentration of free acetaldehyde at the end of fermentation

The sum of the above, in relation to acetaldehyde, gives:

$$\sum = 2a + 5s + 2m + b + h$$

The above calculation shows all the acetaldehyde molecules derived from glyceropyruvic fermentation during fermentation. Because glyceropyruvic fermentation yields an equal number of glycerol and pyruvic acid (for which read acetaldehyde) molecules, the corresponding calculation would be:

$$\sum = 2a + 5s + 2m + 2b + h = G$$

where G is the molar concentration of glycerol. Hence:

$$\sum : G = 1$$

The Σ:G ratio observed in different experimental fermentations lies between 0.82 and 0.95, which appears to demonstrate the presence of small quantities of other secondary products, possibly cell components, produced using acetaldehyde from glyceropyruvic fermentation. This is clearly not the case, however, since under anaerobic conditions, the net Σ is never greater than G, as pyruvic acid (acetaldehyde) would have to be produced in greater quantities than glycerol.

Limitations of Mass Balance

The above mass balance has certain limitations:

1. If fermentation occurs in a medium rich in glutamic acid, this acid could lead to the formation of succinic acid.
2. Musts made from grapes with noble rot are rich in glycerol.
3. Wine made from unhealthy grapes has high levels of acetic acid due to the oxidative metabolism of acetic acid bacteria.

The mass balances performed on these products are useful for determining whether or not a wine has undergone pure alcoholic fermentation. They can also be used to classify wines, perform quality controls, or detect adulteration. The above balance obviously has its limitations, as it should ideally contain all the secondary products of pyruvic acid, including residual pyruvic acid. Furthermore, the calculations are based on the Thunberg pathway, and even though other pathways are probably unlikely to be involved, they should be taken into account. These limitations demonstrate the complexity associated with mass balances, and they raise the possibility that the balance observed is due to compensation of errors. Nonetheless, it is certain that these secondary products are formed. Fermentations conducted

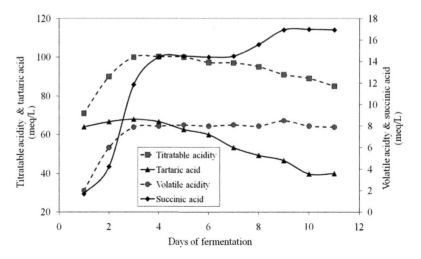

FIGURE 11.13   **Changes in titratable acidity, volatile acidity, and tartaric acid and succinic acid levels during fermentation.**

at different pHs have also shown that, although the Σ:G ratio is always close to 1, the concentrations of the different secondary products of glyceropyruvic fermentation vary considerably.

6. CHANGES IN GRAPE ACIDS DURING FERMENTATION

Between 10 and 30% of the malic acid originally present in grapes is metabolized by yeast during alcoholic fermentation. The proportion varies according to the yeasts that conduct the fermentation. Malic acid, however, is preferentially metabolized by lactic acid bacteria (in malolactic fermentation). Citric acid, in contrast, appears to be metabolized exclusively by bacteria (and more so in white wines than in red wines), in a process that leads to the formation of acetic acid.

Tartaric acid is highly resistant to the action of microorganisms. The production of ethanol during alcoholic fermentation results in a reduction of tartaric acid salts (mainly potassium bitartrate and neutral calcium tartrate); these precipitate, shifting the dissociation constant of tartaric acid and eliminating the bitartrate and tartrate ions at the bottom of the fermentation tanks. Despite this precipitation, however, wine is still oversaturated with these two salts. In other words, their levels in solution are much higher than would be expected given their solubility. As a result, wines can develop haze when there is a drop in temperature.

7. FACTORS AFFECTING ALCOHOLIC FERMENTATION

Wine is produced thanks to the action of yeast flora from grapes on must. During fermentation, the behavior and selection of yeasts depends on the physical and chemical conditions of the medium. Alcoholic fermentation is affected by three main, inter-related, factors:

- Physical factors: temperature, oxygen
- Biological factors: yeast
- Chemical factors: pH levels and sugar, ethanol and SO_2 concentrations

7.1. Physical Factors

Temperature

Temperature is key to the successful production and storage of wine, and this is particularly true for white wines, which are more sensitive to the effects of temperature than red wines. In wine-producing countries with hot climates, the temperatures reached during grape pressing (as high as 40°C) can increase the rate at which the compounds from the grape skins are absorbed into the grape juice. This can result in better quality wines when the process is carefully conducted.

The temperature at which grapes are pressed affects yeast multiplication, fermentation rate, and the formation of secondary products.

INFLUENCE ON YEAST

For all strains of yeast, it is possible to establish:

a. a minimum temperature, under which the yeasts would not multiply (6–9°C);
b. an optimum temperature, at which the yeasts would multiply faster (25–30°C); and
c. a maximum temperature, above which the yeasts would not multiply (33–38°C).

Yeast activity at different stages of fermentation also varies according to temperature:

- At 26 to 28°C, activity intensifies halfway through fermentation and declines considerably at the final stages.
- At 18 to 20°C, activity remains constant during the active fermentation phase.
- At low temperatures, the yeast population is sparser but more stable throughout fermentation.

It should also be noted that higher temperatures increase the toxicity of alcohol for yeast and thus limit its growth, and slow the fermentation.

INFLUENCE ON RATE OF FERMENTATION

Maximum fermentation rates are reached at a temperature of close to 35°C; at 15°C to 25°C, the rate is moderate, and above 37 or 38°C, fermentation slows down and can even become stuck. A temperature of over 35°C is often sufficient to bring fermentation to a halt.

Generally speaking, the higher the temperature of the must, the faster is the fermentation rate. Excessively high temperatures, however, can result in incomplete fermentation, producing wines containing residual sugar and this is a potential source of problems during storage. As shown in Figure 11.14, fermentation conducted at a temperature of 25°C takes longer but is more complete. At 30°C, and especially at 35°C, fermentation is faster but incomplete.

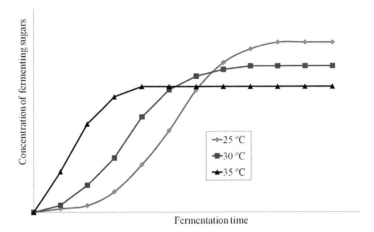

FIGURE 11.14   **Levels of fermented sugars according to temperature.**

The recommended fermentation temperature is 15 to 20°C for white table wines and 20 to 25°C for red wines.

Intermediate temperatures (15–25°C) have a greater impact than high temperatures (30–35°C) on the fermentative capacity of yeasts, i.e., on the amount of alcohol they are capable of producing during fermentation. As long as there are no abrupt drops in temperature (in the range of 5–6°C), yeasts also exhibit greater fermentative capacity at low temperatures, although the process takes much longer.

INFLUENCE ON CHEMICAL COMPOSITION

Temperature also affects the final composition of wine. The average volatilization of alcohol at 26 to 28°C, for example, has been quantified at 0.2% (vol/vol). Glycerol production also increases at higher temperatures. One indirect consequence of this is that musts fermented at higher temperatures need a continuous addition of SO_2 during

TABLE 11.1   Percentage of Ethanol Obtained and Onset of Fermentation According to Temperature

Temperature (°C)	Onset of Fermentation	Ethanol (%)
10	8 days	16.2
15	6 days	15.8
20	4 days	15.2
25	3 days	14.5
30	36 hours	10.2
35	24 hours	6.0

fermentation because the SO_2 added at the beginning of fermentation is more likely to combine with other components of the must and therefore lose its antiseptic properties.

As far as the volatile components of wine are concerned, the general tendency is that, at higher fermentation temperatures the levels of higher alcohols increase while those of fatty acid esters decrease. The volatility of aroma compounds also increases with temperature, explaining why white wines should be fermented at low temperatures (to retain their primary aromas).

Oxygen Concentration

Since yeasts require oxygen to multiply, the oxygen concentration in musts affects yeast activity. Yeast can metabolize sugar by respiration or fermentation and, in theory, both of these catabolic pathways are regulated by the presence or absence of oxygen. In must, however, yeast metabolizes sugar exclusively by fermentation, regardless of the oxygen concentration of the medium (Crabtree or counter-Pasteur effect).

Sterols have an important biological function in yeasts as they participate in membrane synthesis and are the origin of certain hormones. Sterol synthesis is linked to the presence of oxygen. At the beginning of fermentation, yeasts can use their sterol reserves, but these become depleted if fermentation continues under strictly anaerobic conditions. Brief contact with air is therefore important to ensure the synthesis of sterols and consequently the multiplication of yeasts and the completion of alcoholic fermentation. It has been suggested that this can cause the yeast population in a must to double, consequently increasing the rate of fermentation and the amount of sugar converted to alcohol.

The complete clarification of white grape must (performed prior to fermentation) is not recommended because it eliminates part of the microbial population, deprives yeasts of both a physical support and the traces of oxygen present in the highly porous particles removed (essential for the synthesis of sterols), and eliminates nutrients that are essential for yeast growth.

7.2. Biological Factors

Yeasts

Yeasts can be classified according to their ability to produce ethanol (low, moderate, or high). Fermentation is not conducted by a single species of yeast but rather by several species that intervene throughout the process. Indeed, the type of yeast that intervenes varies as the must is gradually converted to wine. The selection of yeasts throughout this process occurs as follows:

1. Grapes are a natural source of different species of yeast.
2. Part of the microflora is eliminated during devatting, and the first selection occurs with the addition of SO_2, with strains that tolerate this compound surviving.
3. Further selection occurs during fermentation depending on tolerance of ethanol and anaerobic conditions. Oxidative yeasts predominate during the early stages and are gradually replaced by fermentative *Saccharomyces* strains.

Apart from their influence on ethanol content, yeasts also affect other compounds such as terpenes, esters, and higher alcohols, which all contribute to the organoleptic characteristics of wine.

7.3. Chemical Factors

Sugar Content

Musts with a sugar concentration of over 300 g/L have a high osmotic pressure, which adversely affects the growth and fermentative ability of yeasts. This effect is accentuated by the formation of alcohol, which inhibits growth, and occasionally, causes yeasts to produce more acetic acid. Tolerance of high sugar concentrations varies from one yeast species to another.

pH

The pH of must generally lies between 3.0 and 3.9. It is erroneously believed that yeasts work better at lower pHs. What actually occurs is that most harmful bacteria grow better in weakly acidic media, explaining why some authors recommend maintaining a pH of approximately 3.2. Nonetheless, acidity is an extremely important organoleptic property of wine, and it is a key determinant of whether a wine is accepted or rejected. Musts made from healthy grapes can be fermented at a pH of over 3.6, as the risk of bacterial contamination is minimal. In such cases, thus, it is the enologist who makes an important contribution to the final character of the wine.

Ethanol Concentration

The final ethanol concentration of wine depends on the sugar content of the must and the fermentative capacity of the yeasts involved. In practice, fermentation is considered to be a success when approximately 1 unit of alcohol is obtained for every 17 grams of sugar.

Ethanol is toxic to yeast, with toxicity increasing with concentration. This toxic effect is reflected by a decrease in the proportion of nitrogen assimilated by yeasts in the presence of ethanol. All species of yeast have an ethanol tolerance limit. Once this is exceeded, they will die. Accordingly, it is desirable that the yeasts responsible for fermentation can tolerate ethanol to ensure that all the sugars in the must are transformed to ethanol. Wines which contain no residual sugars are microbiologically stable.

Sulfites

The addition of sulfites (SO_2) is a widespread practice in wineries, not only to control alcoholic fermentation, but also to preserve the finished product. SO_2 exerts both a biological and chemical effect on fermentation.

The biological effect of SO_2 is to inhibit the proliferation of microorganisms that adversely affect the course of fermentation and the final quality of the wine. Examples of these microorganisms are certain yeasts and in some cases, lactic acid and acetic acid bacteria.

SO_2 also has antioxidant properties (protection against oxygen) and antioxidase properties (protection against enzymatic oxidation). It also has a tendency to combine with carbonyl compounds, in particular acetaldehyde.

The following equilibria can be established for the solubilization of SO_2, although, at the pH of wine, the first reaction predominates:

$$(1)\ SO_2 + H_2O\ \leftrightarrow\ HSO_3^- + H^+$$
$$(2)\ HSO_3^-\ \leftrightarrow\ SO_3^{2-} + H^+$$

There are two main forms of SO_2 in must and wine: free SO_2 (molecular SO_2 and the bisulfite form) and bound SO_2 (mostly involving combinations with carbonyl compounds). Total SO_2 includes all free and bound forms of this compound. Molecular SO_2 is also known as active SO_2 as it is in this form that SO_2 has the greatest antiseptic and antioxidant properties, although levels never exceed 10% of all free forms of SO_2. The bisulfite form also has antiseptic and antioxidant properties but to a lesser degree.

The acetaldehyde produced by yeasts during fermentation combines with SO_2 to form a highly stable compound. Reversible combinations are also formed and their equilibrium depends on numerous factors: the substance to which SO_2 binds, total SO_2 content, and temperature. Unstable, SO_2-bound compounds may be a source of free SO_2 when levels decline.

The addition of low doses of SO_2 can favor the proliferation of undesirable microorganisms and the action of antioxidant enzymes. High doses, in contrast, result in high levels of glycerol and acetaldehyde (glyceropyruvic fermentation). In musts made with healthy grapes, the addition of 50 mg/L of SO_2 is generally sufficient to ensure successful fermentation.

8. FORMATION OF LACTIC ACID BY LACTIC ACID BACTERIA

Small amounts of lactic acid are produced by yeasts during alcoholic fermentation. Red wines contain high levels of lactic acid because they undergo malolactic fermentation. This is not generally desirable in white wines, as it replaces the characteristic fresh character conferred by malic acid with the smoother, warmer mouthfeel associated with lactic acid.

8.1. Lactic Acid Fermentation of Hexoses

Lactic acid bacteria can metabolize glucose via homolactic or heterolactic fermentation.

Homolactic Fermentation

Homolactic fermentation is performed by lactic acid bacteria of the genera *Pediococcus* and *Streptococcus* as well as by certain species of *Lactobacillus*. The first stage of this process is identical to the glycolysis performed by yeasts that leads to the formation of pyruvic acid. The second stage involves the formation of lactic acid from pyruvic acid (Figure 11.15).

Heterolactic Fermentation

Heterolactic fermentation is performed by bacteria from the *Leuconostoc* genus and by certain species of *Lactobacillus*. It is called heterolactic fermentation because in addition to lactic acid, it also yields ethanol, CO_2, and on occasions, acetic acid (Figure 11.16).

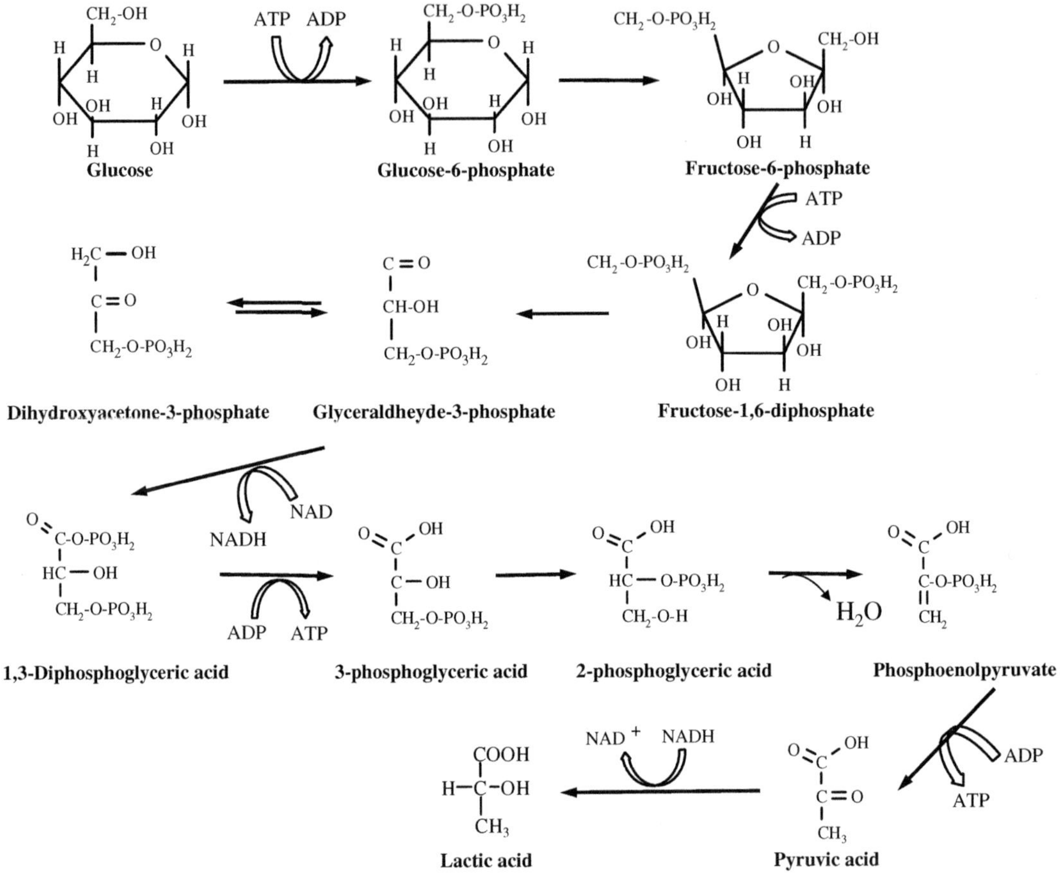

FIGURE 11.15 Homolactic fermentation of glucose. ADP = adenosine diphosphate; ATP = adenosine triphosphate; NAD = nicotinamide adenine dinucleotide.

8.2. Lactic Acid Fermentation of Pentoses

Pentose sugars such as ribose, arabinose, and xylose are fermented by certain strains of *Lactobacillus*, *Pediococcus*, and *Leuconostoc* bacteria, which can be homofermentative or heterofermentative. The conversion mechanism involved is similar to that observed for hexoses but with the production of lactic and acetic acid (Figure 11.17).

8.3. Malolactic Fermentation

Lactic acid bacteria in wine convert L-malic acid exclusively into L-lactic acid, with the release of CO_2.

The general chemical equation for malolactic fermentation is

$$\text{Malic acid} \rightarrow \text{lactic acid} + CO_2$$

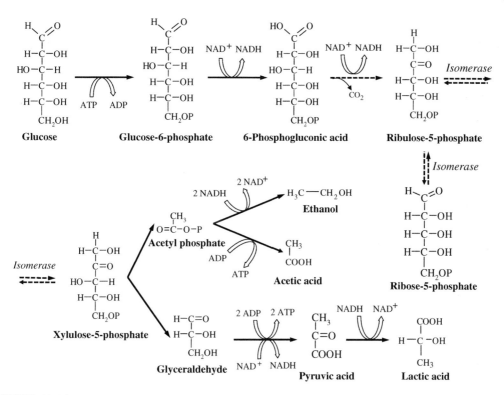

FIGURE 11.16 **Heterolactic fermentation of glucose.** ADP = adenosine diphosphate; ATP = adenosine triphosphate; NAD = nicotinamide adenine dinucleotide.

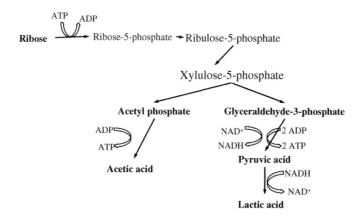

FIGURE 11.17 **Lactic acid fermentation of ribose.** ADP = adenosine diphosphate; ATP = adenosine triphosphate; NAD = nicotinamide adenine dinucleotide.

There are three pathways through which malic acid is converted to lactic acid, each involving a different enzyme:

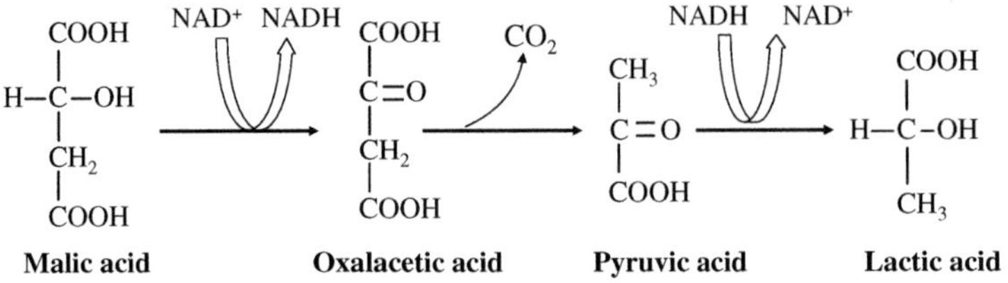

Malate dehydrogenase

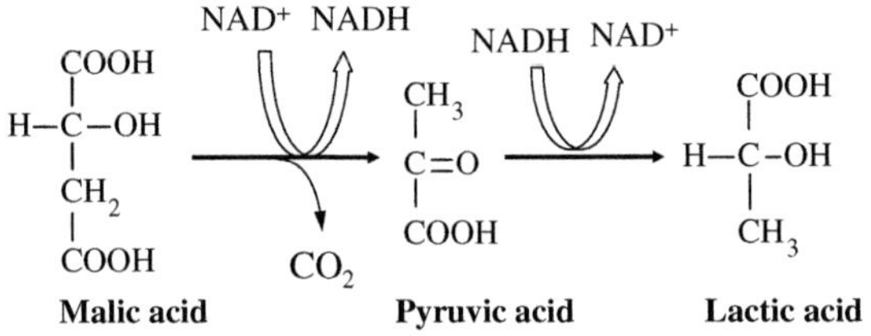

Malic enzyme

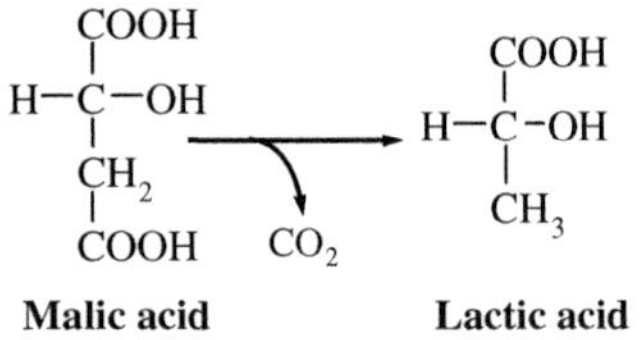

Malolactic enzyme

As can be seen in the above diagrams, no ATP molecules are produced in any of the cases, nor are there any changes in the redox state of the molecules. In other words, there is no production of reduced (or oxidized) coenzymes that subsequently need to be reoxidized, leading to the formation of ATP. What, then, is the purpose of malolactic fermentation? Even though malolactic fermentation is not very exergonic, it does provide an indirect source of real energy for the cell as it increases the pH inside the cell (malic acid is stronger than lactic acid), giving rise to a proton gradient. The entry of malic acid and the exit of lactic acid from the cell contribute to the generation of a proton motive force that allows the formation of ATP.

9. MALOALCOHOLIC FERMENTATION

Highly acidic musts (particularly those with high levels of malic acid) should ideally be deacidified. One possibility is to induce malolactic fermentation as this converts the malic acid present (which has a pK_a of 3.46) into the somewhat less acidic lactic acid

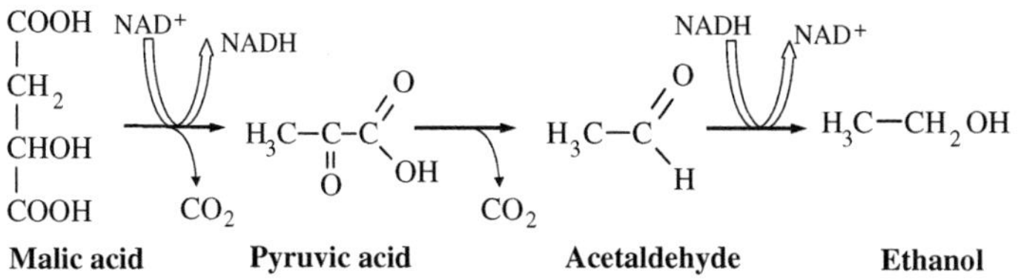

FIGURE 11.18 **Maloalcoholic fermentation.** NAD = nicotinamide adenine dinucleotide

(pK$_a$ of 3.86). It also replaces the sharp taste of malic acid (characteristic of green apples) with the sweet-sour taste of lactic acid. However, very low pH levels inhibit bacterial activity and can also prevent malolactic fermentation from taking place. Furthermore, if the levels of malic acid in the must are very high, excessive levels of lactic acid will be produced, with a detrimental effect on the organoleptic characteristics of the resulting wine.

Another alternative to deacidification with calcium and potassium salts is to use *Schizosaccharomyces* yeasts to convert malic acid to ethanol and CO_2 (maloalcoholic fermentation). This reduces the negative impact that excessive levels of lactic acid can have on the final product.

Schizosaccharomyces pombe has a high tolerance of ethanol in media with low pHs and high levels of SO_2; it also grows well in must. It does, however, have disadvantages, as it can slow down fermentation and, more importantly, generate unpleasant aromas and flavors arising from the metabolism of glucose. While, in theory, this second aspect makes *Schizosaccharomyces* yeasts inappropriate for must fermentation, they can be encapsulated in alginate beads and removed from the medium before they cause undesirable changes (i.e., before they have metabolized all the malic acid in the must). The yeast responsible for alcoholic fermentation is *S. cerevisiae*.

10. RESIDUAL SUGARS AND TYPE OF WINE

All wines contain residual sugar. This is the portion of sugar not metabolized by yeast during fermentation, consisting partly of nonfermentable sugar and partly of reducing sugar (mainly glucose and fructose). The concentration of nonfermentable sugar does not generally exceed 1 g/L in wine made with healthy grapes. Levels of reducing sugar, however, vary according to fermentation conditions and the type of wine being produced. Because the sugar content has an enormous effect on one of the key organoleptic properties of wine — flavor — wines tend to be classified according to the amount of sugar they contain. The classification system used in Spain (established by the *Estatuto de la viña y del vino* (Statute of Vineyards and Wines)) is shown in Table 11.2, along with the approximate English equivalents.

When fermentation comes to a halt because the yeasts are unable to consume all the fermentable sugars in the must due to adverse conditions, immediate measures need to be taken to prevent the growth of bacteria, and particularly that of lactic acid bacteria. These convert fructose to mannitol and any residual glucose to lactic and acetic acid.

Natural sweet wines are made by deliberately halting fermentation before the yeasts consume all the glucose and fructose present in the must, hence the name *natural sweet*. Fructose is largely responsible for the sweet flavor of these wines as it is less readily

TABLE 11.2   Final Concentration of Sugar and Type of Wine

Type of Wine	Sugar (g/L)
Seco (dry)	<5
Abocado (medium-dry)	5–15
Semiseco (semi-dry)	15–30
Semidulce (semi-sweet)	30–50
Dulce (sweet)	>50

metabolized by yeasts than glucose. This has very important technical and commercial implications in terms of wine classification, as the longer the fermentation period, the higher the glucose to fructose (G:F) ratio will be. To produce a natural sweet wine, fermentation is halted by adding grape spirit when the alcohol content of the must reaches approximately 8% (vol/vol). The must is fortified to a strength of approximately 15% (vol/vol). The resulting wine is cooled to 4°C and racked periodically to remove fermentation lees.

The G:F ratio of sweet wines made with a blend of dry wine and concentrated must, or simply stored with an appropriate dose of SO_2, is close to 1.

A similar product to sweet wines is mistelle, which is made by adding ethanol to grape must. The alcohol concentration is similar to that of wine (approximately 15%) as this is sufficient to prevent microbial growth. Because it is not fermented, mistelle contains no secondary fermentation products and has a G:F ratio of approximately 1.

Musts used to make sweet wines that have been heat sterilized contain considerable quantities of 5-hydroxymethylfurfural, a cyclic derivative of fructose, produced when this sugar is heated.

In Spain, cava, a sparkling wine, is also classified according to final sugar content, which is directly related to the dosage (also known by the French term *liqueur d'expedition*) after disgorging and before bottling.

Brut nature cava also exists. This contains no sugar as none is added to the product before bottling.

TABLE 11.3   Types of Cava by Final Sugar Content

Type of Cava	Sugar (g/L)
Extra brut	<6
Brut	6–15
Extra-dry	15–20
Dry	20–35
Semi-dry	35–50
Sweet	>50

Nitrogen Compounds

1. INTRODUCTION

Organic nitrogen-containing compounds have many important roles in nature. They display an enormous structural diversity, in which nitrogen atoms can form part of simple functional groups or complex heterocyclic systems; they also have varying degrees of substitution and oxidation. The most noteworthy of these naturally occurring molecules are proteins, and most vitamins and hormones.

Nitrogen compounds can be classified as mineral or organic. Mineral compounds are essentially formed by the ammonium ion (NH_4^+), which is generated when ammonium salts are dissolved in water. Organic compounds, in contrast, are carbon and hydrogen

TABLE 12.1   Nitrogen Fractions of Interest in Winemaking

Fraction	Compounds	Use by Microorganisms
Inorganic nitrogen	Ammonia and ammonium: NH_3 and NH_4^+	Easily assimilated
Organic nitrogen	Amino acids with a molecular mass of <200 Da	
	Polypeptides with a molecular mass of 200–10,000 Da.	Non-assimilable
	Proteins with a molecular mass of >10,000 Da	

compounds that contain a nitrogen atom. All organic nitrogen-containing compounds can be considered as derivatives of ammonia in which one or more hydrogen atoms are substituted by hydrocarbon radicals. Of particular interest in this group are α-amino acids and their peptide and protein derivatives.

Another classification system, which is of more interest in winemaking, is based on how nitrogen compounds are used by microorganisms; that is, it distinguishes between assimilable and non-assimilable compounds. Examples of the former are the ammonium ion and free amino acids, and examples of the latter are peptides and proteins.

Must made from ripe grapes contains approximately 200 to 500 mg/L of total nitrogen, which is higher than the levels found in wine. Compared to white wine, red wine made by macerating on the skins has higher concentrations of low-molecular-weight nitrogen compounds but lower concentrations of proteins, as these interact with coloring matter, and settle at the bottom of fermentation tanks. The nitrogen content of both red and white wine is equivalent to approximately 20% dry extract, which corresponds to 70 to 700 mg/L of elemental nitrogen and 0.5 to 4 g/L of protein-derived nitrogen. The factor used to convert elemental nitrogen to protein nitrogen is 6.25.

2. TOTAL NITROGEN AND ASSIMILABLE NITROGEN

The total nitrogen content of must and wine is determined using the Kjeldahl method. This consists of subjecting a sample of must or wine to digestion by boiling it in concentrated sulfuric acid containing selenium as a catalyst. All forms of nitrogen present in the sample will be converted to ammonium sulfate. Following the addition of an excess of sodium hydroxide, the ammonium ions will be released in the form of ammonia gas, which is drawn off by steam and collected in a solution of boric or sulfuric acid of a known concentration. Part of the acid is neutralized by the ammonia, and the excess is measured using a solution of sodium hydroxide of a known concentration. The ammonia content of the sample is determined by calculating the difference between the acid present before and after the ammonia is collected. The results can be expressed as % nitrogen, % NH_3, or % protein (% nitrogen × 6.25).

This method is still widely used in winemaking, although other methods exist for quantifying the different forms of nitrogen.

Assimilable nitrogen comprises the ammonium cation and amino acids. Ammonium is the form of nitrogen first consumed by yeasts. The only amino acid that is not assimilable by yeasts during fermentation is proline, although this is consumed during biological aging, a process used to obtain the *Fino* wines of Montilla-Moriles and Jerez. For fermentation to proceed smoothly, a must does not need to have a minimum content of amino acids as yeasts can synthesize all amino acids — it simply needs to have sufficient quantities of ammonium ions. Amino acids, however, are interesting as they have a stimulating effect on fermentation and are precursors of higher alcohols, which give rise to several esters that, together with higher alcohols, contribute to the aroma of wine.

Musts generally contain sufficient levels of nitrogen compounds to ensure yeast growth. In the presence of assimilable nitrogen, yeasts multiply at a faster rate, meaning that fermentation will proceed smoothly. Neither peptides nor proteins contribute to the growth or multiplication of yeasts, as they are not generally assimilable. In musts with an assimilable nitrogen content of less than 120 mg/L, it is desirable to add an ammonium salt to aid fermentation. These musts still need to contain at least 50 mg/L of ammonium, however.

3. NITROGEN COMPOUNDS IN GRAPES AND MUST

Grapes contain both mineral and organic nitrogen compounds distributed throughout the berry. The skin and seeds are the richest in terms of total nitrogen content, while the flesh contains just a quarter or a fifth of the total content.

3.1. Total Nitrogen

The total nitrogen content of ripe berries depends on the variety of grape, the rootstock, the cultivation conditions, and, particularly, the use of nitrogen fertilizers. Levels tend to be low when there is a shortage of water available to the plant. Over-ripening and onset of rot can also cause a decrease in total nitrogen.

3.2. Ammonium

The ammonium cation accounts for 5 to 10% of all nitrogen in must made from ripe grapes (this is the equivalent of a few hundredths of a gram). Ammonium is the form of

TABLE 12.2   Distribution of Total Nitrogen (mg/100 Berries) in Different Parts of Ripe Grapes (Adapted from Blouin and Guimberteau, 2004)

	Sauvignon	Semillon	Cabernet Sauvignon	Merlot
Skin	990	865	530	525
Seeds	465	386	523	630
Must	480	363	214	355

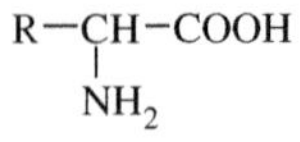

$$R-\underset{\underset{NH_2}{|}}{CH}-COOH$$

FIGURE 12.1   General structure of an α-amino acid.

nitrogen that is most rapidly assimilated by yeasts and its levels determine the fermentability of the must. When they are below 50 mg/L, it is desirable to add diammonium salts (phosphate or sulfate), but when they are below 25 mg/L, nitrogen in the form of ammonium salt should be added. The recommended dose in such cases is 10 g/hL if the aim is to achieve rapid onset of fermentation. The maximum permitted dose in the European Union is 30 g/hL.

The addition of diammonium salts without analytical control (i.e., without checking whether these salts are actually needed) is not recommended as it can lead to a decrease in odorant compounds (particularly higher alcohols) and esters (particularly fatty acid ethyl esters). This is because, in the presence of excess ammonium, yeasts are unable to metabolize amino acids, which are precursors of higher alcohols and other compounds. In turn, a reduction in higher alcohol levels will result in a reduction of their esters.

3.3. Amino Acids

Amino acids are molecules that contain a free carboxyl group and a free amino group. Proteins are formed by α-amino acids.

The R substituent is a feature of all amino acids, and, from a chemical perspective, it varies greatly from one amino acid to another. The α carbon is asymmetric in all amino acids except glycine.

In all cases, the electrical charge of an amino acid in solution is determined by its isoelectric point and the pH of the medium. When the pH is higher than the isoelectric point, the amino acid will carry a negative charge due to the ionization of the carboxyl group. In contrast, when the pH is lower than the isoelectric point, the amino acid will carry a positive charge due to the ionization of the amino group. If the pH of the medium is the same as the isoelectric point, the net charge on the amino acid will be zero.

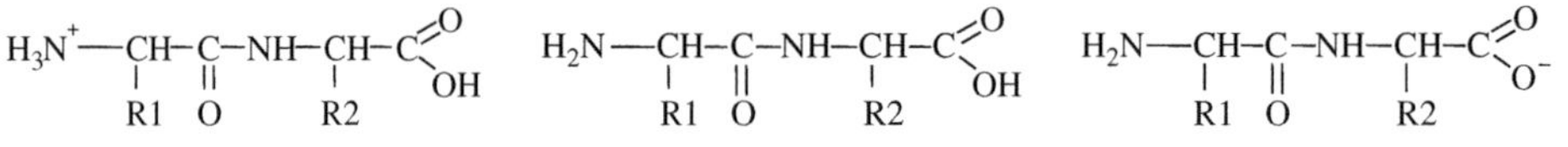

pH < IP   pH = IP   pH > IP

Must contains approximately 20 amino acids, which together account for 30 to 40% of the total nitrogen content. The quantitative and qualitative composition varies with grape variety, type of soil, cultivation practices, and climate. Attempts have been made to establish differences between varieties based on relationships between different amino acids, but with little success as the amino acid content of grape varieties depends on the characteristics of the vintage. Attempts have also been made to link the amino acid profile of wines to particular grape varieties, but the amino acids present in wine do not correspond to those present in the must from which the wine is made.

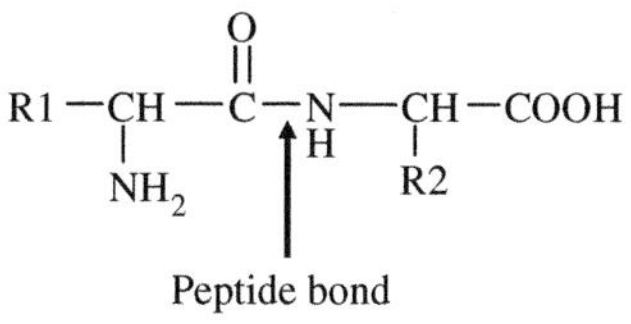

$$R1-CH-C-N-CH-COOH$$

(with O double-bonded to C, NH$_2$ below first CH, H below N, R2 below second CH)

Peptide bond

FIGURE 12.2   **General structure of a dipeptide.**

3.4. Peptides and Proteins

The linking of amino acids via a peptide bond gives rise to peptides and proteins. The structure of a dipeptide is shown in Figure 12.2.

Peptides are amino acid polymers. Peptides were traditionally defined as polymers with a molecular mass of less than 10,000 Da. The current tendency, however, is to distinguish between oligopeptides, which are combinations of at most four amino acids, and polypeptides, which are combinations of more than four amino acids. One particularly noteworthy peptide in winemaking is glutathione, a tripeptide that possesses the sulfur-containing amino acid cysteine and whose presence in must increases resistance to oxidation of coloring matter. Glutathione reacts with quinones resulting from the oxidation of several phenolic compounds to produce a compound that can only be oxidized by the enzyme laccase, which is present in rotten grapes. In contrast, it cannot be oxidized by the enzyme tyrosinase, present in healthy grapes.

Proteins are combinations of amino acids with a molecular weight of over 10,000 Da. Like peptides, they cannot be assimilated by yeast during fermentation. Because of their importance in colloidal processes and the stabilization of wine, they will be examined in a separate section.

4. CHANGES IN NITROGEN CONTENT DURING GRAPE RIPENING

The nitrogen content of grapes increases during ripening; indeed, grape skin and flesh can exhibit a two-fold or higher increase in total nitrogen content during this time.

FIGURE 12.3   **Total nitrogen content of grape juice and skin during the ripening of Tempranillo grapes.**

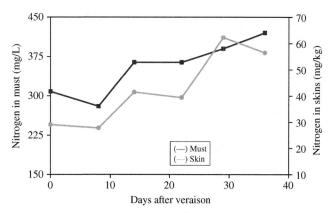

During the herbaceous stage, ammonium ions account for approximately 80% of the total nitrogen content of the berry and are derived from nitrates taken from the soil by the roots of the vine. Ammonium accounts for 50% of the total nitrogen content of grape juice before veraison, but 20% or less at the end of ripening. This is because it is converted to more complex forms of nitrogen after veraison. This conversion occurs via transamination reactions with ketoacids in the Krebs cycle and sugar respiration, in which ammoniacal nitrogen is first converted to amino acids and then to proteins.

The free amino acid content of grape juice increases as the berries ripen; indeed, levels can increase by a factor of between two and five. Free amino acids account for the greatest proportion of total nitrogen by weight, and levels can vary from 1 to 4 g/L from one year to the next. In ripe grapes, free amino acids account for 30 to 40% of the total nitrogen content. Arginine, proline, glutaminic acid, and its aminated form, glutamine, are the main amino acids in grape juice. In most varieties, proline and arginine predominate, although the levels of individual amino acids vary from year to year.

Influence of Cultivation Practices and Climate

Ammonium in grapes comes from nitrates taken from the soil by the roots of the vine. The nitrate content of soil can vary greatly depending on the type of soil, the fertilizers used, water availability, and cultivation practices. Essentially, the more fertile a soil is (higher content of nitrogen or greater use of fertilizers), the higher the total nitrogen content of the grapes will be.

The use of ground cover (vegetation on the vineyard floor) reduces the ammoniacal nitrogen content of grapes. Indeed, must produced from grapes grown in ploughed vineyards contains twice as much ammonium as that made using grapes from vineyards with ground cover. The effect on assimilable nitrogen levels is similar.

The total nitrogen content, like the amino acid content, also depends on climate. In years with heavy rainfall and high temperatures (unfavorable conditions for winemaking), it is higher than in years with a more temperate climate and with less rainfall during the ripening period. Climate has the opposite effect on amino acid content; that is, amino acid levels decrease in years with unfavorable conditions.

5. CHANGES IN NITROGEN CONTENT DURING FERMENTATION

A must needs to contain a minimum amount of ammonium and amino acids if fermentation is to proceed smoothly.

5.1. Ammonium

Ammonium is the most easily assimilable form of nitrogen for yeasts and as such is consumed entirely by these microorganisms. Its level in musts should be approximately 50 mg/L.

5.2. Amino Acids

Most amino acids can be used by yeasts as an exclusive source of nitrogen. In other words, yeasts are capable of conducting fermentation with just a single amino acid, as long as this is present in sufficient quantities. The exceptions are proline, lysine, cysteine, and to a lesser extent, histidine and glycine. Proline is the only amino acid that cannot be assimilated by yeasts during fermentation as the enzyme proline oxidase is inhibited in anaerobic conditions. Indeed, proline can even be excreted during fermentation. Yeasts are also incapable of growing in musts in which lysine or cysteine is the only source of nitrogen because when these compounds are broken down, they produce a toxic intermediate that accumulates in the absence of other nitrogen sources.

Arginine is among the amino acids most widely used by yeasts, and can account for between 30 and 50% of all the nitrogen used by these microorganisms. The other amino acids are consumed in different proportions depending on, among other factors, fermentation conditions and the species or variety of yeast. Indeed it is very difficult to establish a consumption profile.

Fermentation conditions (oxygen levels, temperature, etc.) influence the consumption of amino acids by yeasts, with increased rates seen in musts fermented at higher temperatures and subject to periodic aeration (Figure 12.5).

As already mentioned, successful fermentation depends on the presence of a minimum quantity of assimilable nitrogen compounds. In the case of biological aging, during which *flor* yeasts develop aerobically and form a film over the wine, a minimum amino acid content is necessary to ensure the correct growth of these yeasts. In this case, proline is the main source of nitrogen.

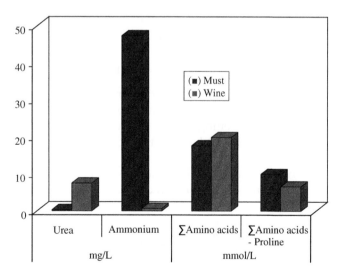

FIGURE 12.4 Ammonium, urea, total amino acid content and assimilable amino acid content in a Pedro Ximénez must fermented at 24°C and in the resulting wine.

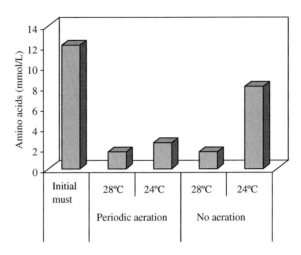

FIGURE 12.5   Consumption of amino acids during the fermentation of a must at two temperatures and with and without aeration.

6. PROTEINS

6.1. General Characteristics

Proteins are macromolecules with a molecular weight of over 10,000 Da. They are formed from linear chains of amino acids linked by peptide bonds. The spatial conformation of the proteins is determined by the sequence of the amino acids.

As occurs with amino acids, the electrical charge of a protein in solution is determined by the isoelectric point of the protein and the pH of the medium. As musts have a low pH, the proteins present will be positively charged.

6.2. Factors That Influence the Protein Content of Musts

Proteins accumulate in the grape throughout ripening, with levels peaking several days before ripeness and decreasing slightly thereafter. Several authors have observed a change in the nature of proteins during ripening, namely a reduction in the level of proteins with a molecular weight of 20,000 to 22,000 Da and the appearance of proteins with a higher molecular weight.

The total protein content of grape juice at harvest time can range between 30 and 300 mg/L. The fractionation of different proteins has shown quantitative but not qualitative differences

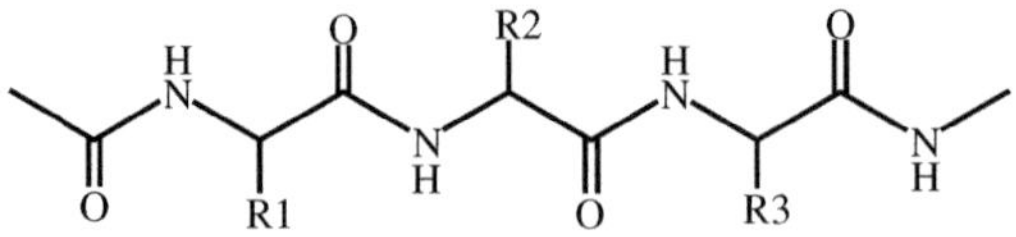

FIGURE 12.6   Structure of a polypeptide chain.

for the same grape variety, regardless of its origin. Nevertheless, both quantitative and qualitative differences are observed when two varieties are compared.

The protein content of must also increases when grapes are macerated on their skins prior to fermentation. The increased extraction of proteins occurs in the first few hours during which the must is in contact with the skins. The addition of SO_2 to grapes during maceration also favors the extraction of proteins, probably because SO_2 weakens the cells of the flesh and the skin, leading to an increased release of proteins into the must. This explains why the addition of SO_2 prior to maceration is not recommended. Indeed, when SO_2 is added, it should be done in an inert CO_2 atmosphere to prevent oxidation.

Musts made from destemmed grapes have a higher protein content than those made from grapes pressed in whole bunches. This is because the stems contain tannins that bind proteins and extract them from the must.

The mechanical processing of grapes weakens the skins and therefore facilitates the secretion of proteins. In such cases, additional bentonite is required to protect the wines against protein degradation.

6.3. Changes in Protein Content during Fermentation

Grape-derived proteins are resistant to the action of yeast proteases and are therefore not assimilated during fermentation. Red wine contains practically no proteins as these are precipitated by tannins. In contrast, the protein content of white and rosé wines is very similar to that of the initial must, and there are even cases when peptide nitrogen levels increase due to the release of thermostable peptides during yeast autolysis.

Grape-derived proteins cause instability in the clarity of white wines. Specifically, they cause protein degradation, which results in haze in bottled wine stored at high temperatures, or in wines enriched with tannins from the cork. There are several methods for testing the protein stability of a wine, but the most practical and reliable method is a test based on the sensitivity of proteins to heat treatment. This consists of heating 100 mL of wine in a water bath at 80°C for 30 minutes. The wine is considered to be stable if the resulting haze is less than 2 NTU (nephelometric turbidity units). If higher, the wine will need to be treated.

Proteins that are naturally present in must are thermally unstable, although there is no direct relationship between the concentration of proteins and the haze caused by heat. This indicates that not all proteins are affected to the same extent by heat, although they are all, to a greater or lesser extent, responsible for haze.

The addition of bentonite is the most common method for eliminating excess proteins. The amount of bentonite needed to stabilize wines has increased considerably in recent years (from 40 to 120 g/hL) because of the increased use of mechanical processes and new production techniques that result in a greater transfer of proteins from the grapes to the must. The problem is that high levels of bentonite have a negative effect on the aroma profile of wine.

One alternative to the use of bentonite is to age the wine on fermentation lees. During this process, the yeast cells release a mannoprotein that provides white wine with increased protein stability and considerably reduces the amount of bentonite needed to stabilize the wine. Wines aged on fermentation lees also appear to exhibit greater stability towards tartrate precipitation.

7. OTHER NITROGEN COMPOUNDS

7.1. Urea and Ethyl Carbamate

Urea can be defined as a derivative of carbonic acid or as the amide of carbamic acid. Its presence in wine is mostly of microbial origin.

| Carbonic acid | Carbamic acid | Urea |

Urea is of interest to winemakers because it is a precursor of ethyl carbamate. While carbamic acid is highly unstable and is easily broken down into carbonic acid and ammonia, its ethyl esters (urethanes) are very stable. The monoethyl ester (ethyl carbamate) is naturally present in fermented beverages as the result of the metabolic activity of yeasts. It has also been found in grapes as a metabolic intermediate. Because ethyl carbamate has been attributed with carcinogenic potential, its levels should be kept at a minimum and under no circumstances exceed 15 µg/L.

Of note among the factors that influence ethyl carbamate levels in wines are:

1. Grapevine variety and use of nitrogen fertilizers. These both influence the final amino acid composition of the must. In general, musts with normal levels of assimilable nitrogen give rise to wines with low urea levels and consequently low carbamate levels. Lower levels have also been detected in wines made from varieties with a lower arginine to proline ratio.
2. Certain prefermentation operations, such as hot maceration, which can give rise to wines with a higher concentration of ethyl carbamate.
3. The strain of yeast.
4. Fermentation at high temperatures; this favors the consumption of nitrogen compounds and the formation of urea.
5. Malolactic fermentation. Wines that undergo this type of fermentation have higher concentrations of ethyl carbamate than other types of wine, but concentrations depend on the strain of bacteria and the moment at which fermentation is ended.
6. Aging on lees. This results in increased levels of ethyl carbamate.
7. High storage temperatures. These favor the formation of ethyl carbamate.

7.2. Biogenic Amines

The substitution of one, two, or three of the hydrogen atoms in NH_3 by aliphatic or aromatic radicals gives rise to primary, secondary, or tertiary amines, respectively. The volatile amines methylamine, ethylamine, diethylamine and phenethylamine are important in winemaking as they have an undesirable effect on aroma. Another group of amines, the biogenic amines (ethanolamine, histamine, tyramine, putrescine, and cadaverine), are found in many products fermented by lactic acid bacteria, or in poorly conserved food. Of note in wine is histamine, which is responsible for allergy and headaches.

Histamine

$$\text{N} \quad \text{CH}_2-\text{CH}_2-\text{NH}_2$$

Most biogenic amines are derived from the decarboxylation of amino acids, and the enzymes that catalyze these transformations are found in numerous strains of *Pediococcus* or *Lactobacillus* bacteria. Of note among the precursors of biogenic amines is arginine, which is directly or indirectly related to the synthesis of at least four biogenic amines.

Histidine is the precursor of histamine, but no association has been found between levels of this amino acid in musts and histamine levels in wine. Histamine appears during alcoholic fermentation and increases during malolactic fermentation.

Biogenic amines will generally be more abundant in red wine as this typically undergoes malolactic fermentation and numerous bacteria have the enzymatic machinery required to produce them. Biogenic amine levels in wine are minimized by adding 50 g/hL of bentonite, which is negatively charged and therefore binds the positively charged amine molecules.

1. INTRODUCTION

The earliest definitions of acids and bases described acids as compounds or substances with a sour taste, and bases as substances that neutralized or reversed the action of acids. These definitions, however, have evolved as knowledge in this area has developed.

Lavoisier believed that all acids contained oxygen, but this theory was disproved with the discovery that hydrochloric acid did not. Later, acidity was attributed to the presence of hydrogen, but many hydrogen-containing compounds do not exhibit acidic behavior. Arrhenius established a definition of acids that is still widely used, especially for aqueous solutions. He essentially defined an acid as a hydrogen-containing substance that produced H^+ ions in solution; accordingly a base was defined as a substance that contained hydroxyl groups and produced OH^- ions in solution. He described neutralization as the combination of H^+ and OH^- ions to form water:

$$H^+ + OH^- \rightarrow H_2O$$

The first difficulties that arose with Arrhenius' theory were related to the fact that H^+ does not exist as such, as it has a high charge density and tends to become hydrated:

$$HCl + H_2O \rightarrow Cl^- + H_3O^+$$

Hence, it was considered necessary to reformulate Arrhenius' theory as it was believed that the actual ion that existed was H_3O^+. Consequently, for a substance to be considered an acid, it did not actually have to dissociate but simply be capable of donating a proton to another molecule. Furthermore, certain substances are considered as bases even though they do not produce OH^- ions:

$$HCl + NH_{3\,(g)}^{\;*} \rightarrow NH_4Cl \rightarrow NH_4^+ + Cl^-$$

*NH_3 is the formula for gaseous ammonia. When it dissolves in water, what really exists is ammonium hydroxide (NH_4OH).

Brönsted and Lowry extended and rectified Arrhenius' theory to produce new definitions of acids and bases, although they were still based on protons:

- Acid: a substance capable of donating a proton
- Base: a substance or species that accepts a proton

According to Brönsted and Lowry's definitions, an acid-base reaction is a proton-transfer reaction:

$$HCl + H_2O \rightarrow Cl^- + H_3O^+$$

This acid-base reaction reflects the competition between the two bases to accept a proton:

$$HCl + H_2O \rightarrow Cl^- + H_3O^+$$
$$\text{acid1 \quad base2} \qquad \text{base1 \quad acid2}$$

$$NH_3 + H_2O \rightleftarrows NH_4^+ + OH^-$$
$$\text{base1 \quad acid2} \qquad \text{acid1 \quad base2}$$

$$HCl + NH_3 \rightarrow NH_4^+ + Cl^-$$
$$\text{acid1 \quad base2} \qquad \text{acid2 \quad base1}$$

$$H_2O + H_2O \rightleftarrows H_3O^+ + OH^-$$
$$\text{acid1 \quad base2} \qquad \text{acid2 \quad base1}$$

The above reactions do not directly show the H^+ ion. The equation shows how an acid donates a proton to a base to form the conjugate base and acid, respectively. According to the Brönsted-Lowry definition, a conjugate acid-base pair or system is established; it is then possible to construct a scale showing the relative strength of acids and bases in which the strongest acid is that which has the greatest tendency to donate a proton. The strongest base, in contrast, is that which has the greatest tendency to accept a proton. It can therefore be deduced that the stronger an acid, the weaker its conjugate base will be, and, conversely, the stronger a base, the weaker its conjugate acid will be. Although this theory is the most widely used for aqueous solutions, one should not rule out an acid-base theory that considers substances that have an acidic or basic nature, even though they do not produce H^+ or OH^- ions in solution. To include these substances, Lewis extended the definition of acidity and basicity by establishing the following:

- Acid: a substance capable of accepting a pair of electrons
- Base: a substance capable of donating a pair of electrons

According to Lewis, acid-base reactions involve the formation of a coordinate or dative covalent bond:

$$H^+ + [:\ddot{O}:H]^- \rightleftarrows H:\ddot{O}:H$$
$$\text{Acid} \quad \text{Base}$$

Lewis' definition is the broadest as it includes and extends all previous definitions, and is particularly useful in non-aqueous solutions. Nonetheless, the most widely used acid-base concept is that which involves the transfer of a proton; that is, the Brönsted-Lowry theory, which was an extension of Arrhenius' work.

2. LAW OF MASS ACTION

When two substances (A and B) react chemically, the progress of the reaction can be followed from the moment the substances are mixed (t_0) to the moment at which there are no longer any changes in the concentrations of the reagents (A and B) or the products (C and D) (t_3). This process is shown in Figure 13.1.

$$A + B \rightarrow C + D$$

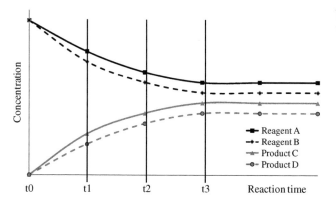

FIGURE 13.1 **Changes in concentration with time in a hypothetical reaction.**

The reaction starts at t_0; at t_1 the concentrations of A and B are decreasing while those of C and D have started to increase; at t_2, A and B continue to decrease while C and D continue to increase. However, the rate at which the concentrations are changing is now slower than at the beginning of the reaction (interval t_0–t_1). By t_3, the concentrations have practically leveled off. The following can thus be established:

- t_0: forward reaction only.

$$A + B \rightarrow C + D$$

- t_1: formation of C and D and initiation of reverse reaction.

$$A + B \rightleftarrows C + D$$

- t_2: decrease in rate of forward reaction and increase in rate of reverse reaction. The reason for this is simple: the concentrations of A and B are decreasing while those of C and D are increasing.
- t_3: no net change in the concentrations of the reagents or the products, as both are formed and consumed at the same rate.

When this point is reached — that is, when there are no changes in the concentrations of either the reagents or the products — the system is said to be at equilibrium and there will be no changes in either direction unless the conditions change.

A chemical system reaches equilibrium when opposing reactions take place at the same rate. This gives the macroscopic appearance of stability and permanence, even though at the molecular level, the reagents are converted to products, which, in turn, are converted to reagents. Equilibrium thus is a dynamic state.

One of the methods used to follow the progress of a reaction and express its current state involves the application of the law of mass action, which, for a general reaction, is expressed as follows:

$$a\,A + b\,B \rightleftarrows c\,C + d\,D$$

$$Q = \frac{[C]^c \cdot [D]^d}{[A]^a \cdot [B]^b}$$

Because the concentrations of C and D increase while those of A and B decrease, the coefficient Q will increase. When equilibrium is reached, there will be no further changes in these concentrations. At this point, Q acquires a specific value, which is constant for each reaction and is denoted K.

At equilibrium:

$$K = Q_{eq} = \frac{[C]_{eq}^c \cdot [D]_{eq}^d}{[A]_{eq}^a \cdot [B]_{eq}^b}$$

[ ] = concentration (mol/L)

The equilibrium constant, K, describes a condition that a system must fulfill at equilibrium. $K \Rightarrow$ describes the "condition of equilibrium".

2.1. Ionic Product of Water

When the above concepts are applied to the reaction in which water is ionized, the following can be expressed:

$$H_2O + H_2O \rightleftarrows OH^- + H_3O^+$$

$$\text{acid1} \quad \text{base2} \quad \text{base1} \quad \text{acid2}$$

This chemical equation is produced because water is amphoteric; that is, it can act as an acid or a base. The equilibrium constant for this reaction is expressed as:

$$K_w = \frac{[H_3O^+] \cdot [OH^-]}{[H_2O]^2} = K_{H_2O} = \text{constant}$$

K_w is the constant of the product of the molar concentration of the protons and hydroxyl ions produced in the self-ionization of water. At 25°C it has a value of 10^{-14} when the concentrations are expressed in moles per liter.

$[H_2O]$, the concentration of water, is considered to be constant as the proportion of water that dissociates is very small compared to the proportion that does not. It can therefore be stated that:

$$K_w = 10^{-14} = [H_3O^+][OH^-]$$

Although the self-ionization of water is an equilibrium reaction that contributes H^+ and OH^- ions to a solution, the concentration of these ions is negligible compared to the H^+ ions produced by an acid or the OH^- ions produced by a base in their dissociation reactions (unless of course the acids or bases are very weak).

2.2. pH

Natural solutions and liquids tend to contain very low concentrations of H^+ ions. Pure water, for example, has $[H^+] = 10^{-7}$ mol/L. To avoid the use of decimals or exponents to express the concentration of protons in aqueous solutions, Sörensen introduced the concept of pH:

$$pH = -\log[H_3O^+] \rightarrow pH = -\log[H^+]$$

Accordingly, pOH was defined as $-\log[OH^-]$.

The equilibrium constant pK can also be expressed as $-\log K$.

For water: $pK_w = pH + pOH \rightarrow pH + pOH = 14$

For pure water: $pH = -\log[H^+] = -\log[10^{-7}] = 7$

$$pOH = 14 - 7 = 7$$

TABLE 13.1   pH Values of Natural Liquids

Liquid	pH
Pure water	7
Water	5 – 8
Beer	4 – 5
Wine	3 – 4
Vinegar	2.5 – 3.5
Sodas and other carbonated drinks	1.8 – 3.0

An acid solution has a higher concentration of H^+ ions than pure water, and accordingly, will have a pH of less than 7. A neutral solution, in turn, has the same concentration of H^+ ions as pure water and will therefore have a pH of 7. Finally, a basic solution will have a pH of more than 7 as it has a lower concentration of H^+ ions than pure water.

The contribution from the self-ionization of water in aqueous solutions is only taken into account for very weak acids or bases; that is, substances that have a concentration of H_3O^+ or OH^- ions less than 10^{-6} M.

3. DISSOCIATION CONSTANTS: STRENGTH OF ACIDS AND BASES

The relative strength of acids and bases is indicated by the value of their dissociation constant (K_a and K_b). Both K_a and K_b are equilibrium constants. Therefore, the higher their value, the more the equilibrium is shifted to the right and the greater their tendency to produce H^+ or OH^- ions.

$$AcOH \leftrightarrows AcO^- + H^+; K_a = \frac{[AcO^-] \cdot [H^+]}{[AcOH]}$$

The strength scale is used only for weak acids and bases as the equilibrium of strong acids and bases is always displaced completely to the right. In such cases, the value of K_a or K_b is ∞, as the molecular species does not exist in solution.

Example for a strong acid:

$$HCl \rightarrow Cl^- + H^+; \qquad K = \frac{[H^+] \cdot [Cl^-]}{[HCl]} = \infty; \text{ as } [HCl] \rightarrow 0$$

Examples of strong acids are $HClO_4$, HCl, HNO_3 and H_2SO_4, and strong bases are $NaOH$ and KOH.

TABLE 13.2   Dissociation Constants and pK_a of a Range of Acids

Acid	K_a (25°C)	pK_a
HSO_4^-	1.2×10^{-2}	1.92
$ClCH_2\text{-}COOH$	1.4×10^{-3}	2.85
HF	6.8×10^{-4}	3.17
HCOOH	1.8×10^{-4}	3.74
$CH_3\text{-}COOH$	1.8×10^{-5}	4.74
CO_2	4.2×10^{-7}	6.38
HS^-	10^{-14}	14

According to the values in the above table, the higher the equilibrium constant (K_a), the lower the pK_a will be. Therefore, acids with a higher K_a (and a lower pK_a) will be stronger. The same occurs with bases (higher K_b and lower pK_b).

3.1. Degree of Dissociation

The degree of dissociation (α) for weak acids and bases is defined as the fraction of 1 mole that dissociates at equilibrium.

$$AcOH \leftrightarrows AcO^- + H^+$$

Initial point:	1 mole	0	0
Equilibrium:	$1 - \alpha$	α	α

$$K_a = \frac{c^2\alpha^2}{c(1 - \alpha)} = \frac{c\alpha^2}{1 - \alpha}$$

If there are c moles: $c(1 - \alpha)$ $c\alpha$ $c\alpha$

By definition, α is a unit factor but it is often also expressed as a percentage.

If absolute concentrations are used instead of the degree of dissociation, the equilibrium constant is expressed as follows:

$$AcOH \leftrightarrows AcO^- + H^+$$

Initial point: c   0   0

$$K_a = \frac{x^2}{c - x}$$

Equilibrium: c − x   x   x

On comparing the expressions at equilibrium, it is seen that:

$$c\alpha = x; \quad \Rightarrow \alpha = \frac{x}{c}$$

Two methods can be used to solve the K_a equation: the first involves resolving the second-degree equation and the second involves performing successive approximations. In the second method, the following approximation is performed:

$c-x \approx c$, which gives the following equation:

$$K_a = \frac{x^2}{c}; \text{ from which } \rightarrow x = \sqrt{K_a \cdot c}$$

The value obtained for x is used to calculate a value for $c-x$, which, in turn, is used to calculate a new value for x. The method ends when there is practically no difference between two successive approximations.

Example

Let us consider a solution of acetic acid: $[AcOH] = 10^{-3}$ M; K_a (AcOH) $= 1.8 \times 10^{-5}$

$$AcOH \leftrightarrows AcO^- + H^+$$
$$10^{-3} - x \quad\quad x \quad\quad x$$

Second-degree equation method:

$$K_a = \frac{x^2}{10^{-3} - x}; \quad x = [H^+] = \frac{-K_a \pm \sqrt{K_a^2 + 4K_aC_0}}{2} = 1.27 \times 10^{-4}M; \Rightarrow pH = 3.90$$

Successive approximation method:

$$K_a = \frac{x^2}{10^{-3}};$$

$$x = \sqrt{10^{-3}K} = 1.36 \times 10^{-4};$$

$$K_a = \frac{x^2}{10^{-3} - 1.36 \times 10^{-4}}; \quad x = \sqrt{8.64 \times 10^{-4}K_a} = 1.26 \times 10^{-4}$$

$$K_a = \frac{x^2}{10^{-3} - 1.26 \times 10^{-4}}; \quad x = \sqrt{8.74 \times 10^{-4}K_a} = 1.27 \times 10^{-4}$$

The successive approximation method is particularly useful when dealing with third- or fourth-degree equations.

Other useful relationships can also be established using K_a equations. Therefore, for a generic weak acid, HA:

$$HA \leftrightarrows H^+ + A^-$$

$$C_0 - x \quad\quad x \quad\quad x$$

$$K_a = \frac{[H^+]^2}{[HA]} = \frac{x^2}{C_0 - x}; \quad K_aC_0 - K_a x = x^2; \quad x^2 + K_ax - K_aC_0 = 0$$

The resolved equation yields:

$$x = \left[H^+\right] = \frac{-K_a \pm \sqrt{K_a^2 + 4K_aC_0}}{2}$$

If a series of conditions are fulfilled, the above expression can be simplified. When $K_a \ll C_0$, for example, the following can be established: $K_a^2 \ll 4\,K_aC_0$

$$\left[H^+\right] = -\frac{K_a}{2} \pm \sqrt{K_aC_0}$$

If the above condition is fulfilled and $K_a \ll C_0$, then:

$$\left[H^+\right] \approx \sqrt{K_aC_0}$$

Therefore, to be able to ignore the concentration of a dissociated acid with respect to the starting concentration of an undissociated acid, the acid must be very weak (low K_a) and rather concentrated (high C_0).

4. ACTIVITY AND THERMODYNAMIC CONSTANT

Acids in wine exist partly as free acids and partly as neutralized acids, forming salts. The relative proportions depend on pH (which tends to be between 3 and 4 for wine) and on the K_a of the acids. In winemaking, it is always interesting to know the pH of a particular wine. Thankfully, this can be determined easily by potentiometry. It is not so easy to determine the ratio of free acid to dissociated acid for a given acid, particularly tartaric acid.

The percentage of dissociated acid and undissociated molecular acid depends on the dissociation constant and on the pH according to the following expression:

$$HA \leftrightarrows A^- + H^+; \quad K_a = \frac{\left[H^+\right] \cdot \left[A^-\right]}{\left[HA\right]}$$

$$\left[H^+\right] = K_a \cdot \frac{\left[HA\right]}{\left[A^-\right]} \quad \rightarrow \quad -\log\left[H^+\right] = -\log K_a - \log \frac{\left[HA\right]}{\left[A^-\right]}$$

$$pH = pK_a - \log \frac{\left[HA\right]}{\left[A^-\right]} \text{ or } pH = pK_a + \log \frac{\left[A^-\right]}{\left[HA\right]}$$

The expression of the acid dissociation constant, K_a, refers to the molar concentrations of the different species that participate in the equilibrium. The validity of this expression depends on the mobility of the ions and species in the solution, and this mobility, in turn, depends on the total number of charged species in the solution.

In the next section, we are going to further explore the concept of the dissociation constant to study the relationship between pH, pK_a, and the amount of free and dissociated acid in real solutions.

The law of mass action as a function of molar concentrations, or in our particular case, the acid dissociation constant, provides an approximation to reality, as it is only valid when activities rather than concentrations are considered. In such cases, the equilibrium constant depends exclusively on temperature and type of solvent.

$$K_t = \frac{(H^+)(A^-)}{(HA)};$$

K_t = thermodynamic constant, ( ) = activity

Activity is related to molar concentration through what is known as the activity coefficient:

$$(A^-) = f_{A^-}[A^-]; \ (HA) = f_{HA}[HA]$$

To help to understand the concept of activity, the next section will look at the Debye-Hückel theory regarding the conductivity of electrolytes, which is based on an extremely simple concept.

4.1. Debye-Hückel Theory

It should be considered that ions with different charges are in constant motion due to the forces of attraction and repulsion. The mobility of ions depends on three factors: size, charge, and concentration.

Debye and Hückel imagined that a cation in solution would be surrounded by anions and cations, but because ions with the same charge repel each other, they believed that this cation

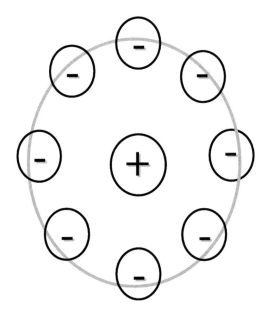

FIGURE 13.2   **Central ion surrounded by a cloud of ions with spherical symmetry, in the absence of an electric field.**

would be mainly surrounded by anions. Likewise, they held that an anion would be surrounded mainly by cations. For simplicity, it can be considered that an ion, whatever its charge, will be surrounded by a cloud of oppositely charged ions. The stronger the electrostatic interaction between any pair of ions, the lower the effective concentration of the ions will be.

The forces of electrostatic interaction (attraction or repulsion) are the main forces responsible for the formation of aggregates and, thus, for the reduction in the mobility of ions. Logically, these forces depend on the size and charge of the ions and their concentration in the medium. Accordingly, they can be considered to be non-existent in infinitely dilute solutions.

Onsager studied the effects of asymmetry and electrophoresis by extending the Debye-Hückel theory. In this model, it is considered that in the presence of an electric field, an ion and its cloud of oppositely charged ions will be facing opposite directions. The effect will be an asymmetric ion cloud, which in the absence of an electric field, will have spherical symmetry.

With respect to the electrophoretic effect, it is considered that the resulting separation of charges reduces the mobility of ions. A positively charged ion moving towards the negative pole will have reduced mobility because its surrounding cloud of ions will be oriented towards the opposite pole, and because it will also be affected by the motion of negatively charged ions, together with their cloud, towards the positive pole. The mobility of ions is also reduced by the fact that these are surrounded by solvent molecules moving towards them.

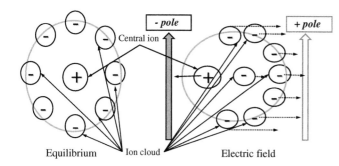

FIGURE 13.3 Effect of asymmetry in the presence of an electric field.

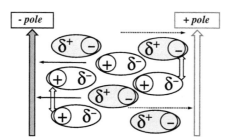

FIGURE 13.4 Electrophoretic effect or drag force.

4.2. Calculating Activity

The effective concentration or activity of an ion in an aqueous solution at 25°C can be calculated using the following expression:

$$\log f = -0.509.[Z_+ Z_-].\sqrt{I}$$

$$I = \frac{1}{2}\sum_{i=1}^{i=n} M_i Z_i^2$$

$$a = f.c$$

Z = charge of the cation and anion
I = ionic strength   a = activity
i = any cation or anion in the solution
f = activity coefficient   c = molar concentration
M_i = molar concentration of ion i

In the case of electrolytes such as NaCl and HCl, which give rise to ions in equal concentrations and with a like charge, the ionic strength is equal to the molar concentration.
Example: Consider a 0.5 M solution of NaCl in water:

$$[NaCl] \rightarrow Na^+ + Cl^-$$

$$0.5 \qquad 0.5$$

$$I = \frac{1}{2}(0.5 \times (-1)^2 + 0.5 \times (1)^2) = 0.5$$

$$\log f = -\frac{Z^2 A\sqrt{I}}{1 + B\sqrt{I}}$$

For solutions in solvents other than pure water, the activity coefficient in the Debye-Hückel expression takes account of the influence of temperature and the dielectric constant of the medium. A and B are dependent on both temperature and the dielectric constant of the medium, and Z is the charge of the ion.

Usseglio-Tomasset and Bosia have proposed the following values for the Debye-Hückel constants A and B as a function of ethanol content:

$$A = 0.5047 + 0.0042 \times (\%EtOH) + 5 \times 10^{-5} \times (\%EtOH)^2$$

$$B = 1.6384 + 0.004597 \times (\%EtOH) + 3.5 \times 10^{-5} \times (\%EtOH)^2$$

Sample Problem

Calculate the activity of Na^+ in a 0.5 M solution of NaCl and 15% EtOH

$$NaCl \rightarrow Na^+ + Cl^-$$

$$0.5 \quad 0.5$$

$$I = \frac{1}{2}\sum M_i Z_i^2 = \frac{1}{2}(0.5 \times (-1)^2 + 0.5 \times (+1)^2) = 0.5$$

$$\log f_i = -\frac{Z^2 A\sqrt{I}}{1 + B\sqrt{I}}$$

$$A = 0.5047 + 0.0042 \times (15) + 5 \times 10^{-5} \times (15)^2 = 0.57895$$

$$B = 1.6384 + 0.004597 \times (15) + 3.5 \times 10^{-5} \times (15)^2 = 1.71523$$

$$\log f_{Na^+} = \frac{1^2 \times 0.57895\sqrt{0.5}}{1 + 1.71523\sqrt{0.5}} = -0.185; \qquad f_{Na^+} = f_{Cl^-} = 10^{-0.185} = 0.653$$

$$(Na^+) = [Na^+] \times f_{Na^+} = 0.5 \times 0.653 = 0.326$$

$$(Cl^-) = [Cl^-] \times f_{Cl^-} \times 0.653 = 0.326.$$

5. MIXED AND THERMODYNAMIC DISSOCIATION CONSTANTS

We have seen the following acid dissociation constants so far:

$K_a = \dfrac{[H^+]\cdot[A^-]}{[HA]}$ Acidity constant as a function of concentration

$K_t = \dfrac{(H^+)(A^-)}{(HA)}$ Acidity constant as a function of activity

In practice, the following mixed dissociation constant is used:

$$K_m = \frac{(H^+)\cdot[A^-]}{[HA]}$$

This constant is calculated as a function of the activity of the H^+ ion and the molar concentration of A^- and HA. This is because the hydrogen ion concentration of a solution is determined by potentiometry using a pH meter, which measures the activity of this ion. It is therefore possible to establish a relationship between K_m and K_t.

$$K_t = \frac{(H^+)(A^-)}{(HA)}; \quad (H^+) = K_t\frac{(HA)}{(A^-)}$$

$$K_m = \frac{(H^+)\cdot[A^-]}{[HA]}; \quad (H^+) = K_m\frac{[HA]}{[A^-]}$$

Bearing in mind that: $(A^-) = f_{A^-}[A^-]$; $(HA) = f_{HA}[HA]$

$$K_m \frac{[HA]}{[A^-]} = K_t \frac{[HA]f_{HA}}{[A^-]f_{A^-}}; \quad K_m = K_t \frac{f_{HA}}{f_{A^-}}$$

$$pK_m = pK_t - \log \frac{f_{HA}}{f_{A^-}}$$

The above expression shows the relationship between the thermodynamic constant and the mixed constant as a function of the activity coefficients of the dissociated and undissociated forms of the acid. Presuming that a second dissociation takes place, that is, that the H^+ ions originate from an ion, the following can be written:

$$A^{Z_A} \rightarrow B^{Z_B} + H^+$$

$$pK_m = pK_t - \log \frac{f_A}{f_B}$$

Relationship between charges: $Z_A = Z_B + 1$

The relationship between the thermodynamic constant and the mixed constant as a function of the ionic strength of the medium is more interesting.

We know that $\log f = -\dfrac{Z^2 A \sqrt{I}}{1 + B \sqrt{I}}$

The activity coefficient f depends on the ionic strength I of the medium; therefore, for each ion:

$$\log f_A = -\frac{Z_A^2 A \sqrt{I}}{1 + B \sqrt{I}}$$

and

$$\log f_B = -\frac{Z_B^2 A \sqrt{I}}{1 + B \sqrt{I}}; \quad \log \frac{f_A}{f_B} = \log f_A - \log f_B$$

$$\log f_A - \log f_B = -\frac{Z_A^2 A \sqrt{I}}{1 + B \sqrt{I}} + \frac{Z_B^2 A \sqrt{I}}{1 + B \sqrt{I}} = (Z_B^2 - Z_A^2) \frac{A \sqrt{I}}{1 + B \sqrt{I}}$$

Bearing in mind that $Z_B = Z_A - 1$, the following is obtained:

$$(Z_B^2 - Z_A^2) = (Z_A - 1)^2 - Z_A^2 = Z_A^2 - 2Z_A + 1 - Z_A^2 = 1 - 2Z_A$$

The value of the mixed dissociation constant can be calculated when the values of the thermodynamic constant, the ionic strength of the medium, and the constants A and B are known.

TABLE 13.3 Relationship Between Thermodynamic Dissociation Constant and Ethanol Content

Acid	Equation
Acetic acid	$pK_t = 4.755 + 7.96 \times 10^{-3}(\%EtOH) + 2.88 \times 10^{-4}(\%EtOH)^2$
Lactic acid	$pK_t = 3.889 + 1.208 \times 10^{-2}(\%EtOH) + 1.50 \times 10^{-4}(\%EtOH)^2$
Gluconic acid	$pK_t = 3.815 + 1.48 \times 10^{-2}(\%EtOH) + 1.7 \times 10^{-5}(\%EtOH)^2$
Tartaric acid	$pK_{t1} = 3.075 + 1.097 \times 10^{-2}(\%EtOH) + 1.64 \times 10^{-4}(\%EtOH)^2$ $pK_{t2} = 4.387 + 1.47 \times 10^{-2}(\%EtOH) + 1.61 \times 10^{-4}(\%EtOH)^2$
Malic acid	$pK_{t1} = 3.474 + 1.187 \times 10^{-2}(\%EtOH) + 1.53 \times 10^{-4}(\%EtOH)^2$ $pK_{t2} = 5.099 + 1.701 \times 10^{-2}(\%EtOH) + 1.09 \times 10^{-4}(\%EtOH)^2$

$$\left. \begin{array}{l} \log \dfrac{f_A}{f_B} = (1 - 2Z_A)\dfrac{A\sqrt{I}}{1 + B\sqrt{I}} \\[3ex] pK_m = pK_t - \log \dfrac{f_A}{f_B} \end{array} \right\} \qquad \boxed{ \begin{array}{l} pK_m = pK_t - (1 - 2Z_A)\dfrac{A\sqrt{I}}{1 + B\sqrt{I}} \\[3ex] pK_m = pK_t + (2Z_A - 1)\dfrac{A\sqrt{I}}{1 + B\sqrt{I}} \end{array} }$$

Sample Calculation for the Mixed Dissociation Constant

Ethanol: 13%			
$pH = 3.23 \rightarrow 10^{-3.23} = |H^+|;$	$\equiv$ 0.59 meq/L		
Ammoniacal nitrogen (NH_4OH):	0.43 meq/L	$\equiv$	6 mgN/L
Potassium:	19.05 meq/L	$\equiv$	745 mg/L
Sodium:	0.87 meq/L	$\equiv$	20 mg/L
Calcium:	4.70 meq/L	$\equiv$	94 mg/L
Magnesium:	9.21 meq/L	$\equiv$	112 mg/L
Σ Cations:	34.85 meq/L		
Alkalinity of ash:			23.2 meq/L
Titratable acidity:			100.9 meq/L
Volatile acidity:			13.0 meq/L
Tartaric acid:			19.67 mmol/L
Malic acid:			Trace levels
Lactic acid:			26.1 mmol/L
Sulfates:		3.87 mmoles/L $\equiv$ 7.74 meq/L	
Chlorides:			1.61 meq/L
Phosphates ($H_2PO_4^-$):			3.02 meq/L

$$\Sigma Anions = \text{alkalinity of ash} + SO_4^{2-} + Cl^- + H_2PO_4^- = 35.57 \text{ meq/L.}$$

$$I = \frac{1}{2}\left[\underbrace{0.59 \cdot 10^{-3} \cdot 1^2}_{H^+} + \underbrace{0.43 \cdot 10^{-3} \cdot 1^2}_{NH_4^+} + \underbrace{19.05 \cdot 10^{-3} \cdot 1^2}_{K^+} + \underbrace{0.87 \cdot 10^{-3} \cdot 1^2}_{Na^+} + \underbrace{\frac{4.70}{2} \cdot 10^{-3} \cdot 2^2}_{Ca^{+2}} + \right.$$

$$\left. + \underbrace{\frac{9.21}{2} \cdot 10^{-3} \cdot 2^2}_{Mg^{+2}} + \underbrace{23.2 \cdot 10^{-3} \cdot 1^2}_{\text{Alkalinity of ash}} + \underbrace{3.87 \cdot 10^{-3} \cdot 2^2}_{SO_4^{2-}} + \underbrace{1.61 \cdot 10^{-3} \cdot 1^2}_{Cl^-} + \underbrace{3.02 \cdot 10^{-3} \cdot 1^2}_{H_2PO_4^-} \right]$$

Hence, $I = 0.046$

In theory, the sum of individual cations and the alkalinity of ash, which is a measure of the total number of cations that neutralize acids weaker than HCl (used to determine alkalinity) should be the same. In the current example:

ΣCations: 34.85 meq/L
Alkalinity of ash: 23.2 meq/L
SO_4^{2-}: 3.87 mmol/L = 7.74 meq/L
Cl^-: 1.61 mmol/L = 1.61 meq/L
$H_2PO_4^-$: 3.02 mmol/L = 3.02 meq/L
ΣAnions: 35.57 meq/L(*)

(*)Anions other than those that can be determined from the alkalinity of ash should also be considered, such as SO_4^{2-}, Cl^-, and $H_2PO_4^-$. As can be seen, the sum of cations (34.85 meq/L) and anions (35.57 meq/L) are quite similar, proving that the analyses were performed correctly.

As can be seen, calculating ionic strength is a very laborious process, as the concentration of each ion must be calculated. An approximate value, however, can be calculated very easily

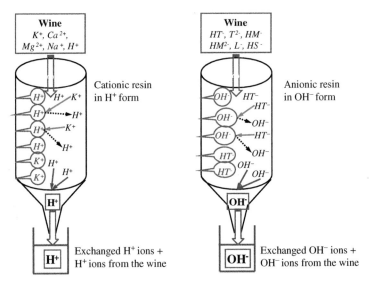

FIGURE 13.5 Analysis of cations and anions via acid (H^+) and base (OH^-) ion-exchange resins.

by analyzing the exchange acidity using a cation-exchange resin (in the H^+ form) (Figure 13.5). The difference in the acidity of a wine before and after it is passed through a resin of this type reflects the sum of all the cations retained; that is, the sum of all the cations originally present in the wine. Because wine is electrically neutral, it contains equal numbers of cations and anions. The concentration of these ions can be used to calculate the approximate ionic strength of the wine.

On passing our wine through a cation-exchange resin, it is seen to have an acidity of 35 meq/L. The ionic strength thus is $I = 0.035$. In the next step, we will calculate the value of the mixed constant as a function of the two ionic strengths calculated.

$A = 0.5678$ and $B = 1.704$, both for 13% EtOH.

It is now necessary to calculate the value of n for each acid:

$n = 2 \cdot Z_A - 1$ where Z_A is the charge of the acid species that dissociates.

For tartaric acid:

$$pK_{t1} = 3.25 \text{ and } pK_{t2} = 4.61 \text{ (13\%EtOH)}$$

$$I = 0.046 \quad pK_{m1} = 3.25 + \frac{(-1) \cdot 0.5678\sqrt{0.046}}{1 + 1.704\sqrt{0.046}} = 3.25 - 0.0892 = 3.16$$

$$pK_{m2} = 4.61 + \frac{(-3) \cdot 0.5678\sqrt{0.046}}{1 + 1.704\sqrt{0.046}} = 4.61 - 3 \times 0.0892 = 4.34$$

$$I = 0.035 \quad pK_{m1} = 3.25 + \frac{(-1) \cdot 0.5678\sqrt{0.035}}{1 + 1.704\sqrt{0.035}} = 3.25 - 0.0805 = 3.17$$

$$pK_{m2} = 4.61 + \frac{(-3) \cdot 0.5678\sqrt{0.035}}{1 + 1.704\sqrt{0.035}} = 4.61 - 3 \times 0.0805 = 4.37$$

As can be seen from the above results, the calculation of ionic strength based on the sum of individual ions (which yielded a value of 0.046) and using the exchange resin (which yielded a value of 0.035) led to an error of 1 to 2 hundredths of a unit for the value calculated for the mixed pK_a. Therefore, the acidity calculated using the cation-exchange resin can be considered adequate for calculating a close approximation of the ionic strength of the wine.

Finally, because lactic acid is the second most abundant acid in the wine analyzed, we will have the following mixed K_a values:

$$\text{Lactic acid: } pK_t = 4.08 \text{ (13\%EtOH)};$$

$$I = 0.046; \quad pK_m = 4.08 + \frac{(-1) \cdot 0.5678\sqrt{0.046}}{1 + 1.704\sqrt{0.046}} = 4.08 - 0.0892 = 3.991$$

$$I = 0.035; \quad pK_m = 4.08 + \frac{(-1) \cdot 0.5678\sqrt{0.035}}{1 + 1.704\sqrt{0.035}} = 4.8 - 0.0805 = 3.999$$

6. DISSOCIATION STATE OF ORGANIC ACIDS

A series of equations can be used to resolve acid-related problems, with the four most relevant shown below:

1. Acidity constant $K_a = \dfrac{(H^+) \cdot [A^-]}{[HA]}$: based on the equilibrium $HA \leftrightarrow H^+ + A^-$

2. Conservation of matter: $[HA]_{initial} = [HA]_{equilibrium} + [A^-]_{equilibrium}$

3. Condition of electroneutrality $\Sigma[+] = \Sigma[-]$

4. Self-ionization of water $K_w = 10^{-14} = [H_3O^+][OH^-]$, based on the equilibrium:

$$2\,H_2O \rightleftarrows H_3O^+ + OH^-$$

6.1. Monoprotic Acids

The most common monoprotic acids found in wine are acetic acid, lactic acid, gluconic acid, glucuronic acid, and galacturonic acid.

In this general study of acids in wine, the influence of ethanol content on the dissociation of the acid is not taken into account, although, generally speaking, pK_a values for organic acids increase with ethanol content by approximately one tenth for a strength of approximately 10% ethanol. It should be recalled at this point that pK_a increases with a decrease in K_a (reduction in acidity).

In the case of an acid HA dissolved in water, if the solution is sufficiently dilute, it can be considered that the activity of water is equal to 1 and hence the following equilibrium can be used:

$$HA + H_2O \rightleftarrows H_3O^+ + A^-$$

As $[H_2O]$ is a constant (1 L of water = 1 kg of water = 55.5 mol/L), the amount of water that dissociates (expressed by K_w) is negligible compared to the amount of acid that dissociates. The following can thus be written:

$K_a = \dfrac{(H_3O^+) \cdot [A^-]}{[HA]}$ (All the concentrations refer to a state of equilibrium.)

Winemakers are essentially interested in knowing the pH of a given acid solution or the proportion of free and dissociated acids.

The K_a and pK_a values of different acids have been determined, and are extremely important as, if you know the K_a and the concentration of the acid at equilibrium, it is possible to calculate the pH of the corresponding solution.

$[H_3O^+] = K_a \dfrac{[HA]_{eq}}{[A^-]_{eq}}$ $[HA]_i = [HA]_{eq} + [A^-]_{eq}$

Conservation of mass

	HA	$\leftrightarrows$ H$^+$	+ A$^-$
Initial	i	0	0
Equilibrium	i − x	x	x
Equilibrium	c(1 − α)	cα	cα;

The degree of dissociation is the fraction of 1 mole that dissociates at equilibrium.

$$\alpha = \frac{[A^-]_{eq}}{[HA]_i} = \frac{[A^-]_{eq}}{[HA]_{eq} + [A^-]_{eq}}; \quad \text{hence}: K_a = \frac{c^2\alpha^2}{c(1-\alpha)} = \frac{c\alpha^2}{(1-\alpha)}$$

The pH can be calculated using K_a, the initial concentration of the acid, and its degree of dissociation.

Given that $c\alpha = [H_3O^+]$, $pH = pK_a - \log\dfrac{1-\alpha}{\alpha}$ or $pH = pK_a + \log\dfrac{\alpha}{1-\alpha}$

Using the above expression, for an initial concentration (c) of acid, it is possible to calculate the proportion of free and dissociated acid using the pH and the pK_a.

$$[HA]_{eq} = c(1-\alpha); \quad [A^-] = c\alpha;$$

$$\text{Hence: } \log\frac{[A^-]_{eq}}{[HA]_{eq}} = pH - pK_a$$

The above equation is extremely practical as pH is easily measured and pK_a values are predefined.

6.2. Diprotic Acids

Diprotic (dicarboxylic or dibasic) acids are the most common acids in wine. In order of abundance, they are tartaric, malic, and succinic acid. Tartaric acid is the most relevant of these acids as it is the strongest and therefore makes the greatest contribution to the pH of wine.

Considering the diacid acid H_2A, there will be two acid-base ionization equilibria and we can therefore refer to two ionization steps. To better understand the mathematics involved, it is useful to analyze the equilibria in successive steps. In other words, it is considered that the second dissociation of the acid does not take place until the first one is complete.

First ionization: $H_2A + H_2O \leftrightarrows HA^- + H_3O^+$

$$K_{a1} = \frac{(H_3O^+)[HA^-]}{[H_2A]} \quad (H_3O^+) = K_{a1}\frac{[H_2A]}{[HA^-]}$$

$$pH = pK_{a1} + \log\frac{[HA^-]}{[H_2A]}; \quad pH - pK_{a1} = \log\frac{[HA^-]}{[H_2A]}$$

TABLE 13.4 K_a and pK_a Values of Diprotic Acids in Wine

	K_{a1}	pK_{a1}	K_{a2}	pK_{a2}
Tartaric acid	9.77×10^{-4}	3.01	8.9×10^{-5}	4.37
Malic acid	3.47×10^{-4}	3.46	7.47×10^{-6}	5.13
Succinic acid	6.61×10^{-5}	4.18	5.89×10^{-6}	5.23

Second ionization: $HA^- + H_2O \leftrightarrows A^{2-} + H_3O^+$

$$K_{a2} = \frac{[H_3O^+][A^{2-}]}{[HA^-]} \quad (H_3O^+) = K_{a2}\frac{[HA^-]}{[A^{2-}]}$$

The total concentration of (H_3O^+) of the solution will be the sum of the contributions of the first and second ionization steps. Nonetheless, it is standard practice to use approximations, which basically consist of ignoring the second ionization when K_{a1} and K_{a2} are very different (approximately if $K_{a1} > 10K_{a2}$). Another option is to use the successive approximation method, which consists of calculating (H_3O^+) from the first dissociation, placing this in the second equation, and calculating a value for $[HA^-]$, which, in turn, is placed in the first equation and used to calculate a new value for (H_3O^+), and so on, successively. The method is complete when two successive approximations yield identical or practically identical results.

It is also possible to calculate the proportion of each of the species at equilibrium as a function of the concentration of protons and K_a values. This is done using the dissociation equations of a diprotic acid and the mass conservation equation:

First equation: $H_2A + H_2O \leftrightarrows HA^- + H_3O^+$

$$K_{a1} = \frac{[H_3O^+][HA^-]}{[H_2A]} \qquad [HA^-] = \frac{[H_2A] \cdot K_{a1}}{[H_3O^+]}$$

Second equation: $HA^- + H_2O \leftrightarrows A^{2-} + H_3O^+$

$$K_{a2} = \frac{[H_3O^+][A^{2-}]}{[HA^-]} \qquad [A^{2-}] = \frac{[HA^-] \cdot K_{a2}}{[H_3O^+]} = \frac{[H_2A] \cdot K_{a1} \cdot K_{a2}}{[H_3O^+]^2}$$

Third equation: law of conservation of mass

$$c_i = [H_2A]_i = [H_2A]_{eq} + [HA^-]_{eq} + [A^{2-}]_{eq}$$

$$[H_2A]_i = [H_2A]_{eq} + \frac{[H_2A]_{eq} \cdot K_{a1}}{[H_3O^+]} + \frac{[H_2A]_{eq} \cdot K_{a1} \cdot K_{a2}}{[H_3O^+]^2}$$

$$[H_2A]_i = \frac{[H_2A]_{eq} \cdot [H_3O^+]^2 + [H_2A]_{eq} \cdot K_{a1} \cdot [H_3O^+] + [H_2A]_{eq} \cdot K_{a1} \cdot K_{a2}}{[H_3O^+]^2}$$

For simplicity, if we take the common factor $[H_2A]_{eq}$:

$$[H_2A]_i = [H_2A]_{eq} \cdot \left(\frac{[H_3O^+]^2 + K_{a1} \cdot [H_3O^+] + K_{a1} \cdot K_{a2}}{[H_3O^+]^2} \right)$$

This equation can be simplified by referring to the second-degree polynomial as D, as a function of $[H_3O^+]$, such that:

$$c_i = [H_2A]_i = [H_2A]_{eq} \cdot \frac{D}{[H_3O^+]^2}$$

In general, we would obtain a polynomial of n+1 terms for an H_nA acid:

$$[H_3O^+]^n + [H_3O^+]^{n-1}K_1 + \cdots + K_1K_2...K_n$$

The denominator would correspond to the concentration of protons raised to the number of dissociable protons in the acid molecule.

6.3. Calculating the Degree of Dissociation

Recalling the definition of the degree of dissociation (α), in the case of dibasic acids, there are three chemical species in solution, meaning that there will be three degrees of dissociation.

$$\alpha_0 = \frac{[H_2A]_{eq}}{[H_2A]_i} = \frac{[H_2A]_{eq}}{\dfrac{[H_2A]_{eq} \cdot D}{[H_3O^+]^2}} = \frac{[H_3O^+]^2}{D} = \frac{[H_3O^+]^2}{[H_3O^+]^2 + K_1 \cdot [H_3O^+] + K_1 \cdot K_2}$$

$$\alpha_1 = \frac{[HA^-]}{[H_2A]_i} = \frac{\dfrac{[H_2A] \cdot K_1}{[H_3O^+]}}{\dfrac{[H_2A] \cdot D}{[H_3O^+]^2}} = \frac{[H_3O^+] \cdot K_1}{[H_3O^+]^2 + K_1 \cdot [H_3O^+] + K_1 \cdot K_2}$$

$$\alpha_2 = \frac{[A^{2-}]}{[H_2A]_i} = \frac{\dfrac{[H_2A] \cdot K_1 \cdot K_2}{[H_3O^+]^2}}{\dfrac{[H_2A] \cdot D}{[H_3O^+]^2}} = \frac{K_1 \cdot K_2}{[H_3O^+]^2 + K_1 \cdot [H_3O^+] + K_1 \cdot K_2}$$

In summary, the molar fractions of the different species α_0, α_1, α_2,...α_n present when an H_nA acid containing n protons is in solution can be calculated using a fraction whose denominator is a polynomial containing $n + 1$ $[H_3O^+]$ terms and whose numerator consists of each of the terms of the following polynomial:

$$D = [H_3O^+]^n + [H_3O^+]^{n-1}K_{a1} + [H_3O^+]^{n-2}K_{a1}K_{a2} + \cdots + K_{a1}K_{a2} ... K_{an}$$

These expressions can be used to calculate the different degrees of dissociation or molar fractions of each species in solution for all weak polyprotic acids. Using the α values obtained, it is then easy to calculate the absolute concentration of each species if the total concentration of acid added (initial concentration) is known.

As can be deduced from Table 13.5, at a given pH, as an acid becomes weaker, its degree of ionization decreases and the molecular species starts to predominate. It is also clear that the

TABLE 13.5 Ionization Percentage of the Main Acids Found in Wine at Two pH Levels

pH	Species at Equilibrium	Tartaric Acid	Malic Acid	Citric Acid	Succinic Acid
3.0	H_2A	51.3	74.1	57.6	94.3
	HA^-	46.7	25.7	41.7	5.7
	A^{2-}	2	0.2	0.7	0.01
3.5	H_2A	23.4	47.1	29.3	83.9
	HA^-	67.5	51.7	67.1	16.0
	A^{2-}	9.1	1.2	3.6	0.1

pH of the medium is determined by the strongest acids, as these have a higher dissociation percentage.

7. CHANGES IN DEGREE OF DISSOCIATION AND MOLAR FRACTION WITH pH

Figure 13.6 shows the degree of dissociation for tartaric acid and malic acid, considering the first and second ionization steps. The graph shows how each of the acids behaves at different pHs. As can be seen, for the same pH, the proportion of dissociated acids (α) is increased when K_a increase (pK_a decrease). As the pH increases, so does the degree of dissociation.

Another type of graph, which provides very valuable information, is that which shows the molar fraction of the different species of an acid that are present as a function of pH. These graphs are interesting because they show what are known as equivalence points, which are the points at which the concentrations of the different species that participate in the equilibrium are identical.

FIGURE 13.6 Degree of dissociation of acids in wine by pH.

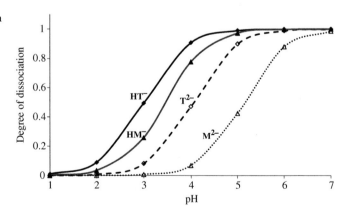

First equivalence point:

$$H_2T \leftrightarrows HT^- + H^+$$

$$K_1 = \frac{[HT^-] \cdot [H^+]}{[H_2T]} \quad \text{if } [HT^-] = [H_2T] \quad [H^+] = K_1 \Rightarrow pH = pK_1$$

Second equivalence point:

$$HT^- \leftrightarrows T^{2-} + H^+$$

$$K_2 = \frac{[T^{2-}] \cdot [H^+]}{[HT^-]} \quad \text{if } [T^{2-}] = [HT^-] \quad [H^+] = K_2 \Rightarrow pH = pK_2$$

It is noteworthy that the proportion of HT^- reaches a peak at a value which can be easily calculated using the expression α_{HT^-}.

$$\alpha_{HT^-} = \frac{[H_3O^+] \cdot K_{a1}}{[H_3O^+]^2 + [H_3O^+] \cdot K_{a1} + K_{a1} \cdot K_{a2}}$$

The boundary condition is that the first derivative of the function must be zero at this point. It is therefore only necessary to calculate the first derivative of the function and to establish the pH at which HT^- reaches its peak concentration (i.e., when α_{HT^-} is maximum).

$$\alpha_{HT^-(max)} \left[[H_3O^+]^2 + [H_3O^+] \cdot K_{a1} + K_{a1} \cdot K_{a2} \right] = [H_3O^+] \cdot K_{a1}$$

$$\alpha_{HT^-(max)} \left[\frac{[H_3O^+]}{K_{a1}} + 1 + \frac{K_{a2}}{[H_3O^+]} \right] = 1$$

For α_{HT^-} to be maximum, the derivative of its function must be $d(f_x) = 0$.

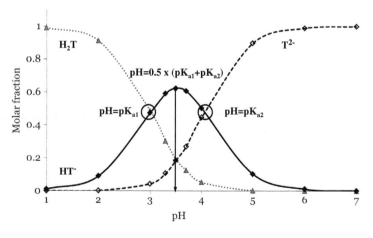

FIGURE 13.7 Equivalence points formed by the species produced by tartaric acid in solution.

The derivative of

$$\left[\frac{[H_3O^+]}{K_{a1}} + 1 + \frac{K_{a2}}{[H_3O^+]}\right]$$

is

$$d(f_x) = \frac{1}{K_{a1}} - \frac{K_{a2}}{[H_3O^+]^2}$$

and as

$$d(f_x) = 0 \Rightarrow \frac{1}{K_{a1}} = \frac{K_{a2}}{[H_3O^+]^2} \Rightarrow [H_3O^+]^2 = K_{a2} \cdot K_{a1}$$

$$\Rightarrow pH = \frac{pK_{a2} + pK_{a1}}{2}$$

For tartaric acid:

$$pH = \frac{pK_{a2} + pK_{a1}}{2} = \frac{4.37 + 3.01}{2} = 3.69$$

At maximum α_{HT^-}, the concentrations of H_2T and T^{2-} are identical, and one equilibrium or the other will predominate if the pH of the maximum point is shifted to the right or the left. This is important, above all, in equilibria in which HT^- precipitates out of the medium following tartaric stabilization.

At pH values above the maximum, the following equilibrium predominates:

$$HT^- \rightleftarrows T^{2-} + H^+$$

At values below the maximum, in contrast, the following equilibrium predominates:

$$HT^- + H^+ \rightleftarrows H_2T$$

8. PROPORTION OF NEUTRALIZED ACIDS IN WINE

When acids are neutralized by bases in wine, they form salts. The best approximation to the true concentrations of the different chemical species that each acid contributes to wine is obtained by using the mixed dissociation constants of each acid. As already discussed, to calculate a mixed dissociation constant, it is necessary to know the ionic strength of the solution and the value of the A and B constants that appear in the expression for the Debye-Hückel activity coefficient.

Once the mixed dissociation constants are known (these incorporate the ionic strength of wine), the true concentrations of acids associated with the pH of wine can be easily calculated using the equations described in this chapter.

Continuing with the previous example:

The combined percentage of tartaric acid and lactic acid in wine can be calculated as follows:

Tartaric acid: $pK_{m1} = 3.16$; $pK_{m2} = 4.34$

$$c_i = 19.67 \text{ mmol/L} \equiv 2.95 \text{ g/L}$$

$$pH = 3.23$$

$$\alpha_0 = \frac{[H_2T]_{eq}}{[H_2T]_i} = \frac{[H_3O^+]^2}{[H_3O^+]^2 + K_1[H_3O^+] + K_1K_2} = \frac{(10^{-3.23})^2}{(10^{-3.23})^2 + 10^{-3.16}10^{-3.23} + 10^{-7.393}}$$

$$\alpha_0 = \frac{3.467 \times 10^{-7}}{8.658 \times 10^{-7}} = 0.400 \qquad \%H_2T_{eq} = 40.0\%$$

$$[H_2T]_{eq} = 0.400 \times v19.67 \text{ mmol} = 7.87 \text{ mmol/L}$$

$$\alpha_1 = \frac{[H_3O^+] \cdot K_1}{D} = \frac{10^{-3.23} \cdot 10^{-3.16}}{8.658 \times 10^{-7}} = \frac{4.775 \times 10^{-7}}{8.658 \times 10^{-7}} = 0.5515$$

$$\%HT^-_{eq} = 55.15\%$$

$$[HT^-]_{eq} = 0.5515 \times v\,19.67 \text{ mmol} = 1085 \text{ mmol/L}$$

$$\alpha_2 = \frac{K_1 \cdot K_2}{D} = \frac{10^{-7.393}}{8.658 \times 10^{-7}} = \frac{4.046 \times 10^{-8}}{8.658 \times 10^{-7}} = 0.0467$$

$$\%T^{2-}_{eq} = 4.67\%$$

$$[T^{2-}]_{eq} = 0.0467 \times 19.67 \text{ mmol} = 0.92 \text{ mmol/L}$$

It is particularly interesting to know the fraction of combined (neutralized) acid in the case of acids whose salts precipitate when an aqueous medium changes to an aqueous alcoholic solution (e.g., formation of potassium salts of tartaric acid in wine).

Lactic acid:

$$pK_m = 3.991$$
$$c_i = 26.1 \text{ mmol/L} \equiv 2.35 \text{ g/L}$$
$$pH = 3.23$$
$$HL \leftrightarrows H^+ + L^-$$
$$c(1 - \alpha) \qquad c\alpha \qquad c\alpha$$

The degree of dissociation can be calculated using the following expressions:

$$pH = pK_m - \log \frac{1 - \alpha}{\alpha}; \text{ or } pH = pK_m + \log \frac{\alpha}{1 - \alpha}$$

Or using the equilibrium constant:

$$K_m = \frac{[H^+] \cdot [L^-]}{[HL]}; \quad \frac{[L^-]}{[HL]} = \frac{K_m}{[H^+]} = \frac{c \cdot \alpha}{c(1-\alpha)} = \frac{10^{-3.991}}{10^{-3.23}} = 10^{-0.761} = 0.173$$

$$\frac{\alpha}{1-\alpha} = 0.173 \Rightarrow \alpha = 0.173 - 0.173\alpha;$$

$$\alpha = \frac{0.173}{1.173} = 0.1478; \quad \alpha = 14.78\%$$

$$[L^-] = 0.1478 \times 26.1 = 3.86 \, mmol/L$$

$$[HL] = 26.1 - 3.86 = 22.24 \, mmol/L$$

On applying the equations obtained for diprotic acids to the case of lactic acid, we get:

$$\alpha = \frac{K_m}{[H_3O^+] + K_m} \qquad \alpha = \frac{10^{-3.991}}{10^{-3.23} + 10^{-3.991}} = 0.1478$$

The data obtained can then be used to calculate the proportions of neutralized acids:
Neutralized tartaric acid: 10.85 meq/L as HT^-
0.92(x2) meq/L as T^{2-}
Neutralized lactic acid: 3.86 meq/L as L^-
16.55 meq/L of tartaric acid and lactic acid are neutralized.
On adding total acidity (sum of free acids) and alkalinity of ash (total combined acids), we get:

$$100.9 + 23.2 = 124.1 \, meq/L$$

This figure of 124.1 meq/L reflects the sum of all the acids in the wine, irrespective of their state of dissociation. If we then subtract all the acids whose individual concentrations are known, we would get the following:

$$\sum Acids(\neq H_2T, HL, H_2M, and \, HAc) = 124.1 - \underset{\text{Acetic acid}}{13.0 \, meq/L} - \underset{\text{Tartaric acid}}{19.67(x2)meq/L} - \underset{\text{Lactic acid}}{26.1 \, meq/L}$$

$$\sum Acids(\neq H_2T, HL, H_2M, and \, HAc) = 124.1 - 78.44 = 45.66 \, meq/L$$

On the other hand, the alkalinity of ash value reflects the total concentration of neutralized acids. Because we know that tartaric acid and lactic acid account for 16.55 meq of neutralized acids, the remaining value (alkalinity of ash value minus 16.55 meq) corresponds to neutralized acids other than lactic acid, tartaric acid, malic acid, and, in this case, acetic acid.

$$\text{Alkalinity of ash} = 23.2 \, meq/L$$
$$\sum \text{neutralized } H_2T \text{ and } HL = 16.55 \, meq/L$$
$$\text{Difference: } 6.65 \, meq/L$$

Therefore, the proportion of neutralized acids other than tartaric acid and lactic acid (which were calculated previously) is:

$$\%\text{Neutralization} = \frac{6.65 \ \text{meq/L}}{45.66 \ \text{meq/L}} \times 100 = 14.56\%$$

In other words, 14.56% of all the weak acids in the wine (including succinic acid, gluconic acid, and galacturonic acid) are neutralized.

14

Buffering Capacity of Wines

1. INTRODUCTION

The acidity of must and wine is due to the presence of a significant quantity of weak organic acids in solution. These acids are partially neutralized by cations (particularly potassium, calcium, sodium, and magnesium) extracted from the soil by the roots of the vine. If we represent a generic acid in wine as HA, the fundamental acid-base equilibrium can be presented as follows:

$$HA \rightleftarrows H^+ + A^-$$

The pH of wine changes little when a moderate quantity of a strong acid or a strong base is added. This ability of wine to resist large changes in pH is an indication of its buffering capacity, a phenomenon common to all solutions that contain a weak acid and its salt.

Buffering capacity is explained by an equilibrium that is established between the molecular form of an acid and its conjugate base. The addition of a strong acid to a wine introduces H^+ ions that react with the A^- anion, leading to a reduction in its concentration and an increase in the molecular form (HA) of the acid. When the equilibrium between the molecular and dissociated forms of the acid is re-established in the wine, a new concentration of H^+ is obtained that is derived solely from this dissociation and that is responsible for the new pH value. Similarly, the addition of a strong base to a wine leads to a neutralization reaction with the molecular form of the acid, which reduces the concentration of HA and increases the concentration of A^-. The re-establishment of the acid-base equilibrium generates a new concentration of H^+ and therefore a new pH value in the wine.

2. BUFFERING CAPACITY OF WEAK ACID SOLUTIONS

The titration curve of a weak acid with a strong base contains a region in which the pH rises sharply during the addition of the first few milliliters of the base. This is followed by a period during which the pH remains relatively constant despite continued addition of the base. A similar situation occurs in the titration curve for a weak base with a strong acid, except that in this case there is an initial sharp reduction in pH followed by a period of relative stability.

In each case, the slow change in pH upon addition of a base or acid is indicative of the buffering capacity of the solution. Hence, a buffer solution is any solution that undergoes only limited changes in pH when H^+ or OH^- ions are added. The equilibrium that governs these solutions is the dissociation of a weak acid or base. The following would occur in the case of acetic acid:

$$AcOH \rightleftarrows AcO^- + H^+$$

$$K_a = \frac{[AcO^-] \cdot [H^+]}{[AcOH]}; \quad [H^+] = K_a \cdot \frac{[AcOH]}{[AcO^-]}; \quad pH = pK_a - \frac{[AcOH]}{[AcO^-]}$$

In general, the pH of a buffering solution of a weak acid is determined by the ratio of the dissociated to the undissociated acid, in other words the ratio of acid to salt:

$$pH = pK_a - \log\frac{[acid]}{[salt]}$$

From the above equation — known as the Henderson-Hasselbach equation — we can deduce that maximum buffering capacity is achieved when the concentration of acid is equal to that of its salt. In these cases, the equilibrium can be easily displaced in one direction or the other.

$$\text{If } [acid] = [salt]; \quad \log\frac{[acid]}{[salt]} = 0; \quad \text{and} \quad pH = pK_a$$

The behavior of a buffering solution is explained in the example below, where the addition of H^+ ions will shift the equilibrium between AcOH and AcO^- to the left whereas the addition of OH^- ions will shift it to the right. The buffering capacity of a solution will be greater the higher the concentration of the molecular form of the acid and of its dissociated form.

Example: Let a buffer solution be composed of acetic acid and its acetate anion, both at a concentration of 1 mol/L. Calculate the pH before and after the addition of 0.1 moles of strong acid to 1 L of the buffer and following the addition of 0.1 moles of strong base to 1 L of buffer.

$$[AcOH] = 1M = [AcONa];$$
$$AcOH \rightleftarrows AcO^- + H^+$$

Initial equilibrium $\quad 1 \quad\quad 1 \quad\quad x; \; pH = pK_a - \log\frac{1}{1} = 4.74$

Equilibrium following the addition of 0.1 mol H^+/L :
$$1 + 0.1 \quad\quad\quad 1 - 0.1 \quad x'$$

Following the addition of 0.1 mol/L of strong acid:

$$pH = pK_a - \log\frac{1 + 0.1}{1 - 0.1} = 4.74 - 0.09 = 4.65$$

If we add 0.1 moles of base to 1 L of the buffer solution, we obtain a pH of 4.83.

Therefore, the addition of 0.1 moles of a strong acid or base to 1 L of a solution of AcOH (1M) and AcO^- (1M) would lead to a pH change of only ± 0.09 units.

In contrast, if we were to add 0.1 moles of the strong acid or base to 1 L of water:

0.1 moles $H^+ \rightarrow 1L\ H_2O \Rightarrow pH = -\log 0.1 = 1$
0.1 moles $OH^- \rightarrow 1L\ H_2O \Rightarrow pOH = -\log 0.1 = 1; \; pH = 14 - 1 = 13$

In the same way that an expression can be calculated for the pH of a buffer solution prepared from a weak acid and its conjugate base, one can also be deduced for the value of the pH in buffer solutions prepared from a weak base and its conjugate acid.

$$pH = 14 - pK_b + \log\frac{[Base]}{[Salt]}$$

According to the ionization curves for acids at the normal pH of wine, tartaric acid exhibits the highest degree of dissociation, followed by malic acid, and finally succinic acid. A good approach to the analysis of the buffering capacity of wines is to treat all the acids present as monocarboxylic acids whenever small variations in pH are to be considered around standard pH values for wine (3.3–3.7) and at normal concentrations of the acid. Based on these assumptions, titratable acidity can be considered as a measure of the concentration of free acids in wine and the alkalinity of ash as a measure of the concentration of salts.

3. BUFFERING CAPACITY OF WINE

The buffering capacity of wine is defined as the quantity of acid or base that must be added to the wine in order to reduce or increase the pH value by one unit. It is expressed as milli-equivalents of acid or base per liter of wine (meq H^+/L or meq OH^-/L, respectively).

According to the Henderson-Hasselbach equation

$$pH = pK_a - \log \frac{[HA]}{[A^-]}$$

$$pH = pK_a - \log [HA] + \log [A^-] \quad \text{as } \log_{10} x = \frac{1}{2.303} \ln x$$

$$pH = pK_a - \frac{1}{2.303} \ln|HA| + \frac{1}{2.303} \ln|A^-|$$

To calculate the buffering capacity of a weak acid in response to the addition of a base (B) or an acid (HA), we must first obtain the derivatives of pH and B:

$$dpH = -\frac{1}{2.303} \times \frac{1}{[HA]} d[HA] + \frac{1}{2.303} \times \frac{1}{[A^-]} d[A^-]$$

When the differential of a quantity of strong base (dB) is added to a buffering solution, there is an equal reduction in the quantity of the molecular acid and in the quantity of the acid salt, as per the following expression:

$$dB = d[A^-] = -d[HA]$$

Substituting in the expression for dpH:

$$dpH = \frac{dB}{2.303} \left(\frac{1}{[HA]} + \frac{1}{[A^-]} \right)$$

$$dpH = \frac{dB}{2.303} \left(\frac{[HA] + [A^-]}{[HA] \times [A^-]} \right)$$

$$\frac{dB}{dpH} = 2.303 \times \left(\frac{[AH] \times [A^-]}{[HA] + [A^-]} \right)$$

The previous equation yields the buffering capacity of a solution in response to the addition of a base. In the case of wine, we have seen that titratable acidity represents the sum of the acids in molecular (free) form, whereas the alkalinity of ash represents the sum of the acids in the anionic form (salts). This is only true when the acids present carry a single proton. As the pH variations in wine are extremely small (dpH), this approximation is perfectly valid. Consequently, the buffering capacity of a given wine can be calculated as follows:

$$\frac{dB}{dpH} = 2.303 \times \frac{TA \times AA}{TA + AA}$$

Where TA = titratable acidity and AA = alkalinity of ash.
Example:

A wine has the following values: TA = 100.9 meq/L
                                        AA = 23.2 meq/L
                                        pH = 3.23

The buffering capacity will be:

$$\frac{dB}{dpH} = 2.303 \times \frac{100.9 \times 23.2}{100.9 + 23.2} = 43.44 \text{ meq/L}$$

In other words, the addition of 43.44 meq of a strong base to a liter of this wine will cause the pH to increase to a value of 4.23. If we only want to increase the pH by 0.1 units, it will be sufficient to add a tenth of the volume of the strong base (4.34 meq/L); in this case the titratable acidity will be reduced:

$$TA = 100.9 - 4.34 = 96.56 \text{ meq/L}$$

In contrast, the alkalinity of ash will increase:

$$AA = 23.2 + 4.34 = 27.54 \text{ meq/L}$$

If now we want to reduce the pH of a wine by adding a strong acid, the following expression is obtained:

$$-\frac{dAc}{dpH} = 2.303 \times \left(\frac{[HA] \times [A^-]}{[HA] + [A^-]} \right)$$

The expression links a reduction in the pH of a buffering solution containing a single acid to the addition of a strong acid. The reduction in pH is indicated by a minus sign.

In the case of a mixture of weak monoprotic acids, such as those found in wine, we obtain the following expression:

$$-\frac{dAc}{dpH} = 2.303 \times \frac{TA \times AA}{TA + AA}$$

These expressions are valid for wines in which the appropriate analyses have been carried out (principally titratable acidity, pH, and alkalinity of ash). They are useful to maintain the titratable acidity and alkalinity of ash at appropriate values according to specific criteria or production requirements.

In practice, if one wishes to alter the buffered pH of a must, there is often insufficient time or material to analyze titratable acidity or alkalinity of ash. In this case, a quantity of acid or base is gradually added to a liter of must with a known pH value until the target pH is reached.

4. EFFECT OF MALOLACTIC AND MALOALCOHOLIC FERMENTATION ON THE ACID-BASE EQUILIBRIUM OF WINE

The elimination or transformation of one of the three main acids found in wine affects the equilibria of the other two acids. This occurs when malic acid is transformed into lactic acid or ethanol during malolactic or maloalcoholic fermentation, respectively.

The effect of these transformations of malic acid on the remaining acids can be studied using a model solution with an initial pH of 3.0 that contains 40.0 mmol/L tartaric acid (6.0 g/L) and 40.0 mmol/L malic acid (5.36 g/L). Two assumptions are made: 1) that malic acid is transformed into lactic acid in an equimolar (1:1) reaction, and 2) that maloalcoholic fermentation leads to the complete disappearance of malic acid.

TABLE 14.1 Model Solution to Study the Effects of Malolactic and Maloalcoholic Fermentation on pH

Acids in Wine	pK_{a1}	pK_{a2}	Initial	After Malolactic Fermentation	After Maloalcoholic Fermentation
Tartaric	3.04	4.37	40	40	40
Malic	3.46	5.13	40	0	0
Lactic	3.86		0	40	0

It is advisable to first calculate the fraction of dissociated acids under the initial conditions to analyze the effect of malolactic or maloalcoholic fermentation.

Fraction of dissociated tartaric acid under initial conditions:

$$C_i = 40 \text{ mmol/L}; \quad pH = 3.0; \quad pK_{a1} = 3.04; \quad pK_{a2} = 4.37$$

$$\alpha_{H_2T} = \frac{[H_3O^+]^2}{[H_3O^+]^2 + [H_3O^+]K_{a1} + K_{a1}K_{a2}} = \frac{(10^{-3})^2}{(10^{-3})^2 + (10^{-3} \times 10^{-3.04}) + (10^{-3.04} \times 10^{-4.37})}$$

$$= 0.5126$$

$$\alpha_{H_2T} = 0.5126 \Rightarrow 51.26\% \text{ of } H_2T$$

$$\alpha_{HT^-} = \frac{[H_3O^+]K_{a1}}{[H_3O^+]^2 + [H_3O^+]K_{a1} + K_{a1}K_{a2}} = \frac{(10^{-3}) \times 10^{-3.04}}{(10^{-3})^2 + (10^{-3} \times 10^{-3.04}) + (10^{-3.04} \times 10^{-4.37})}$$
$$= 0.4675$$

$$\alpha_{HT^-} = 0.4675 \Rightarrow 46.75\% \text{ of } HT^-$$

$$\alpha_{T^{2-}} = \frac{K_{a2}K_{a1}}{[H_3O^+]^2+[H_3O^+]K_{a1} + K_{a1}K_{a2}} = \frac{10^{-7.41}}{1.95 \times 10^{-6}} = 0.01994$$

$$\alpha_{T^{2-}} = 0.0199 \Rightarrow 1.99\% \text{ of } T^{2-}$$

The absolute quantity of each of the species that gives rise to a 40 mmol/L solution of tartaric acid at pH 3 is:

$$[H_2T]_{eq} = 0.5126 \times 40.0 = 20.504 \text{ mmol/L}$$
$$[HT^-]_{eq} = 0.4675 \times 40.0 = 18.70 \text{ mmol/L}$$
$$[T^{2-}]_{eq} = 0.0199 \times 40.0 = v\ 0.796 \text{ mmol/L}$$

The sum of these three concentrations is 40 mmol/L
Fraction of dissociated malic acid under initial conditions:

$$C_i = 40 \text{ mmol/L}; \ pH = 3.0; \ pK_{a1} = 3.46; \ pK_{a2} = 5.13$$

$$\alpha_{H_2M} = \frac{10^{-6}}{D} = \frac{10^{-6}}{1.349 \times 10^{-6}} = 0.7411 \Rightarrow [H_2M]_{eq} = 0.7411 \times 40 = 29.645 \text{ mmol/L}$$

$$\alpha_{HM^-} = \frac{10^{-6.46}}{1.349 \times 10^{-6}} = \frac{3.46 \times 10^{-7}}{1.349 \times 10^{-6}} = 0.2570 \Rightarrow [HM^-]_{eq} = 0.2570 \times 40 = 10.28 \text{ mmol/L}$$

$$\alpha_{M^{2-}} = \frac{10^{-8.59}}{1.349 \times 10^{-6}} = \frac{2.57 \times 10^{-9}}{1.349 \times 10^{-6}} = 1.905 \times 10^{-3} \Rightarrow [M^{2-}]_{eq}$$

$$= 1.905 \times 10^{-3} \times 40 = 0.076 \text{ mmol/L}$$

4.1. Malolactic Fermentation

If we assume that all the malic acid present is transformed into lactic acid (although in practice only around 75% is transformed), the pH of the solution would change, as would the degree of dissociation of the acids present under the new equilibrium conditions. Accordingly, the buffering capacity would also change.

To analyze the effect of malolactic fermentation on the pH and buffering capacity of wine, we must take into account the following:

1. Lactic acid formed: 40 mmol, from the initial 40 mmol of malic acid.
2. Lactic acid salt: In the absence of tartaric acid, the quantity of lactic acid salt would be equivalent to that of malic acid salt in the original solution. This is explained by the requirement to maintain the electrochemical neutrality of the wine. The equivalent of the

positive charges that compensated the negative charge on the M^{2-} and HM^- ions would not disappear from the medium even if malic acid itself disappeared, and consequently, the following is required to maintain the electrochemical neutrality of the solution:

$$[L^-] = [HM^-] + 2[M^{2-}] = 10.28 + 2 \times 0.076 = 10.432 \text{ moles}$$

The undissociated lactic acid will be: $[HL] = 40 - 10.432 = 29.568$ moles

However, since tartaric acid is present, and its first dissociation is stronger than that of lactic acid, a cation-displacement reaction will occur, affecting the concentration of free and dissociated lactic acid according to the following equation:

$$H_2T + KL \rightleftarrows KHT + HL$$

If we apply the previously calculated concentrations of the different species to this equilibrium, we obtain the following:

$$
\begin{array}{ccccccc}
H_2T & + & L^- & \rightleftarrows & HT^- & + & HL \\
20.504 - x & & 10.432 - x & & 18.70 + x & & 29.568 + x
\end{array}
$$

Effect of Malolactic Fermentation on pH

To calculate the pH of the solution containing tartaric and lactic acid, the acid dissociation constants of both are used:

For tartaric acid: $H_2T \rightleftarrows HT^- + H^+$
(the second dissociation can be effectively ignored at the pH normally found in wine)

$$K_{aT} = \frac{[HT^-][H^+]}{[H_2T]} \qquad [H^+] = K_{aT}\frac{[H_2T]}{[HT^-]}$$

For lactic acid: $\quad HL \rightleftarrows L^- + H^+$

$$K_{aL} = \frac{[H^+][L^-]}{[HL]}$$

$$[H^+] = K_{aL}\frac{[HL]}{[L^-]}$$

Given that the two acids are present in the same solution, the pH will be the same and therefore:

$$[H^+] = K_{aT}\frac{[H_2T]}{[HT^-]} = K_{aL}\frac{[HL]}{[L^-]}$$

The values of the molecular and dissociated species are calculated from the expression of the acid dissociation constant for each acid under the conditions of the new equilibrium corresponding to the mixture of the two.

$$K_{a1T}\frac{20.504 - x}{18.70 + x} = K_{aL}\frac{29.568 + x}{10.432 - x};$$

From this, a second-degree equation can be obtained:

$$(20.504 - x)(10.432 - x)K_{a1T} = K_{aL}(29.568 + x)(18.70 + x)$$
$$K_{a1T}(x^2 - 30.936x + 213.90) = K_{aL}(x^2 + v48.268x + 552.92)$$
$$(10^{-3.04} - 10^{-3.86})x^2 - (30.936K_{a1t} + 48.268K_{aL-})x + (213.9K_{a1T} - 552.92K_{aL}) = 0$$
$$7.74 \times 10^{-4}x^2 - 0.0349x + 0.119 = 0, \quad x_1 = 41.35, \quad x_2 = 3.71$$

If $x = 3.71$ mmol/L, we can now calculate the fractions of free and dissociated forms of tartaric and lactic acid and use these to calculate the concentration of $[H^+]$ and the pH:

$$\frac{[H_2T]_{eq}}{[HT^-]_{eq}} = \frac{20.504 - 3.71}{18.7 + 3.71} = \frac{16.79}{22.41}$$

$$[H^+] = 9.12 \times 10^{-4} \frac{16.79}{22.41} = 6.83 \times 10^{-4} \Rightarrow pH = 3.165$$

$$\frac{[HL]_{eq}}{|L^-|_{eq}} = \frac{29.568 + 3.71}{10.432 - 3.71} = \frac{33.278}{6.722}$$

$$[H^+] = 1.38 \times 10^{-4} \frac{33.278}{6.722} = 6.83 \times 10^{-4} \Rightarrow pH = 3.165$$

In summary, the transformation of malic into lactic acid during malolactic fermentation leads to an increase in pH.

We must take into account that the alkalinity of ash for the solution before and after malolactic fermentation remains constant.

1. The alkalinity of ash prior to this transformation, expressed in meq/L, can be obtained by adding the millimoles of the combined acids:

$$AA = [HT^-] + [HM^-] + 2[T^{2-}] + 2[M^{2-}] = 18.70 + 10.28 + 2(0.796) + 2(0.076)$$
$$= 30.724 \text{ meq/L}.$$

2. Following the transformation of malic acid into lactic acid, the alkalinity of ash corresponds to the sum of the acids as follows:

$AA = [HT^-] + [L^-] + 2[T^{2-}]$, and at the pH that has been calculated (3.165) these concentrations correspond to the sum:

$$AA = \left(\frac{K_{a1}[10^{-3.165}]}{(10^{-3.165})^2 + K_{a1} \times 10^{-3.165} + K_{a1}K_{a2}} \times 40\right) + 6.722 + 2 \times \left(\frac{K_{a1}K_{a2}}{D} \times 40\right)$$
$$= 22.08 + 6.722 + 2 \times 1.377 = 31.56 \text{ meq/L}$$

The difference observed in the alkalinity of ash (31.56 versus 30.724) reveals that the initially calculated pH (3.165) is, in theory, slightly high.

The error in the estimate used to calculate the pH is around 0.01 to 0.02, as demonstrated by the calculation of the different acid species present in the solution at a pH of 3.15.

$$AA = 40 \times \left(\frac{K_a + 10^{-3.15}}{D}\right) + 6.527 + 40 \times \left(\frac{K_{a1}K_{a2}}{D}\right)$$

$$HL \rightleftarrows L^- + H^+$$

$$40 - x \times 10^{-3.15} \quad K_a = \frac{10^{-3.15} + x}{40 - x} = 10^{-3.86} \quad x = 6.527 \text{ mmol/L}$$

$$AA = 21.78 + 6.527 + 2(1.312) = 30.93$$

In other words, at a pH of 3.15, the value obtained for the alkalinity of ash essentially coincides with that obtained prior to malolactic fermentation. This suggests that the method described yields a theoretical pH value that is sufficiently close to the true value to allow monitoring of malolactic fermentation through measurements of pH alone during the transformation.

Effects of Malolactic Fermentation on the Buffering Capacity of Wine

The buffering capacity of wine as a function of titratable acidity and the alkalinity of ash corresponds to the following expression:

$$\frac{dB}{dpH} = 2.303 \times \frac{TA \times AA}{TA + AA}$$

The alkalinity of ash in the example shown is 30.724 meq/L and the titratable acidity can be calculated by subtracting the alkalinity of ash from the total acid content:

$$TA = (40 \text{ mmol/L} \times 2) \text{ meq} + (40 \text{ mmol/L} \times 2) \text{ meq} - 30.724 = 129.276 \text{ meq/L}$$

Therefore, prior to malolactic fermentation, the buffering capacity was:

$$\frac{dB}{dpH} = 2.303 \times \frac{129.276 \times 30.724}{129.276 + 30.724} = 57.17 \text{ meq/L}$$

After malolactic fermentation, the buffering capacity was:

$$TA = (2 \times 40 \text{ mmol/L}) + (1 \times 40 \text{ mmol/L}) - 30.724v = 89.276 \text{ meq/L}$$

$$\qquad\quad H_2T \qquad\qquad HL$$

$$\frac{dB}{dpH} = 2.303 \times \frac{89.276 \times 30.724}{89.276 + 30.724} = 52.64 \text{ meq/L}$$

In other words, the buffering capacity of a solution of tartaric and malic acid is reduced by approximately 5 meq/L when 100% of the 40 mmol/L of malic acid ($\equiv$ 5.36 g/L) is converted by malolactic fermentation into lactic acid.

4.2. Maloalcoholic Fermentation

During maloalcoholic fermentation, malic acid is transformed into ethanol. In other words, an acidic compound is replaced by a non-acidic compound. In this case, the alkalinity of ash will remain constant, as in malolactic fermentation, but the titratable acidity will decrease with the disappearance of malic acid.

Effect of Maloalcoholic Fermentation on pH

1) *Initial values.* Prior to maloalcoholic fermentation the solution contains the following concentrations of acids:

40 mmol/L H_2T
40 mmol/L H_2M
pH = 3.0

At a pH of 3.0, the alkalinity of ash and the titratable acidity have the following values:

$$AA = [HT^-] + [HM^-] + 2 \times [T^{2-}] + 2 \times [M^{2-}] = 18.7 + 10.28 + 2 \times 0.796 + 2 \times 0.076$$
$$= 30.724 \text{ meq/L}$$

$$TA = [H_2T] \times 2 + [H_2M] \times 2 - AA = 80 + 80 - 30.726 = 129.274 \text{ meq/L}$$

2) *Final values.* When the acid-base equilibria are re-established following the complete conversion of malic acid to ethanol, only tartaric acid remains:

$$H_2T \rightleftarrows HT^- + H^+ \qquad K_{a1} = \frac{[H^+] \times [HT^-]}{[H_2T]}$$

$$HT^- \rightleftarrows T^{2-} + H^+ \qquad K_{a2} = \frac{[H^+] \times [T^{2-}]}{[HT^-]}$$

The concentration of H^+ in the solution of the different species in equilibrium remains the same, allowing us to state that:

$$[H^+] = K_{a1} \times \frac{[H_2T]}{[HT^-]} = [H^+] = K_{a2} \times \frac{[HT^-]}{[T^{2-}]}$$

$$K_{a1} \times [H_2T] = K_{a2} \times \frac{[HT^-]^2}{[T^{2-}]} \qquad K_{a1} \times [H_2T] \times [T^{2-}] = K_{a2} \times [HT^-]^2$$

The equation obtained is: $K_{a2} \times [HT^-]^2 - K_{a1} \times [H_2T] \times [T^{2-}] = 0$ (Equation 1)

Following maloalcoholic fermentation the alkalinity of ash and the total quantity of tartaric acid remain constant. Consequently, we can establish the following:

$$[H_2T]_{\text{initial}} = [H_2T]_{\text{final}} = v \ 40 \text{ mmols}$$
$$AA_{\text{initial}} = AA_{\text{final}} = 30.724$$
$$TA = [H_2T]_{\text{initial}} \times 2 - AA = 80 - 30.724 = 49.276$$

The following relationship between acid content, alkalinity of ash, and titratable acidity is established following maloalcoholic fermentation:

$$TA = 2 \times [H_2T] + [HT^-][H_2T] = \frac{TA - [HT^-]}{2}$$

$$AA = 2 \times [T^{2-}] + [HT^-][T^{2-}] = \frac{AA - [HT^-]}{2}$$

Substituting the values of $[H_2T]$ and $[T^{2-}]$ in equation 1 we obtain the following:

$$K_{a2} \times [HT^-]^2 - K_{a1} \times \left(\frac{TA - [HT^-]}{2}\right) \times \left(\frac{AA - [HT^-]}{2}\right) = 0$$

$$4K_{a2} \times [HT^-]^2 - \left(K_{a1}\left[(TA \times AA) - (TA + AA) \times [HT^-] + [HT^-]^2\right]\right) = 0$$

$$(4K_{a2} - K_{a1})[HT^-]^2 + K_{a1}(TA + AA)[HT^-] - K_{a1} \times TA \times AA = 0$$

Changing the sign, because $K_{a1} > 4K_{a2}$ and the coefficient "a" is negative, we get:

$$(K_{a1} - 4K_{a2})[HT^-]^2 - K_{a1}(TA + AA)[HT^-] + K_{a1} \times TA \times AA = 0$$

Then:

$$[HT^-] = \frac{+K_{a1}(TA + AA) \pm \sqrt{[K_{a1}(TA + AA)]^2 - 4(K_{a1} - 4K_{a2})K_{a1} \times TA \times AA}}{2(K_{a1} - 4K_{a2})}$$

We know that $K_{a1} = 10^{-3.04}$ $K_{a2} = 10^{-4.37}$

$$TA = 49.276 \text{ meq/L} \qquad A = 30.724 \text{ meq/L}$$

The result of the equation:
$[HT^-] = 25.57$ (the alternative value of 72.83 meq/l is not logical)
Using the calculated concentration of the HT^- ion, we can calculate the concentration of the remaining species in the equilibrium:

$$[T^{2-}] = \frac{30.724 - 25.57}{2} = 2.577 \text{ mmol/L}$$

$$[H_2T] = \frac{49.276 - 25.57}{2} = 11.853 \text{ mmol/L}$$

Once the concentrations of the free form and the salt are known, the pH can be calculated:

$$pH = pK_{a1} - \log\frac{[H_2T]}{[HT^-]} = 3.04 - \log\frac{11.853}{25.57} = 3.374$$

The pH can also be calculated from the equation for K_{a2}:

$$pH = pK_{a2} - \log\frac{[HT^-]}{[T^{2-}]} = 4.37 - \log\frac{25.57}{2.577} = 3.373$$

The increase in pH, from 3.0 to 3.37, is much greater than that observed with malolactic fermentation.

Effect of Maloalcoholic Fermentation on the Buffering Capacity of Wine

The variation in buffering capacity is calculated from the titratable acidity and the alkalinity of ash after conversion of malic acid into ethanol.

$$\frac{dB}{dpH} = 2.303 \times \frac{TA \times AA}{TA + AA} = 2.303 \times \frac{49.276 \times 30.724}{80} = 43.58$$

The initial buffering capacity is 57.17 meq/L, and the difference is 57.17 − 43.58 = 13.59 meq/L. This is approximately 3 times the reduction generated by malolactic fermentation.

5. DRY EXTRACT DETERMINED BY DENSITOMETRY

In addition to affecting pH, buffering capacity, and the equilibria derived from the acidification and deacidification of wine, the amount of acid present as a salt influences the physical property of density and, consequently, the value of the dry extract of wine. The total dry extract corresponds to all substances that remain following evaporation under physical conditions specifically chosen to prevent changes in their properties. It is expressed as sucrose content and is made up of non-volatile organic substances and mineral constituents. The mineral components are represented by the ash, which corresponds to all of the products of incineration of the residue following evaporation of wine carried out in such a way that the cations are transformed into carbonates and other non-volatile mineral salts.

The total dry material is influenced by processes such as crushing, destemming, maceration, or time in the vat. It also depends on the conditions used for the concentration and evaporation of the liquid phase of the wine. It is important that the constituents of the wine do not undergo decarboxylation or oxidation reactions, or other reactions in which they are transformed into volatile substances while the extract is being prepared.

Wine contains various types of dry extract:

- Total dry extract, which is obtained by evaporation at 70°C under an air current of 40 L/h. This is expressed in g/L. Red wines contain between 25 and 30 g/L, whereas white wines do not exceed 25 g/L, and in the case of sweet wines the total dry extract depends on the sugar content.
- Non-reducing extract, which is calculated by subtracting the total sugar content from the total dry extract.

TABLE 14.2   Density Values (20°/20°) for a 1 g/L Solution of Some of the Compounds Present in Wine

Compound	Extract in g/L Obtained by Density (20/20)
Tartaric acid	1.190
Potassium bitartrate	1.533
Neutral potassium tartrate	1.762
Malic acid	1.030
Potassium bimalate	1.412
Neutral potassium malate	1.671
Lactic acid	0.639
Potassium lactate	1.370
Succinic acid	0.775
Monopotassium phosphate	1.837
Potassium sulfate	2.200
Glutamic acid	1.050
Arginine	0.900
Proline	0.650
Glycerol	0.603
Sucrose	1.000

- Reduced extract, which is obtained from the total dry extract by subtracting those sugars that exceed 1 g/L, potassium sulfate that exceeds 1 g/L, and any substances that may have been added to the wine.

A more precise method for the determination of dry extract is based on relative density obtained by hydrometry. The volumetric mass of the wine (VM_v) and of an aqueous alcohol solution (VM_d) containing an equal concentration of alcohol to that found in the wine are measured. The relative density is calculated using the following equation, and this allows the equivalent dry extract to be calculated using conversion tables. The relative density must be calculated to at least three decimal places.

$$rd = 1.0018 \, (VM_v - VM_d) + 1.000$$

The dry extract and ethanol content of the wine are used to establish equations that can be used to discover fraudulent activities such as the addition of alcohol or sugar.

The acids and their salts affect the density of the solution and therefore the dry extract calculated using this method. This effect is different for each compound; thus, 1 g of K_2SO_4 changes the density of 1 L of aqueous solution by the equivalent of 2.2 g of sucrose. The dry extract calculated by densitometry would have a value of 2.2 g/L.

Acids increase the dry extract to a lesser extent than their salts. Must contains many compounds that increase the dry extract obtained by densitometry, and therefore the extract calculated in this way is greater than that calculated using gravimetric methods. Nevertheless, in wines, the presence of glycerol compensates for the increase in dry extract calculated by densitometry as a result of acid salts and other compounds, and there is a good correspondence between the values obtained for the dry extract using both methods.

6. RELATIONSHIP BETWEEN TOTAL AND TITRATABLE ACIDITY

The relationship between pH and titratable acidity described by Boulton is based on the premise that the acidity obtained by analyzing a wine using a base up to a pH of 7, added to the potassium and sodium content of the wine, is equal to total acidity.

$$[K^+] + [Na^+] + [TA] = \sum [acids]$$

All of the concentrations in this equation should be expressed as meq/L.

The above equation approximates to the law of conservation of mass, which has been used previously to show that the total acid content is equal to the sum of the neutralized acids (dissolved as anions) and the acids present in the solution as free or molecular forms.

According to Boulton:

Titratable $[K^+] + [Na^+] + [H^+] = [H^+]$ expected for the acid composition

This equation considers the pH of wine to be a function of acid content and the concentration of potassium and sodium ions.

$$pH = f(\text{concentration of acids, } K^+, Na^+)$$

This equation is an approximation to reality, since it does not take into account the concentration of other highly abundant cations such as calcium and magnesium that are present in musts and wines. Nevertheless, it yields acceptable levels of correlation.

The equation of the curve (y = ax + b) is as follows:

$$a = \text{curve} = 0.742 \qquad\qquad b = \text{ordinate at the origin} = 1.872$$

Equation (I) :   Titratable acidity $(meq/L) = 0.742 \times$ total $[H^+]$ (meq/L)

$$(pH = 7) \qquad\qquad (\text{total acidity})$$

Correlation coefficient $(P < 0.001)$: 0.943

For the model solution containing 40 mmol H_2T, 40 mmol H_2M, and 30.726 mmol alkalinity of ash in each liter of solution:

According to equation (I): $TA = (40 \times 2 + 40 \times 2) \times 0.742 + 1.872 = 120.592$ meq/L

According to the model solution:

$$TA = A_{Total} - AA = 80 + v80 - 30.726 = 129.274 \text{ meq/L}$$

Other equations of interest relate titratable acidity plus potassium and sodium content to total acidity. Yet another equation includes only potassium content and titratable acidity. Through a combination of the three equations it is possible to determine $[K^+]$, $[Na^+]$, and titratable acidity:

Equation (II): Titratable acidity $+ [K^+] + [Na^+] = 0.940 \times$ [Total acidity] $+ 4.610$
All concentrations should be expressed as millimoles of positive charge per liter.
Correlation coefficient ($P < 0.001$): 0.996

Equation (III): Titratable acidity(mmol/L) $+ [K^+] = v0.911 \times$ [Total acidity] $+ 4.166$
All concentrations should be expressed as millimoles of positive charge per litre.
Correlation coefficient ($P < 0.001$): 0.993

Using equation (III) and equation (I), the potassium content $[K^+]$ can be calculated:

$$[K^+] = (III) - (I) = [(0.911 \times 160) + 4.166] - [(0.742 \times 160) + 1.872] = 149.926 - 120.592$$

$$= 29.334$$

$$[K^+] = 29.334 \text{ mmol } H^+/L \approx AA = 30.726 \text{ mmol } H^+/L$$

Note: millimoles of positive charge per liter is an expression of the concentration of positive charges in a solution and is equal to the expression of meq/L of acid or base.

The Boulton equations are useful for obtaining an approximate measure of total acid content and the total potassium and sodium cation content that neutralizes a substantial fraction of these acids. In some ways they are a variation of the equations derived in the previous section, which relate titratable acidity to the total acid content, the alkalinity of ash, and the buffering capacity of must and wine.

7. ACIDITY AND ORGANOLEPTIC CHARACTERISTICS

7.1. Flavor

There are four fundamental flavors: acidic, bitter, sweet, and salty. The sensation of acidity in a drink is determined by two factors: the pH and the buffering capacity. The pH is responsible for the intensity of the acidic flavor that is perceived on the palate, whereas the buffering capacity influences the duration of the sensation. A low pH and a low buffering capacity, as found in a strong acid solution, produces a strong sensation of acidity that disappears quickly. In contrast, strongly buffered solutions of weak acids produce a longer lasting acidic flavor with an intensity that increases with decreasing pH. The acidic flavor of wine is due to the contribution of the six principal organic acids that are present, since the mineral acids found in wine are present entirely as salts.

The three acids derived from the grape (tartaric, malic, and citric acid) give rise to the main acidic flavor. Nevertheless, if tartaric acid predominates, the acidic flavor has a metallic character that is more apparent at higher concentrations of this acid. If malic acid predominates, a fresh, "green" sensation reminiscent of green apples is obtained, and if citric acid predominates, the common sensation of acids, which on its own is irritant and corrosive, is modified and softened, producing a flavor reminiscent of the acidic, but not aggressive, taste of lemon. When tartaric acid is present at high concentrations, it is responsible for hardness and a sensation of astringency. It should be remembered that, in general, the best aged red and white wines contain low concentrations of tartaric acid and have a pH of around 3.5 to 3.8.

Acetic, succinic, and lactic acid derived from alcoholic and malolactic fermentation contribute complementary flavors. Acetic acid gives a sharp, vinegary flavor but ethyl acetate is the first species to be perceived during tasting. Succinic acid contributes a bitter, salty flavor and therefore appears to be responsible for the so-called "vinous" character. Finally, lactic acid contributes a special flavor that is both sharp and warm, and resembles the taste of yoghurt.

Sensory analyses using the duo-trio test have shown that, when assessed in 10% aqueous alcohol solutions with an identical pH and buffering capacity, individual solutions of the acids found in wine do not differ significantly in their organoleptic properties. Only lactic acid has a specific identifiable flavor that is different from that of tartaric, malic, citric, glucuronic, or gluconic acid.

In summary, the organic acids contained in wine contribute to acidic flavor or sensation in a manner proportional to their concentration and acidic strength, whereas lactic acid makes a specific contribution. In general terms, the "warm" sensation in the mouth that is desirable in red wines is contributed by a high pH. In contrast, in white table wines, particularly young, fruity ones, the aim is to generate a sensation of "freshness" in the mouth, which is partly due to a low pH.

7.2. Aroma

Organic acids containing 4 and 5 carbon atoms — mainly butyric, isobutyric, and isovalerianic acids — have powerful aromas reminiscent of butter and cheese. They have a negative effect on the aroma of wine, despite being found at low concentrations (parts per million). The remaining C_6-C_{12} medium-chain fatty acids do not make a significant contribution to the aroma of wines, since they have high perception thresholds and relatively neutral aromas. Nevertheless, their esters can increase the quality of wines by contributing pleasant fruity aromas.

Acetic acid, which could in principle be assumed to contribute to vinegary aromas, has a much higher perception threshold than its derivative ethyl acetate. Consequently, the aroma commonly associated with vinegar is mainly due to the contribution of ethyl acetate rather than to that of acetic acid.

7.3. Color (Clarity and Brilliance)

Phenols, specifically anthocyans in red wines, are responsible for the color of wine. These anthocyans are initially present in the skin of the grapes and are transferred more easily to the aqueous alcoholic solution of the must during fermentation at lower pH, since under those conditions the flavylium cation, which is more soluble in aqueous alcohol solutions, is formed more easily. In addition, the color contributed by polyphenols varies with pH, which is the main factor responsible for the equilibrium between the ionic and molecular forms. At a pH of less than 3, the purple form predominates, and at a pH close to neutral, red wines acquire a reddish-blue color. The pH thus influences the intensity and hue of the color of red wines, as well as their brightness (brilliance).

In white wines, a low pH increases the brightness and brilliance of the yellow color. When the pH approaches neutral, brownish-black colors appear. Indirectly, the pH influences the color of white wines by favoring redox reactions of certain compounds with the oxygen in the air. The chemical reactions that cause browning are less likely to occur at low pH; products with an antioxidant effect such as sulfite are also more active at low pH. It is important to remember that polyphenol oxidases usually have optimal pH values that are not especially low.

7.4. Haze

The presence of substances in suspension reduces the quality of wine. These substances have various origins and cause the precipitation that occurs to a greater or lesser extent depending on the pH of the wine. The physicochemical stability of wine is greater at a low pH, which also favors clarity and brilliance and the effect of fining. Table 14.3 shows some examples of how pH affects certain "accidents" in wine.

Finally, pH influences the activity of enzymes and microorganisms found in wine and leads to selection of microorganisms according to their adaptive capacity. In general, the microorganisms of interest in winemaking (yeasts, lactic acid bacteria, and acetic acid bacteria) are more resistant to high concentrations of alcohol than they are to low pH. Consequently, the following factors — which also serve as a reminder of the importance of pH in the winemaking and storage process — are particularly important:

- From a technological point of view, higher acidity (lower pH) offers greater protection against microbial spoilage and at the same time offers advantages during the winemaking process that are reflected in the more rapid natural clarification of a new wine, greater effectiveness of stabilization treatments (clarification and lower doses of sulfite), and a longer shelf life of the bottled wine.
- Nevertheless, from a sensory perspective, excessive acidity can have a negative effect on the flavor of wine, and the equilibrium between substances that produce pleasant flavors (sweet), such as sugars and alcohols, or unpleasant flavors, such as the astringent, acidic flavor contributed by tannins (higher concentrations of tannins should coincide with lower acidity and vice versa), must be taken into consideration.

TABLE 14.3   Poorly Soluble Compounds Found in Wine and the pH at Which They are Formed.

Tartrate precipitates	Greater risk at pH >3.6
White casse	Greatest risk at pH = 3.3
Blue casse	Greatest risk at pH >3.3
Protein casse	Greatest risk at pH ≥4 (which is the isoelectric point for proteins in wine)

8. CORRECTION OF ACIDITY

8.1. Calculating the Acid Dissociation Constant of Wine

In practical terms, it is worthwhile considering all of the acids present in wine as a single acid represented as HV. This acid will have an acid dissociation constant or pK_a value that can be easily calculated from the pH, titratable acidity, and the alkalinity of ash obtained by chemical analysis in an enology laboratory.

The equilibrium that is established in wine can be represented as:

$$HV \rightleftarrows H^+ + V^-$$

As shown in the above equilibrium, the acids present in the must and wine are partially dissociated. The anions formed in this equilibrium are neutralized by the cations (M^+) of the must and wine (principle of electrochemical neutrality).

It can therefore be established that: $[V^-] = [M^+] =$ alkalinity of ash
$[HV] =$ titratable acidity

The value of the acid dissociation constant K_v for the above equilibrium would be:

$$K_v = \frac{[V^-] \times [H^+]}{[HV]} \qquad pK_v = pH - \log\frac{[V^-]}{[HV]} \qquad \text{or alternatively}: pK_v = pH - \log\frac{[AA]}{[TA]}$$

The alkalinity of ash is determined using a very painstaking analytic technique, but it can be calculated indirectly based on the buffering capacity:

$$\frac{dB}{dpH} = 2.303 \frac{TA \times AA}{TA + AA}$$

In order to apply this equation, it is best to determine the buffering capacity over the normal pH range used. Thus, to increase the value of the pH of wine by 0.3 units, the buffering capacity is calculated over this range, taking 100 mL of wine and adding 0.1 N sodium hydroxide until the pH of the wine is increased by approximately 0.3 units. If the aim is to reduce the pH of the wine by 0.3 units, the buffering capacity is determined by adding 0.1 N hydrochloric acid to 100 mL of wine until the pH is reduced by approximately 0.3 units.

Example: Calculate the acid dissociation content of a wine with a pH of 2.96 and a titratable acidity of 118.5 meq/L. In addition, 100 mL of wine requires 16.55 mL of 0.1 N sodium hydroxide to increase its pH by 0.33 units.

$$\frac{16.55 \times 0.1 \text{ meq NaOH}}{100 \text{ mL wine}} \times 1000 \text{ mL}/\text{L} = 16.55 \text{ meq NaOH}/\text{L wine}$$

Then, the buffering capacity of the wine for a range of 0.3 units is:

$$\frac{16.55}{0.33} = 50.15 \text{ meq/L} = \frac{dB}{dpH} \text{ (buffering capacity determined)}$$

The alkalinity of ash, or the acid content present as salt, is calculated using the equation for the buffering capacity:

Characteristics of the wine:

$$TA = 118.5 \text{ meq/L} \qquad pH_i = 2.96 \qquad \Pi_{calculated} = 50.15 \text{ meq/L}$$

$$\Pi = \frac{dB}{dpH} = 2.303 \, \frac{TA \times AA}{TA + AA}$$

$$\Pi \times (TA + AA) = 2.303 \, (TA \times AA)$$

$$2.303 \times TA \times AA - \Pi \times AA = \Pi \times TA$$

$$AA = \frac{\Pi \times TA}{2.303 \times TA - \Pi}; \qquad AA = \frac{50.15 \times 118.5}{2.303 \times 118.5 - 50.15} = 26.7 \text{ meq/L}$$

The value of the pK_v for the hypothetical monocarboxylic acid in the wine (equivalent to the combination of individual acids present) can now be calculated.

As we saw: $pK_v = pH - \log \dfrac{[AA]}{[TA]}$

Then the wine has the following value: $pK_v = 2.96 - \log \dfrac{26.7}{118.5} = 3.61$

8.2. Deacidification

Once the acid dissociation constant of the wine is known, it is easy to predict the pH following addition of deacidifying agents. We should bear in mind that potassium salts are the most commonly used deacidifying agents and their addition leads not only to neutralization of the acid but also to the precipitation of the potassium bitartrate that is formed, particularly following cold stabilization. This effect influences the final titratable acidity and the alkalinity of ash. The effect of adding K_2CO_3 should therefore be considered in two steps.

Example: Calculate the effect of adding 16.7 meq/L of K_2CO_3 on the pH of the wine used in the previous example. (The value of 16.7 meq/L has been chosen because it is essentially the same as the amount of sodium hydroxide required to increase the pH by 0.33 units in the previous example.)

(1) Effects of neutralization by addition of 16.7 meq/L of K_2CO_3 to a wine with a pH of 2.96

The reaction that occurs following addition of the potassium salts is a cation-displacement reaction:

$$2 \, HV + K_2CO_3 \rightarrow 2 \, KV + H_2CO_3$$

The carbonic acid formed breaks down at the pH of wine:

$$H_2CO_3 \rightarrow CO_2 \text{ gas} \uparrow + H_2O \text{ liquid}$$

Considering that the acidity is generated by a single monoprotic acid:

$$HV \rightleftarrows V^- + H^+$$

The addition of K_2CO_3 causes displacement of the equilibrium towards the right to compensate for the loss of H^+, and there is therefore a reduction in the quantity of HV (titratable acidity) and an increase in the quantity of V^- (alkalinity of ash).

$$HV \rightleftarrows V^- + H^+$$

a) Resulting reduction of titratable acidity: 16.7 meq/L. Then, the new titratable acidity is:

$$118.5 - 16.7 = 101.8 \text{ meq/L}$$

b) Increase in the alkalinity of ash (or neutralized acid): 16.7 meq/L. Then, the new alkalinity of ash is:

$$26.7 + 16.7 = 43.4 \text{ meq/L}$$

c) New pH: $pH = pK_v + \log\dfrac{43.361}{101.8} = 3.61 + (-0.37) = 3.24$

d) Increase in pH: 0.3 units compared with the initial pH

e) New buffering capacity: $\dfrac{dB}{dpH} = 2.303 \dfrac{101.8 \times 43.4}{101.8 + 43.4} = 70.0 \text{ meq/L}$

(2) Effect of potassium bitartrate precipitation on pH

Although in practical terms the acidity of wine is hypothetically attributed to a monoprotic acid, it should not be forgotten that wine is saturated with potassium bitartrate, such that the addition of potassium salts will cause the wine to become supersaturated with potassium bitartrate, which will precipitate during cold stabilization. In this respect, we can consider that all potassium precipitates in the form of potassium bitartrate. This will affect the alkalinity of ash and the titratable acidity in relation to the values achieved during neutralization. The precipitation of potassium bitartrate reduces the titratable acidity, since the hydrogen ions in the precipitated material are not included in the determination. Likewise, the disappearance of HT^- when it precipitates as bitartrate will reduce the alkalinity of ash.

$$\text{Final AA} = \text{Initial AA} + \text{meq } K_2CO_3 - \text{meq } K^+$$
$$\text{meq } K_2CO_3 = \text{meq } K^+ \rightarrow \text{Final AA} = \text{Initial AA}$$
$$\text{Final TA} = \text{Initial TA} - \text{meq } K_2CO_3 - \text{meq } K^+$$

The titratable acidity and the alkalinity of ash become:
Δ Titratable acidity: $(101.8 - 16.7) = 85.1 \text{ meq/L}$
Δ Alkalinity of ash: $(43.4 - 16.7) = 26.7 \text{ meq/L}$
Then, the new pH will be:

$$pH = pK_v + \log\dfrac{[AA]}{[TA]} = 3.61 + \log\dfrac{26.7}{85.1} = 3.10$$

Consequently, the pH has been reduced by 0.14 units. Following deacidification and cold stabilization, the new buffering capacity will be:

$$\frac{dB}{dpH} = \Pi = 2.303\frac{TA \times AA}{TA + AA} = 2.303\frac{85.1 \times 26.7}{85.1 + 26.7} = 46.8 \ \ meq/L$$

In summary, the addition of a deacidifying agent causes:

1. A reduction in titratable acidity and an increase in the pH and the alkalinity of ash during deacidification treatment.
2. A new reduction in the titratable acidity and a reduction in the alkalinity of ash and the pH as a consequence of precipitation of potassium bitartrate, either spontaneously or during stabilization.
3. The equivalents of sodium hydroxide used to calculate the buffering capacity, which correspond to an increase of 0.33 units in the pH, are very similar to those of the deacidifying agent added (in this case, K_2CO_3). However, the pH change obtained with the deacidifying agent is less ($pH_{final} = 3.13$) than with sodium hydroxide ($pH_{final} = 3.30$). This is due to the precipitation of the poorly soluble salts formed by the potassium cation and the bitartrate anion.

The deacidifying agents most commonly used are K_2CO_3, $KHCO_3$, and $CaCO_3$. However, the precipitation equilibrium for $CaCO_3$ involves tartrate rather than bitartrate.

$$T^{2-} + Ca^{2+} \rightarrow CaT\downarrow$$

The precipitation of tartrate has the same effect on the alkalinity of ash and the titratable acidity as observed in the case of potassium carbonate; that is, it causes a reduction in titratable acidity and pH. In general terms, the following equation can be described:

$$H_2T + CaCO_3 \rightarrow CaT\downarrow + H_2CO_3$$
$$H_2CO_3 \rightarrow CO_{2(gas)}\uparrow + H_2O$$

$CaCO_3$ is usually less effective as a deacidifying agent than $KHCO_3$, since Ca^{2+} can be linked to other equilibria. Nevertheless, the effect on the pH takes place faster than with potassium salts, since the solubility of calcium tartrate is lower than that of potassium bitartrate. Finally, it should be remembered that this form of deacidification only affects tartaric acid content.

Calculating the Quantity of Deacidifying Agent Required

Once the effect of adding a deacidifying agent on the titratable acidity and the alkalinity of ash is known, we can calculate the quantity of deacidifying agent that needs to be added to the must or wine to achieve the desired pH.

Let x = meq K_2CO_3 to be added

$$\text{Final TA} = \text{Initial TAv} - 2x$$
$$\text{Final AA} = \text{Initial AA}$$

Using the equation $pH = pK_v - \log\dfrac{TA}{AA}$ we can calculate the value of x to achieve a given pH in the wine:

$$pH = pK_v - \log\frac{TA - 2x}{AA} \qquad pH - pK_v = -\log\frac{TA - 2x}{AA}$$

$$pK_v - pH = \log\frac{TA - 2x}{AA} \qquad 10^{pK_v - pH} = \frac{TA - 2x}{AA}$$

In the example that we are using, if the intention is to change the pH from an original value of 2.96 to 3.30, knowing that the pK_v is 3.61, the following quantity of deacidifying agents is obtained:

$$10^{3.61-3.30} = \frac{118.5 - 2x}{26.7} \qquad -2x = (10^{0.31} \times 26.7) - 118.5$$

$$x = 32.0 \text{ meq/L of deacidifying agent}$$

As mentioned above, more equivalents of deacidifying agent are required to achieve the desired pH than those obtained through calculation of the buffering capacity with sodium hydroxide (32.0 meq/L versus 16.55 meq/L).

Of course, if 32 meq/L (2.4 g/L) of molecular tartaric acid is not present in the wine and therefore cannot be precipitated by potassium ions, the pH cannot be increased to the desired value.

8.3. Acidification

Addition of Tartaric Acid as an Acidifying Agent

Tartaric acid is commonly used to correct acidity. The procedure used to calculate the dose required to reduce the pH to a given value is similar to that described for deacidification.

As in the case of deacidification, it is necessary to determine the pH, the titratable acidity, and the buffering capacity. The buffering capacity is determined based on addition of 0.1 N hydrochloric acid to 100 mL of wine until a pH close to the desired value is obtained. Using this information, the alkalinity of ash and finally the acid dissociation constant of the hypothetical monoprotic acid are calculated.

Example: Let the wine have the following characteristics determined by chemical analysis:

Initial characteristics:

TA = 54 meq/L.
pH = 3.92
Volume of 0.1 N hydrochloric acid required to achieve a pH of 3.43 in 100 mL of wine: 22.2 mL

Calculated values:

$$\Pi = \frac{22.2 \text{ mL} \times 0.1 \frac{\text{meq}}{\text{mL}}}{\dfrac{0.1L}{3.92 - 3,43}} = 45.3 \frac{\text{meq}}{\text{L pH}}$$

$$AC = \frac{\Pi \times AT}{2.303 \times AT - \Pi} = 30.95 \frac{\text{meq}}{\text{L}}$$

$$pK_v = pH - \log \frac{[AA]}{[TA]} = 3.92 - \log \frac{30.95}{54.0} = 4.16$$

We can consider the addition of tartaric acid in two stages: the first corresponds to the acidification reaction itself and the second to the precipitation of potassium bitartrate that occurs during natural or induced stabilization of the wine.

(1) Acidification with x meq/L of tartaric acid

Tartaric acid is assumed to act as a strong monoprotic acid at the pH of wine. This implies that all of the tartaric acid added will dissociate during its first dissociation and that, in order to calculate the number of grams to be added, the equivalent mass is assumed to be equal to the molecular mass.

It is worth first considering the effect of the proton derived from the first dissociation of H_2T on the equilibrium of the monoprotic acid of the wine, followed by the effect of the anion (HT^-), which is also derived from the first dissociation of H_2T.

1a. The protons derived from the first dissociation of H_2T displace the equilibrium of the acid in the wine towards the left, such that the titratable acidity increases by the same number of milliequivalents as the equivalents of acid added. The alkalinity of ash is reduced to the same degree as the titratable acidity is increased:

$$H_2T \rightleftarrows H^+ + HT^-$$

$$\overleftarrow{HV \rightleftarrows H^+ + V^-}$$

$$TA = TA_{initial} + x; \text{ and also: } AA = AA_{initial} - x$$

1b. The presence of x milliequivalents of the anion HT^- derived from the first dissociation of the added H_2T contributes to the titratable acidity, since it contains a proton that is included in the determination up to a pH of 7, and also to the alkalinity of ash, since it is an anion derived from the dissociation of an acid.

Consequently, taking into account these two effects, the final balance is:

$$TA_{final} = TA_{initial} + x + x = TA_{initial} + 2x$$
$$AA_{final} = AA_{initial} - x + x = AA_{initial}$$

If we assume that 22.2 meq/L of tartaric acid has been added:

$$TA_{final} = TA_{initial} + 2x = 54 + 2 \times 22.2 = 98.4 \text{ meq/L}$$
$$AA_{final} = AA_{initial} = 30.95 \text{ meq/L}$$

The following pH buffering capacity would be obtained following the acidification step:

$$pH = pK_v + \log\frac{[AA]}{[TA]} = 4.16 + \log\frac{30.95}{98.4} = 3.66$$

$$\Pi = \frac{dB}{dpH} = 2.303 \times \frac{98.4 \times 30.95}{98.4 + 30.95} = 54.21 \ meq/L$$

(2) Bitartrate precipitation stage

Given that wine is considered to be a saturated solution of potassium bitartrate, the addition of tartaric acid as an acidifying agent will introduce the common ion HT^-, and this will precipitate naturally or during cold stabilization as potassium bitartrate. As with the addition of potassium salts as deacidifying agents, we can consider that all of the HT^- added will precipitate as potassium bitartrate.

Again, precipitation of bitartrate has an effect on the alkalinity of ash and the titratable acidity.

$$AA_{final} = AA_{initial} - x \ meq, \ H_2T = 30.95 - 22.2 = 8.75 \ meq/L$$
$$TA_{final} = TA_{initial} + 2x - x = 54 + 2 \times 22.2 - 22.2 = 76.2 \ meq/L$$

The pH and the buffering capacity following the precipitation phase are:

$$pH = pK_v + \log\frac{[V^-]}{[HV]} = 4.16 + \log\frac{8.75}{76.2} = 3.22$$

$$\Pi = \frac{dB}{dpH} = 2.303 \times \frac{76.2 \times 8.75}{76.2 + 8.75} = 18.07 \ meq/L$$

The increase in pH obtained in each step is:

1. Acidification step: Δ pH $= 3.92 - 3.66 = 0.26$
2. Precipitation step: Δ pH $= 3.66 - 3.22 = 0.44$
3. Total: Δ pH $= 3.92 - 3.22 = 0.70$

As can be seen, the reduction in the pH generated in the wine by the addition of 22.2 meq/L of tartaric acid and the subsequent precipitation of the bitartrate formed is greater than that obtained through the addition of the same milliequivalents of hydrochloric acid when determining the buffering capacity (0.70 versus 0.49 units). This should be taken into account during the acidification of musts and wines in order to prevent miscalculating the dose of tartaric acid required and hence obtaining wines with a much lower pH than desired.

Calculating the Quantity of Tartaric Acid to be Added

As in the case of deacidification, once the effect of adding tartaric acid on the titratable acidity and the alkalinity of ash is known, we can calculate the quantity of tartaric acid that must be added to the must or wine to obtain the desired pH.

Example: Using the same wine as in the previous section, calculate the quantity of tartaric acid that must be added to reduce the pH to a value of 3.43.

Initial characteristics:

TA = 54 meq/L
pH = 3.92
pK_v = 4.16
Π = 45.3 meq/L pH
AA = 30.95 meq/L

Considering the two steps established in the previous section, the final effect on the titratable acidity and the alkalinity of ash caused by acidification and subsequent stabilization is:

Final characteristics:

Final pH desired = 3.43

$$TA_{final} = TA_{initial} + x$$

$$AA_{final} = AA_{initial} - x$$

$$pK_v - pH = \log\frac{TA + x}{AA - x}; \qquad 10^{(4.16 - 3.43)} = \frac{54 + x}{30.95 - x}$$

$$5.37(30.95 - x) = 54 + x; \quad x = 17.24 \text{ meq/L}$$

The outcome of adding 17.24 meq/L tartaric acid is a reduction in the pH of the wine to 3.43. The quantity of tartaric acid, expressed in grams, is obtained by considering that tartaric acid behaves as a monoprotic acid at the pH of wine. Therefore:

$$17.24\frac{meq}{L} \times 150\frac{mg\ H_2T}{meq} \times 1\frac{g\ H_2T}{1000\ mg} = 2.6\ g/L$$

$$\frac{3.92 - 3.43\ pH}{2.6\ g/L} \Rightarrow 1\ g/L \text{ reduces } 0.19\ pH \text{ units}$$

The effect on the buffering capacity is:

$$\Pi = \frac{dB}{dpH} = 2.303 \times \frac{71.24 \times 13.71}{71.24 + 13.71} = 26.48\ meq/L$$

According to these calculations, the addition of 1 g of tartaric acid to the wine causes a reduction of approximately 0.2 pH units and a reduction of approximately 7 meq/L in the buffering capacity, due mainly to a reduction in the alkalinity of ash. However, the wine retains a substantial "acid reserve" since the titratable acidity increases as a consequence of the addition of the acid.

Addition of Calcium Sulfate as an Acidifying Agent

In some regions, the use of $CaSO_4$ (Plaster of Paris) as an acidifying agent is permitted. The equilibria that govern this form of acidification are the following:

$$CaSO_4 \rightarrow Ca^{2+} + SO_4^{2-}$$

$$Ca^{2+} + T^{2-} \rightarrow CaT\downarrow$$

$$H_2T \rightleftarrows HT^- + H^+ \rightleftarrows T^{2-} + H^+$$

The Ca^{2+} obtained from dissociation of $CaSO_4$ binds the T^{2-} ion present in the must or wine, producing a poorly soluble salt that deposits at the bottom of the tank and causes displacement of the dissociation equilibrium for tartaric acid towards the right, releasing protons and therefore reducing the pH. The complete reaction is:

$$H_2T + Ca^{2+} \rightarrow TCa\downarrow + 2\,H^+$$

Addition of x milliequivalents of calcium sulfate causes elimination of x milliequivalents of tartrate through precipitation as calcium tartrate, and the titratable acidity and the alkalinity of ash are reduced in the same way as during precipitation in wine deacidifed with $CaCO_3$.

$$AA_{final} = AA_{initial} - x \text{ meq } CaSO_4$$
$$TA_{final} = TA_{initial} - x \text{ meq } CaSO_4$$

Example: Calculate the quantity of $CaSO_4$ necessary to reduce the pH of the wine used in the previous example to a pH of 3.43.

Initial pH $= 3.92$
$TA_{initial} = 54.0$
$AA_{initial} = 30.95$
$pK_v = 4.16$
$\Pi = 45.3$ meq/L
Target pH $= 3.43$

$$10^{(4.16-3.43)} = \frac{54 - x}{30.95 - x}$$
$$5.37(30.95 - x) = 54 - x; \quad x = 25.67 \text{ meq/L}$$

The required quantity of $CaSO_4$, expressed in grams, is calculated from its equivalent weight.

$$25.67\,\frac{meq}{L} \times \frac{136\ mg}{2\ meq} \times \frac{1\ g}{1000\ mg} = 1.75\,\frac{g\ CaSO_4}{L}$$

The new buffering capacity is:

$$TA_{final} = 54 - 25.67 = 28.33$$

$$AA_{final} = 5.28$$

$$\Pi = \frac{dB}{dpH} = 2.303 \times \frac{28.33 \times 5.28}{28.33 + 5.28} = 10.25 \text{ meq/L}$$

The buffering capacity is considerably reduced compared to that of the initial wine (45.31 meq/L) and it is also lower than that obtained by acidification with H_2T (26.22 meq/L). The titratable acidity is also lower, since in this case the reduction in the pH was obtained through a reduction in titratable acidity, whereas with the addition of tartaric acid the titratable acidity is increased. Consequently, the addition of calcium sulfate causes an imbalance in the wine since it considerably reduces the undissociated acid reserve and, as it has a low buffering capacity, the pH of the wine is more susceptible to change.

9. FINAL CONSIDERATIONS

As described for the analysis of acids in wine, the bitartrate ion has a maximum concentration that depends upon its dissociation constants.

In an aqueous alcohol solution, the values of the dissociation constants are reduced (the pK_a increases) and the maximum is displaced to the right. Thus, in aqueous solution, the maximum value is achieved at a pH of 3.69, whereas in aqueous alcohol solution (11% vol/vol ethanol), the maximum is observed at a pH of approximately 3.9.

The pH of wine is usually lower than 3.9, meaning that it is also lower than the dissociation maximum for tartaric acid. Under these conditions, the primary dissociation of tartaric acid predominates, and therefore the precipitation of potassium bitartrate during tartaric stabilization causes a reduction in pH. However, if the pH is greater than 3.9, the second dissociation of tartaric acid predominates. In these cases precipitation of potassium bitartrate causes the previous equilibrium to shift to the left, towards the formation of HT^-, and this reduces the concentration of protons in the medium with a consequent increase in the pH.

It is preferable to carry out corrections in must rather than wine. When correcting the pH of must, it is important to take into account the pH that you want to achieve in the wine after the precipitation of potassium bitartrate that occurs naturally during the winter or is induced by cold stabilization. This helps to reduce the metallic flavor caused by addition of H_2T to a finished wine. In some regions, rather than correct for a given pH, corrections are carried out to achieve a titratable acidity established as an optimal value of 5 g/L H_2T. This, however, does not take into account subsequent precipitation of potassium bitartrate and the changes that this causes in the pH, and it is therefore always preferable to correct the acidity to a pH value.

The correction of pH is based on the following fundamental considerations:

1. Analysis of pH and titratable acidity.
2. Analysis of the buffering capacity for the range within which the pH is to be reduced or increased (or determination of the alkalinity of ash).

3. Calculation of the pK_a value of a hypothetical monocarboxylic acid in the wine.
4. Calculation of the quantity of acidifying or deacidifying agents required to obtain the desired pH after tartaric stabilization of the wine.

In the case of deacidification, it should be remembered that the calculations are valid so long as the wine contains sufficient quantities of tartaric acid and bitartrate to ensure that all the potassium or calcium added in the form of carbonate will be precipitated as potassium bitartrate or calcium tartrate. In the case of acidification with tartaric acid, the reduction in pH caused by the precipitation of potassium bitartrate is influenced by the quantity of potassium ions. It is also important to take into consideration legislation regarding permissible quantities of added tartaric acid and the negative effect that large quantities of this acid can have on the flavor of wine.

15

Precipitation Equilibria in Wine

1. INTRODUCTION

Wine is constantly changing, from the moment it is obtained from the fermentation of must to the moment it is served. The appearance of ethanol during alcoholic fermentation alters the physical and chemical properties of the medium, which changes from water to an aqueous alcoholic solution. One of the main changes that occurs is that the polarity of the medium decreases with increasing concentrations of ethanol. This decrease in polarity reduces the solubility of polar compounds, giving rise to a haze caused by particles that gradually settle at the bottom of the vessel. These phenomena, which are mostly chemical, affect the clarity of wine and are influenced by both chemical factors (such as pH) and physical factors (such as temperature).

Enological Chemistry. DOI: 10.1016/B978-0-12-388438-1.00015-7

253

Haze can develop naturally during the cold winter months following fermentation, or it can be induced by deliberately reducing the temperature of the wine. The organoleptic properties of wine can also be affected by biological factors, which need to be taken into account to ensure that a wine remains stable up to the moment it is served.

Stabilization processes consist of eliminating or reducing the impact of factors that can upset the natural development of wine or alter its organoleptic properties. The purpose of stabilization is to ensure that the wine retains all its qualities (color, aroma, and flavor) for as long as possible and that it deviates as little as possible from the characteristics sought by the winemaker or the expectations of the consumer.

The main causes of instability are:

1. The formation of tartaric acid salts with potassium and calcium ions. These crystalline salts cause wine haze and lead to the formation of crystals at the bottom of the vessel.
2. Excess levels of iron, copper, and protein. These also cause wine haze and the precipitation of coloring matter.
3. The abnormal development of certain microorganisms that can alter the composition and organoleptic properties of wine.

This chapter will analyze the ionic equilibria responsible for the development of haze and the precipitation of crystals in wine.

2. TARTRATE SOLUBILITY

Tartaric acid is a weak acid that, at the pH of wine, is partially dissociated according to the following equilibrium:

$$H_2T \leftrightarrow HT^- + H^+ \leftrightarrow T^{2-} + H^+$$

The proportions of the three species that arise from tartaric acid when it dissolves (H_2T, HT^-, and T^{2-}) depend on pH, temperature, the concentration of other ions (ionic strength), and ethanol content. Of the three species, the bitartrate ion (HT^-) is of particular interest as, together with the potassium ion, it forms the potassium bitartrate salt (KHT), which is sparingly soluble in aqueous alcoholic solutions and, therefore, largely responsible for the precipitations that occur during fermentation and in finished wine.

The solubility of a salt at a given temperature is determined by its solubility product, which is an equilibrium constant. The solubility product is calculated by determining the effective concentrations of products and reagents, expressed in moles per liter, at equilibrium.

The equilibrium that governs the precipitation of potassium bitartrate is a heterogeneous equilibrium as it involves two phases: a solid phase, formed by the precipitate (which settles at the bottom of the vessel) and a liquid phase comprising the ions in the solution. The equilibrium constant is thus expressed by K_{SP} or simply SP (solubility product) and is written as follows:

$$H_2T \leftrightarrow HT^- + H^+$$
$$HT^- + K^+ \leftrightarrow KHT \qquad KHT_{(solution)} \leftrightarrow KHT_{(solid)}$$

The heterogeneous equilibrium in solution is expressed as:

$$KHT_{(solid)} \leftrightarrow HT^-_{(solution)} + K^+_{(solution)}$$

The solubility product is: $SP = [HT^-]_{solution} \times [K^+]_{solution}$

When the product of the effective concentrations of the two ions is the same as the solubility product, the system is at equilibrium and no precipitates are formed. If it is higher than the solubility product, precipitation occurs, with the formation of larger or smaller crystals that are deposited on the walls and bottom of the vessel.

The solubility (S) of a sparingly soluble substance refers to the maximum amount of substance that can dissolve in a given solvent at a given temperature. When a sparingly soluble substance is ionic (i.e. when it forms ions in solution), the solubility product is related to the concentration of dissolved ions. For a generic substance with the formula A_mB_n, the following expression holds:

$$A_mB_n \leftrightarrow m\ A^{n-} + n\ B^{m+} ;$$
$$S \qquad mS \qquad nS$$

$$SP = [A^{-n}]^m_{equilibrium} \times [B^{+m}]^n_{equilibrium}; \quad SP = (mS)^m \times (nS)^n = m^m n^n S^{(m+n)}$$

S is an equilibrium concentration, which, when expressed as a function of SP, is:

$$S_{(mol/L)} = \sqrt[m+n]{\frac{SP}{m^m n^n}}$$

Important examples in wine include:

$$KHT(solid) \leftrightarrow HT^-(solution) + K^+(solution)$$
$$S \qquad\qquad S \qquad\qquad S$$

$$SP = [HT^-][K^+] = S^2; \quad S = \sqrt{SP}$$
$$SCu_{2(solid)} \leftrightarrow 2\,Cu^+_{(solution)} + S^{2-}_{(solution)}$$
$$S \qquad\qquad 2S \qquad\qquad S$$

$$SP = [S^{2-}][Cu^+] = S \times (2S)^2 = 4S^3; \quad S = \sqrt[3]{\frac{SP}{4}}$$

If we refer to the product of ion concentrations at a given temperature as Q, we get the following expression: $Q = [A^{n-}]^m [B^{m+}]^n$, and a relationship can be established between the product of concentrations, their solubility product, and the stability of wine towards the precipitation of sparingly soluble substances:

1. If $Q = SP \rightarrow$ the system is at equilibrium. The solution is saturated and there is a risk of precipitation if the equilibrium conditions change.
2. If $Q < SP \rightarrow$ the solution is not saturated and there will be no precipitation.
3. If $Q > SP \rightarrow$ the solution is supersaturated and precipitation will occur until $Q = SP$.

The solute concentration of a saturated solution is identical to the solubility of the solute. That of an unsaturated solution will be lower than its solubility and the solution will therefore be able to dissolve more solute. A supersaturated solution, in contrast, has a higher concentration of solute than its solubility and will therefore be unstable; accordingly, it will eliminate excess solute by forming precipitates that are deposited at the bottom of the vessel. In winemaking, the activity of ions in solution rather than the concentration of these ions is calculated.

$$AP = f_{A^{n-}}^{m} [A^{n-}]^{m} f_{B^{m+}}^{n} [B^{m+}]^{n}$$

3. FACTORS THAT INFLUENCE SOLUBILITY

Generally speaking, the presence of other ions in solution affects the solubility-precipitation equilibrium, such that there are two possible situations:

1. The solution contains dissolved ions other than those that participate in the precipitation equilibrium of the sparingly soluble compound.
2. Within the solution there are other salts that contain the same ions as those that participate in the precipitation equilibrium.

Other factors that can influence precipitation equilibria in wine are pH, ethanol content, and temperature.

3.1. The Uncommon Ion Effect

The uncommon ion effect, or salt effect, refers to a situation in which the solution contains ions other than those involved in the precipitation equilibrium. These ions, known as uncommon ions, influence the activity of the ions that participate in the equilibrium through their ionic strength.

$$\log f = \frac{-0.5 \, x Z_A \, Z_B \sqrt{I}}{1 + \sqrt{I}}; \quad I = \frac{1}{2} \sum_{i=1}^{i=n} C_i Z_i^2$$

The activity coefficient is less than 1 in all solutions except infinitely dilute solutions, in which the ionic activity is the same as ionic concentration; in such cases, the activity product (AP) is equal to the solubility product.

Considering the above ($f \leq 1$):

$$AP = f_{A^{n-}}^{m} [A^{n-}]^{m} f_{B^{m+}}^{n} [B^{m+}]^{n} = f_{A^{n-}}^{m} f_{B^{m+}}^{n} SP;$$

$$\text{Hence}: \; AP \leq SP$$

The uncommon ion effect results in an increase in the solubility of a sparingly soluble compound. This occurs because the concentration of ions that form the sparingly soluble salt that is needed to achieve the value of the solubility product has to be higher than it would be in the absence of uncommon ions. The uncommon ion effect is directly related to the charge and concentration of all the ions in the solution.

3.2. The Common Ion Effect

The common ion effect refers to a situation in which a solution of ions that form an insoluble salt contains ions that participate in the precipitation equilibrium (common ions). In such cases, solubility decreases as the concentration of the common ion increases.

$$AgCl \rightleftarrows Ag^+ + Cl^-; \quad SP = [Ag^+][Cl^-] = S^2 = 10^{-10} \ S = 10^{-5}$$

If 0.1 mol/L of NaCl is added to a solution containing Cl^- and Ag^+, the common ion Cl^- is being added at a concentration of: $[Cl^-] = 0.1$.

$SP = [Ag^+][Cl^-] = S'(S' + 0.1) \approx S' \times 0.1 = 10^{-10} \rightarrow S' = 10^{-9}$. (The value of $S' + 0.1$ has been approximated to 0.1 as S' will be smaller than S, which is much smaller than 0.1.)

The solubility of sparingly soluble compounds is also affected by factors such as pH, the presence of ligands that form complexes with ions in solution, and the redox potential, which can change the oxidation state of the ions that participate in the precipitation equilibrium. In winemaking, the ethanol concentration is also an important factor to consider.

4. FACTORS THAT AFFECT TARTRATE PRECIPITATION

Bitartrate and potassium ions participate in potassium bitartrate precipitation, the most important ionic precipitation that occurs in wine. Potassium bitartrate is the main component of cream of tartar, which appears in wine when the temperature is sufficiently low. We will therefore first examine the different factors that affect bitartrate ion concentrations.

4.1. pH

pH influences the concentration of all the species of tartaric acid in solution according to the following equilibrium:

$$H_2T \leftrightarrow HT^- + H^+ \leftrightarrow T^{2-} + H^+$$

At point 1, $pH = pK_{a1}$. The concentration of undissociated tartaric acid is equal to that of the bitartrate ion and much higher than that of the tartrate ion.

If $pH = pK_{a1} \rightarrow [HT^-] = [H_2T] >> [T^{2-}]$.

At point 2, $pH = pK_{a2}$. The concentration of the bitartrate ion is the same as that of the tartrate ion, and both are much higher than that of undissociated tartaric acid.

If $pH = pK_{a2} \rightarrow [HT^-] = [T^{2-}] >> [H_2T]$.

Another interesting point is the maximum height of the mole fraction curve for the bitartrate ion, which corresponds to the maximum concentration of this ion. The pH at this point is:

$$pH_{HT^-_{max}} = \frac{pK_{a2} + pK_{a1}}{2} = \frac{4.37 + 3.01}{2} = 3.69$$

$$[HT^-] > [T^{2-}] = [H_2T]$$

FIGURE 15.1 Effect of pH on the proportion of tartaric acid species.

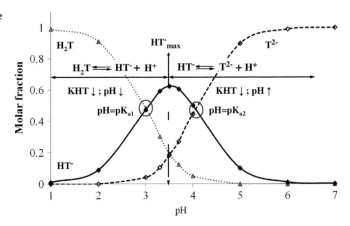

   In an aqueous solution, the maximum precipitation of potassium bitartrate will occur at this pH of 3.69, as this is when the concentration of the bitartrate ion is at its maximum. Nonetheless, it should be recalled that this pH was calculated using pK_{a1} and pK_{a2} values in an aqueous solution. Because pK_a increases with ethanol content, the pH at which the bitartrate ion will reach its maximum concentration will also increase with ethanol content. For example, in a wine with an ethanol content of 11% (vol/vol), the bitartrate ion will reach its maximum concentration at a pH of 3.9.
   The precipitation of potassium bitartrate will increase or decrease the pH of the solution depending on whether this pH is higher or lower than the pH at which the bitartrate ion reaches its maximum concentration.

$$KHT_{(solid)} \leftrightarrow HT^-_{(solution)} + K^+_{(solution)}$$

1. If $pH_{solution} < pH_{maximum}$, the precipitation of potassium bitartrate will cause a decrease in bitartrate ion concentration, which is compensated by a shift in the dominant equilibrium $(H_2T \leftrightarrow HT^- + H^+)$ towards the right. This shift causes an increase in the concentration of H^+ ions in solution, and therefore a decrease in pH.
2. If $pH_{solution} > pH_{maximum}$, the precipitation of potassium bitartrate will cause a decrease in bitartrate ion concentration, which is compensated by a shift in the dominant equilibrium $(HT^- \leftrightarrow T^{2-} + H^+)$ to the left. This shift causes a decrease in the concentration of H^+ ions in solution, and therefore an increase in pH.
3. If $pH_{solution} = pH_{maximum}$, the precipitation of potassium bitartrate will not cause any changes in pH.

4.2. Ethanol

   Ethanol influences the dissociation constants of tartaric acid and the solubility of the ionic salts relative to their solubility in pure water. The effect is due to a reduction in the dielectric constant (ε) of the aqueous medium, which, in turn, also causes a reduction in the solubility of the ionic and polar compounds present.

TABLE 15.1 Solubility of Calcium Sulfate ($CaSO_4$) in Aqueous Alcoholic Solutions

% Ethanol (vol/vol)	Dielectric Constant of Medium	Solubility of $CaSO_4$ (g/L)
0	80	2.084
3.9	78	1.314
10.0	73	0.970
13.6	71	0.436

The ethanol content has a greater effect on the solubility of potassium bitartrate than on that of neutral calcium tartrate, as can be seen in Table 15.2.

TABLE 15.2 Solubility (S) and Solubility Product (SP) of Tartrate Salts

	MM	Water			12% (vol/vol) Ethanol		
		S (g/L)	S (mol/L)	SP	S (g/L)	S (mol/L)	SP
KHT	188.1	4.91	26.1×10^{-3}	6.81×10^{-4}	2.76	14.7×10^{-3}	2.15×10^{-4}
CaT	188.1	0.206	1.095×10^{-3}	1.20×10^{-6}	0.103	5.48×10^{-4}	3.0×10^{-7}

KHT = Potassium bitartrate; MM = molar mass; CaT = neutral calcium tartrate

4.3. Temperature

Generally speaking, an increase in temperature increases the solubility of potassium and calcium salts, with a greater increase observed for the solubility of potassium bitartrate than for that of neutral calcium tartrate. Likewise, a decrease in temperature reduces the solubility of these solutes, and particularly so in the case of potassium bitartrate. This explains why cold stabilization (chillproofing) is used to protect wine against the precipitation of this salt.

5. TARTRATE STABILITY

5.1. Potassium Bitartrate Stability

The chemical principle that governs the stability of a solution has already been established. In the case of potassium bitartrate:

1. If $Q = [HT^-][K^+] > SP \rightarrow$ potassium bitartrate will precipitate.
2. If $Q < SP \rightarrow$ potassium bitartrate will not precipitate.
3. If $Q = SP \rightarrow$ the solution will be saturated and at equilibrium, and will not accept any more HT^- or K^+ ions. Any change in the equilibrium conditions will cause the equilibrium to shift to the right or left.

In winemaking, ionic activity rather than the ionic concentration should be determined. This involves calculating an activity product and a solubility based on the activity of the ions involved.

$$(K^+) = f_{K^+} [K^+]$$
$$(HT^-) = f_{HT^-} [HT^-]$$
$$AP = (K^+)(HT^-) = f_{K^+} [K^+] f_{HT^-} [HT^-] = S_{active}^2$$
$$S_{(active)}(g/L) = MW_{KHT} \times \sqrt{AP}$$

Active solubility refers to the concentration of potassium bitartrate, in grams per liter, calculated using the activity product for a given wine. The solubility of potassium bitartrate in aqueous alcoholic solutions has been established at different temperatures.

It is possible to determine whether or not a wine is saturated with potassium bitartrate (and therefore to determine if it is prone to the precipitation of this sparingly soluble salt) by comparing the active solubility of the wine with the solubility of a solution with the same alcohol content.

If we replace Q with the activity product in the expression that governs the stability of a wine relative to the precipitation of potassium bitartrate, the following conditions can be established:

1. If $AP = (HT^-)(K^+) > SP \rightarrow$ potassium bitartrate will precipitate.
2. If $AP < SP \rightarrow$ potassium bitartrate will not precipitate.
3. If $AP = SP \rightarrow$ the solution is saturated and at equilibrium. This equilibrium will shift in one direction or the other depending on what changes occur in the wine.

Sample problem: Determine whether a wine is stable at a temperature of 15°C.
Calculation of active solubility
Characteristics of wine:

pH = 3.3
11% (vol/vol) ethanol
$[K^+] = 0.90$ g/L Atomic mass = 39 g/mol
$[H_2T] = 3.02$ g/L Molar mass = 150 g/mol

Calculate the activity product and then the active solubility of potassium bitartrate (g/L) at 15°C.

Table 15.4 in the Appendix at the end of this chapter shows the degree of dissociation of tartaric acid over a range of pHs and ethanol contents, and can be used to calculate the concentration of the HT^- ion. Table 15.5 shows the activity coefficient of the HT^- and K^+ ions in aqueous alcoholic solutions at different temperatures.

Calculation of the activity of the HT^- and K^+ ions in the given wine conditions:

$$[H_2T] \times \alpha = [HT^-] = \frac{3.02 \text{ g/L}}{150 \text{ g/mol}} \times 0.6029 = 0.01214 \text{ mol/L}$$

$$(HT^-) = [HT^-] \times f_{HT^-} = 0.01214 \times 0.8675 = 0.01053$$

$$(K^+) = [K^+] \times f_{K^+} = \frac{0.9}{39} \times 0.8675 = 0.02$$

The molar mass of potassium bitartrate is 188.1 g/mol, hence:

$$AP = (HT^-)(K^+) = 0.01053 \times 0.02 = 2.11 \times 10^{-4}$$

The active solubility is $S_A = \sqrt{AP} = \sqrt{2.11 \times 10^{-4}} \times 188.1$ g KHT/mol $= 2.73$ g KHT/L

According to Table 15.6, the solubility of potassium bitartrate in a stable 11% (vol/vol) aqueous alcoholic solution at 15°C is $S = 2.35$ g/L.

Because the value obtained for the wine analyzed is higher than the established value ($S_A > S$), it can be concluded that the wine is unstable and that potassium bitartrate precipitation will occur.

5.2. Relative Saturation

Before drawing any conclusions regarding the stability of a wine towards potassium bitartrate precipitation, it should be recalled that wine is an aqueous alcoholic solution, with a given pH, and that in addition to potassium bitartrate it contains other substances that can affect the activity of the ions in solution. This means that even though the active solubility calculated for a given wine is higher than the predefined value for an ideal aqueous alcoholic solution, the wine will not necessarily be unstable in terms of bitartrate precipitation. It is therefore necessary to correct the criterion used to define an unstable wine.

As previously indicated, wine can retain higher quantities of potassium bitartrate in solution than the corresponding aqueous alcoholic solution. The excess that the wine will accept at a given temperature is related to the activity coefficient at this temperature; this coefficient takes into account the ionic strength of the medium. One means of calculating the percentage of excess potassium bitartrate that a wine will accept is to consider the inverse of the activity coefficient as follows:

$$\%_{\text{Excess KHT}} = 100 \times \left(\frac{1}{f} - 1\right)$$

Sample problem: Calculate the percentage of excess potassium bitartrate that a wine with an ethanol concentration of 12% (vol/vol) can contain and still be stable at 4°C.

The activity coefficient at 4°C and 12% (vol/vol) ethanol is $f = 0.920$.

The percentage of excess potassium bitartrate would thus be:

$$\%_{\text{excess}} = 100\left(\frac{1}{0.92} - 1\right) = 8.7$$

In other words, the wine could accept 8.7% more potassium bitartrate than the corresponding aqueous alcoholic solution:

For a wine with an alcoholic strength of 11% (vol/vol) at a temperature of 15°C:

The activity coefficient at 15°C and 11% (vol/vol) of ethanol is $f = 0.8675$.

$$\%_{\text{excess}} = 100\left(\frac{1}{0.8675} - 1\right) = 15.27$$

The percentage obtained using this method ranges between 7 and 20%, depending on the temperature and the concentration of ethanol. This method is more restrictive than others, but it ensures that if the wine contains a lower percentage of potassium bitartrate than that calculated, it will be stable and precipitation will not occur.

Another less restrictive method involves calculating relative saturation (R), which is defined as:

$$R = \frac{S_A - S}{S}$$

S_A = g/L of potassium bitartrate in the wine
S = g/L of potassium bitartrate in an aqueous alcoholic solution with an ethanol content equal to that of the wine.

The temperature chosen is normally that at which you want the wine to remain stable. Active solubility and solubility are therefore normally calculated at $-4°C$, which is generally the lowest temperature at which wine is stored.

Relative solubility can be:

Positive ($S_A > S$); this does not mean that the wine will undergo potassium bitartrate precipitation, however, as that depends on the specific value of R.
Zero ($S_A = S$); in this case, there is no risk of precipitation.
Negative ($S_A < S$); in this case, there is also no risk of precipitation.

Relative saturation is expressed as a percentage that is used to classify wines as stable or not (Table 15.3).

TABLE 15.3 Relative Saturation and Wine Stability

R	Crystal Formation	Stability of Wine
R < 0.2	No risk of crystal formation	Stable
0.2 < R < 0.6	Crystal formation in the long term	Stable
R > 0.6	Formation of crystals due to natural phenomena (decrease in temperatures in winter) or during stabilization	Unstable, treatment required
R > 2.5	Spontaneous and very rapid formation of crystals	These values are only seen in wine subjected to reverse osmosis

The relative saturation method is less restrictive than the previous method as it considers that a wine will remain stable as long as the excess percentage of potassium bitartrate does not exceed 20% relative to the corresponding aqueous alcoholic solution with the same ethanol content (R < 0.2). This is because it is considered that nucleation (the formation of crystal nuclei) requires high potential energy and will therefore only occur when there is a high level of relative saturation. Wines with a relative saturation of less than 0.2 can be considered stable, even though they have a medium saturation. When the level of saturation exceeds 0.2, the risk of precipitation increases with storage or aging due to the transformation of several substances that prevent potassium bitartrate precipitation. These substances are mainly proteins and coloring matter, which, with time, cease to exert their stabilizing or protective effect.

Using the data from the previous exercise to calculate R, we get:

$$S_A = 2.73 \text{ g/L};$$

$$S = 2.35 \text{ g/L}$$

$$R = \frac{2.73 - 2.35}{2.35} = 0.162$$

According to this method, thus, the wine will be stable. Nonetheless, the value obtained using the activity coefficient method established an excess percentage of:

$$\%_{\text{excess KHT}} = 100\left(\frac{1}{0.8675} - 1\right) = 15.27$$

This means that a wine with an ethanol content of 11% (vol/vol) will not be stable at 15°C as it will only accept an excess of 15.27% potassium bitartrate, and according to the relative solubility method, the wine in this example will have a relative saturation level of 16.2% (R = 0.162).

5.3. Calculating the Amount of Potassium Bitartrate to Eliminate from a Wine

The amount of potassium bitartrate that should be eliminated for a wine to be stable towards the precipitation of this salt is calculated using the active solubility value that corresponds to a relative saturation of less than 0.2, or less than the percentage calculated using the activity coefficient (depending on which method is chosen by the enologist).

Let us take a wine with the following characteristics, which should be stable at 4°C:

Ethanol = 10% (vol/vol)
pH = 3.2; $[H_2T] = 2.83$ g/L; $[K^+] = 0.92$ g/L
$f = 0.8875 \rightarrow 1/f = 1.1267$
$S_A = 2.65$
$S = 1.520$

$$R = \frac{2.65 - 1.52}{1.52} = 0.74; \text{ the wine is unstable.}$$

According to the criterion of relative saturation (R):

$0.2 = \dfrac{S_A - 1.52}{1.52} \rightarrow S_A = 1.824$ g/L $\rightarrow$ Hence, there will be an excess of: $2.65 - 1.824 = 0.826$ g/L.

According to the activity coefficient method, the wine can accept an excess of 12.7% potassium bitartrate.

$$S_A = S + S \times 0.127 = 1.52 + 1.52 \times 0.127 = 1.713 \text{ g/L. Excess}: 2.65 - 1.713 = 0.937 \text{ g/L}$$

5.4. Protection Against Calcium Tartrate Precipitation

The precipitation of neutral calcium tartrate is more likely in white wine than in red wine as the former contains approximately twice as much calcium as the latter. This is because the grapes used to make red wines undergo maceration, which favors the extraction of not only color compounds but also compounds such as pectins and racemic tartaric acid, which form insoluble compounds with calcium. These are precipitated as ethanol is produced during alcoholic fermentation.

Normal calcium levels in wine range between 50 and 60 mg/L. Higher levels may be found in wine stored in uncoated cement tanks.

The solubility product of neutral calcium tartrate decreases to a lesser extent than that of potassium bitartrate with a decrease in temperature, and several authors have indicated that cold stabilization treatment actually favors the precipitation of neutral calcium tartrate.

The likelihood of precipitation of neutral calcium tartrate increases at high pH values due to a higher proportion of tartrate ions. The pH values at which the bitartrate-tartrate equilibrium dominates over the bitartrate-tartaric acid equilibrium should be kept in mind.

6. PROTECTION AGAINST PRECIPITATION

6.1. Protection Against Potassium Bitartrate Precipitation

The traditional method for stabilizing wine and preventing the appearance of potassium bitartrate crystals when the temperature of bottled wine decreases used to consist of storing the wine in tanks exposed to winter temperatures. In such cases, the lower the temperature, the greater the spontaneous precipitation of potassium bitartrate. At the end of winter, the wines were filtered to remove the crystals responsible for haze.

Nowadays, wines are artificially chilled to a temperature of $-4°C$ to ensure the precipitation of potassium bitartrate and prevent hazing. Nonetheless, incomplete precipitation (and therefore insufficient stabilization) has been observed in wines stored at $-5°C$ for 8 to 10 days. The chilling of wine has a negative impact on quality as it triggers the precipitation of proteins, coloring matter, and other components that become insoluble at low temperatures.

Another factor to take into account is that potassium bitartrate crystals may not necessarily form in wine that becomes supersaturated at a given temperature, meaning that the wine will not be stable. This is due to the presence of compounds that prevent or slow down the precipitation of potassium bitartrate. This effect has been seen in model solutions in which precipitation was decreased by 12% and 18% in the presence of 10% glucose and 5% glycerol, respectively. Cold stabilization has also been seen to be ineffective in wines in which a large proportion of the coloring matter is colloidal. This suggests that potassium bitartrate somehow binds via hydrogen bonds that are probably formed between the oxygen atoms of carboxyl groups of tartaric acid and the hydrogen atoms of hydroxyl groups of phenols in the coloring matter.

The above phenomenon can be explained by the mechanism of bitartrate crystal formation. In order for dissolved bitartrate and potassium ions to form a crystal lattice and change to the solid state, the wine must contain crystal nuclei (on which the ions are deposited) and the crystal must grow in both size and weight. When the bitartrate and potassium ions are near this nucleus, they are simultaneously attracted by the forces of the lattice and cause the crystal to grow. This explains the appearance of so-called active sites during the formation of the crystal. These sites are areas with a high concentration of electrical charge (free valence). They are preferentially located at the edges and angles of the crystal and attract oppositely charged ions, thus ensuring the growth of the crystal. The two ions have to be bound simultaneously, as if only one is captured, there will be a chemical imbalance between the crystal and the solution, which will be resolved by the capture of the missing ion, or the release of the extra ion. The likelihood that two ions will coincide at an active site of the crystal is directly proportional to the concentration of the ions in the solution.

Also important are colloidal particles, which are naturally present in wine. These are large particles (polymers) that often carry a negative charge (polyphenols) or a positive charge (high-molecular-weight proteins and peptides). When attracted by the active sites of the crystal, they inhibit crystal growth and consequently the precipitation of bitartrate crystals. They are known as protective colloids.

The fact that proteins are naturally present in wine greatly helps to explain why the potassium bitartrate content of wine is generally higher than the theoretical content dictated by the solubility product. Proteins have been shown to exert a protective effect against precipitation in wines aged on lees. Specifically, these wines become stable towards tartaric precipitation after months of aging. In contrast, wines not aged using this method need to be treated. Mannoproteins may be responsible for this protective effect, as wine aged on lees is systematically enriched with these compounds.

It used to be thought that polyphenols also exerted a certain protective effect against precipitation, but it is now believed that they may actually favor precipitation as sufficiently large polyphenols can interact with proteins, causing them to precipitate and therefore reducing their protective effect.

Tartrate stability can be induced via the addition of certain stabilizing agents that exert a protective effect. Of note among these substances are:

- Metatartaric acid (a polyester that results from the intermolecular esterification of tartaric acid)
- Carboxymethyl cellulose (a structure composed of a variety of complex products that are poorly defined and therefore of variable efficacy)
- Industrial yeast mannoproteins (strongly glycosylated mannoproteins with a molecular weight of approximately 40 kDa).

6.2. Protection Against Calcium Tartrate Precipitation

Calcium tartrate is a very poorly soluble salt (it is twenty times less soluble than potassium bitartrate). Calcium in wine can come from a variety of sources.

- Addition of calcium bentonite to musts
- Addition of calcium carbonate as a deacidifer

- Use of calcium sulfate as an acidifying agent
- Chaptalization with sucrose contaminated with calcium
- Accidental contamination

The risk of precipitation becomes significant at calcium levels of 60 mg/L in red wine and 80 mg/L in white wine. Both potassium bitartrate and calcium tartrate have the same crystalline structure. Although, logically, the crystallization of the former should induce that of the latter, the opposite is in fact the case. Once the nucleus on which the calcium tartrate crystal grows is formed, the kinetics of crystallization are very fast (faster than in the case of potassium bitartrate), but nucleation is very slow and can take several years. This explains why calcium tartrate crystal deposits are found in wines that have been aged for long periods of time.

Cold stabilization is not very effective in the case of calcium tartrate as the solubility of this salt is not strongly dependent on temperature. The use of metatartaric acid, however, can prevent precipitation in wine stored at 4°C for several months. Racemic tartaric acid and the L-isomer of calcium tartrate have also been proposed as methods for eliminating calcium via the precipitation of the calcium racemate. The effectiveness of this treatment depends on the absence of protective colloids.

It is therefore recommendable to eliminate protective colloids (high-molecular-weight proteins and peptides) prior to cold stabilizing wine against tartrate precipitation.

7. APPENDIX

TABLE 15.4 Degree of Dissociation of Tartaric Acid by pH and Ethanol Content

pH	8	8.5	9	9.5	10	10.5	11	11.5	12	12.5	13	13.5	14
					% (vol/vol) Ethanol								
2.80	0.3762	0.3751	0.3740	0.3730	0.3718	0.3674	0.3644	0.3633	0.3622	0.3611	0.3600	0.3563	0.3526
2.85	0.4022	0.4010	0.4000	0.3990	0.3979	0.3934	0.3904	0.3893	0.3882	0.3871	0.3860	0.3823	0.3786
2.90	0.4277	0.4269	0.4260	0.4250	0.4240	0.4194	0.4164	0.4153	0.4142	0.4127	0.4112	0.4079	0.4046
2.95	0.4514	0.4502	0.4484	0.4491	0.4498	0.4454	0.4424	0.4413	0.4402	0.4391	0.4380	0.4343	0.4306
3.00	0.4763	0.4735	0.4707	0.4733	0.4758	0.4714	0.4684	0.4673	0.4662	0.4651	0.4640	0.4603	0.4566
3.05	0.5058	0.5027	0.5002	0.5004	0.5006	0.4965	0.4938	0.4929	0.4919	0.4910	0.4900	0.4863	0.4826
3.10	0.5338	0.5318	0.5297	0.5274	0.5251	0.5210	0.5183	0.5174	0.5164	0.5155	0.5145	0.5111	0.5077
3.15	0.5558	0.5536	0.5517	0.5538	0.5476	0.5441	0.5418	0.5411	0.5404	0.5397	0.5390	0.5356	0.5322
3.20	0.5771	0.5754	0.5737	0.5758	0.5696	0.5661	0.5638	0.5631	0.5624	0.5617	0.5610	0.5581	0.5552
3.25	0.5956	0.5937	0.5922	0.5905	0.5888	0.5862	0.5844	0.5841	0.5837	0.5834	0.5830	0.5801	0.5772
3.30	0.6132	0.6120	0.6107	0.6090	0.6073	0.6047	0.6029	0.6026	0.6022	0.6016	0.6010	0.5989	0.5968
3.35	0.6272	0.6252	0.6247	0.6235	0.6222	0.6209	0.6196	0.6196	0.6197	0.6194	0.6190	0.6169	0.6148
3.40	0.6401	0.6395	0.6387	0.6375	0.6362	0.6349	0.6336	0.6334	0.6332	0.6331	0.6330	0.6317	0.6304
3.45	0.6486	0.6477	0.6472	0.6465	0.6458	0.6456	0.6454	0.6456	0.6458	0.6664	0.6470	0.6457	0.6444
3.50	0.6560	0.6559	0.6557	0.6550	0.6543	0.6541	0.6539	0.6544	0.6548	0.6552	0.6555	0.6553	0.6551
3.55	0.6590	0.6587	0.6587	0.6585	0.6584	0.6588	0.6592	0.6607	0.6621	0.6630	0.6640	0.6638	0.6636
3.60	0.6608	0.6612	0.6617	0.6615	0.6614	0.6605	0.6597	0.6612	0.6626	0.6636	0.6645	0.6659	0.6673
3.65	0.6578	0.6578	0.6687	0.6592	0.6596	0.6597	0.6598	0.6613	0.6628	0.6639	0.6650	0.6664	0.6678
3.70	0.6537	0.6545	0.6557	0.6562	0.6566	0.6579	0.6593	0.6606	0.6618	0.6632	0.6645	0.6661	0.6677
3.75	0.6452	0.6456	0.6472	0.6482	0.6492	0.6487	0.6482	0.6538	0.6593	0.6617	0.6640	0.6656	0.6672
3.80	0.6356	0.6368	0.6387	0.6397	0.6407	0.6425	0.6443	0.6476	0.6508	0.6532	0.6555	0.6587	0.6619
3.85	0.6216	0.6228	0.6247	0.6263	0.6278	0.6316	0.6354	0.6389	0.6423	0.6447	0.6470	0.6502	0.6534

Adapted from Llaguno, 1982.

TABLE 15.5 Activity Coefficients for Calculating Activity of K^+ and HT^- Ions in Aqueous Alcoholic Solutions

°C	8	8.5	9	9.5	10	10.5	11	11.5	12	12.5	13	13.5	14
					% (vol/vol) Ethanol								
−4	0.8935	0.8993	0.9050	0.9063	0.9075	0.9110	0.9145	0.9183	0.9220	0.9255	0.9290	0.9320	0.9350
−3	0.8910	0.8962	0.9014	0.9044	0.9073	0.9097	0.9120	0.9158	0.9195	0.9230	0.9265	0.9295	0.9325
−2	0.8885	0.8931	0.8977	0.9007	0.9036	0.9066	0.9095	0.9133	0.9170	0.9205	0.9240	0.9270	0.9300
−1	0.8860	0.8901	0.8941	0.8971	0.9000	0.9035	0.9070	0.9108	0.9145	0.9180	0.9215	0.9245	0.9275
0	0.8835	0.8870	0.8905	0.8940	0.8975	0.9010	0.9045	0.9083	0.9120	0.9155	0.9190	0.9220	0.9250
1	0.8810	0.8845	0.8880	0.8915	0.8950	0.8986	0.9021	0.9058	0.9095	0.9130	0.9165	0.9195	0.9225
2	0.8785	0.8820	0.8855	0.8890	0.8925	0.8961	0.8997	0.9034	0.9070	0.9105	0.9140	0.9170	0.9200
3	0.8760	0.8795	0.8830	0.8865	0.8900	0.8937	0.8973	0.9009	0.9045	0.9080	0.9115	0.9145	0.9175
4	0.8735	0.8770	0.8805	0.8840	0.8875	0.8913	0.8950	0.8985	0.9020	0.9055	0.9090	0.9120	0.9150
5	0.8710	0.8745	0.8780	0.8815	0.8850	0.8888	0.8925	0.8960	0.8995	0.9030	0.9065	0.9095	0.9125
6	0.8685	0.8720	0.8755	0.8791	0.8826	0.8863	0.8900	0.8935	0.8970	0.9005	0.9039	0.9069	0.9099
7	0.8660	0.8695	0.8730	0.8766	0.8802	0.8839	0.8875	0.8910	0.8945	0.8979	0.9013	0.9043	0.9073
8	0.8635	0.8670	0.8705	0.8742	0.8778	0.8814	0.8850	0.8873	0.8920	0.8954	0.8987	0.9017	0.9047
9	0.8610	0.8645	0.8680	0.8717	0.8754	0.8790	0.8825	0.8860	0.8895	0.8928	0.8961	0.8991	0.9021
10	0.8585	0.8620	0.8655	0.8693	0.8730	0.8765	0.8800	0.8835	0.8870	0.8903	0.8935	0.8965	0.8995
11	0.8560	0.8595	0.8630	0.8668	0.8705	0.8740	0.8775	0.8810	0.8845	0.8878	0.8910	0.8940	0.8969
12	0.8535	0.8570	0.8605	0.8643	0.8680	0.8715	0.8750	0.8785	0.8820	0.8853	0.8985	0.8914	0.8943
13	0.8510	0.8545	0.8580	0.8603	0.8625	0.8675	0.8725	0.8760	0.8795	0.8840	0.8860	0.8889	0.8917
14	0.8485	0.8520	0.8555	0.8593	0.8630	0.8665	0.8700	0.8735	0.8770	0.8803	0.8835	0.8863	0.8891
15	0.8460	0.8495	0.8530	0.8568	0.8605	0.8640	0.8675	0.8710	0.8745	0.8778	0.8810	10.8839	0.8865
16	0.8435	0.8470	0.8505	0.8543	0.8580	0.8615	0.8650	0.8685	0.8720	0.8753	0.8785	0.8813	0.8840
17	0.8410	0.8445	0.8480	0.8518	0.8555	0.8590	0.8625	0.8660	0.8695	0.8728	0.8760	0.8788	0.8815
18	0.8385	0.8420	0.8455	0.8493	0.8530	0.8565	0.8600	0.8635	0.8670	0.8703	0.8735	0.8763	0.8790
19	0.8360	0.8395	0.8543	0.8468	0.8505	0.8540	0.8575	0.8610	0.8645	0.8678	0.8710	0.8738	0.8765
20	0.8335	0.8370	0.8405	0.8443	0.8480	0.8515	0.8550	0.8585	0.8620	0.8653	0.8685	0.8713	0.8740

Adapted from Llaguno, 1982.

TABLE 15.6 Solubility (g/L) of Potassium Bitartrate in Aqueous Alcoholic Solutions

°C	% (vol/vol) Ethanol												
	8	8.5	9	9.5	10	10.5	11	11.5	12	12.5	13	13.5	14
−4	1.190	1.155	1.120	1.085	1.050	1.015	0.980	0.940	0.910	0.880	0.860	0.86	0.810
−3	1.250	1.210	1.170	1.140	1.100	1.070	1.030	0.990	0.960	0.930	0.905	0.875	0.850
−2	1.310	1.270	1.230	1.190	1.155	1.120	1.075	1.040	1.010	0.980	0.950	0.920	0.895
−1	1.370	1.320	1.280	1.250	1.210	1.170	1.120	1.090	1.060	1.030	0.995	0.965	0.940
0	1.430	1.380	1.335	1.300	1.260	1.220	1.170	1.135	1.110	1.075	1.040	1.010	0.980
1	1.500	1.450	1.400	1.365	1.320	1.280	1.230	1.200	1.170	1.130	1.110	1.060	1.030
2	1.570	1.520	1.470	1.430	1.390	1.350	1.300	1.260	1.230	1.190	1.150	1.120	1.080
3	1.640	1.590	1.540	1.495	1.450	1.410	1.360	1.320	1.280	1.250	1.210	1.170	1.140
4	1.710	1.660	1.610	1.560	1.520	1.470	1.430	1.380	1.340	1.300	1.260	1.230	1.190
5	1.780	1.730	1.680	1.625	1.580	1.535	1.490	1.445	1.400	1.360	1.320	1.280	1.240
6	1.870	1.820	1.770	1.715	1.670	1.620	1.570	1.530	1.480	1.440	1.400	1.360	1.320
7	1.965	1.910	1.860	1.805	1.760	1.710	1.660	1.610	1.560	1.520	1.480	1.440	1.400
8	2.060	2.000	1.950	1.895	1.840	1.790	1.740	1.690	1.650	1.600	1.550	1.510	1.470
9	2.150	2.090	2.040	1.985	1.930	1.880	1.830	1.780	1.730	1.680	1.630	1.590	1.550
10	2.240	2.185	2.130	2.075	2.020	1.965	1.910	1.860	1.810	1.760	1.710	1.670	1.630
11	2.350	2.290	2.230	2.170	2.110	2.050	2.000	1.950	1.900	1.850	1.790	1.750	1.710
12	2.460	2.390	2.330	2.265	2.200	2.140	2.090	2.040	1.990	1.930	1.880	1.830	1.790
13	2.560	2.490	2.420	2.360	2.290	2.230	2.170	2.120	2.070	2.020	1.960	1.920	1.870
14	2.670	2.600	2.520	2.455	2.380	2.320	2.260	2.210	2.160	2.100	2.050	2.000	1.950
15	2.780	2.700	2.620	2.550	2.475	2.410	2.350	2.300	2.250	2.190	2.130	2.080	2.030
16	2.900	2.820	2.740	2.670	2.590	2.530	2.460	2.410	2.350	2.290	2.230	2.180	2.130
17	3.020	2.940	2.860	2.790	2.710	2.640	2.580	2.520	2.460	2.390	2.330	2.280	2.220
18	3.140	3.070	2.990	2.910	2.830	2.760	2.690	2.630	2.560	2.500	2.430	2.380	2.320
19	3.265	3.190	3.110	3.030	2.950	2.880	2.810	2.740	2.670	2.600	2.530	2.470	2.410
20	3.385	3.310	3.230	3.150	3.070	2.995	2.920	2.845	2.770	2.700	2.630	2.570	2.510

Adapted from Llaguno, 1982.

1. INTRODUCTION

The appearance of precipitated salts after a wine has been bottled and placed on the market can cause consumer rejection. It is therefore necessary to stabilize the salts prior to bottling. Traditional stabilization techniques tend to have a limited effectiveness, and improvements are therefore constantly being proposed and introduced. Cold treatments are most commonly used. Although other wine-stabilization techniques are available, their use is limited in countries belonging to the International Organization of Vine and Wine (OIV).

2. PHYSICAL TREATMENTS FOR TARTRATE STABILIZATION OF WINE

Stabilization of wine against the precipitation of tartaric acid salts can be achieved using a variety of treatments.

2.1. Cold Treatment

It has long been known that cold winter temperatures have a favorable effect on wines. Attempts have been made to control and enhance this effect by subjecting wines to temperatures close to their freezing point — below 0°C — for a given period of time. Under these conditions, precipitation of potassium and calcium salts of tartaric acid causes haze in the wine. Then, as the precipitate gradually settles at the bottom of the vessel, the wine clears again. Once this precipitate is separated from the wine by decanting or filtration, the wine should, in principle, have stable color and clarity.

Cold treatment is preferred for:

1. Inducing precipitation of tartaric acid salts
2. Inducing the precipitation of colloidal substances such as unstable pigments in red wines or proteins in white wines

Cold is principally used to stabilize wines against precipitation of the potassium and calcium salts of tartaric acid. Cheaper, highly effective treatments are available to induce the flocculation and sedimentation of non-ionic compounds. It should be noted, however, that cold has no effect at all on microbiological stability.

The following equation is used to determine the treatment temperature:

$$T_{treatment} = -\left[\left(\frac{\%EtOH(v/v)}{2}\right) - 1\right] \text{ (temperatures} < 0°C)$$

This equation is based on the relationship between the freezing point of an aqueous alcohol solution and its ethanol content:

$$T_{freezing} = -\left[\left(\frac{\%EtOH(v/v) - 1}{2}\right)\right] \text{ (temperatures} < 0°C)$$

The treatment involves maintaining the wine at a temperature close to its freezing point for a number of days (at least a week for white wines and a number of weeks for red wines). The temperature used in practice is usually slightly higher than that calculated using the equation above (−4°C).

Effect on Crystalline Precipitates

Although the solubility of tartaric acid salts is reduced in the presence of ethanol, precipitation is partially inhibited by substances that interact with the crystal nuclei and impede their growth. Consequently, crystalline precipitates can appear in wines some months after fermentation. This phenomenon is most apparent in red wines.

The solubility of potassium bitartrate is highly dependent upon temperature, and following cold treatment the wine remains stable against new precipitations of this salt, so long as it is not subjected to temperatures lower than those used in the treatment, and only if the colloidal structure of the wine does not change substantially. The solubility of the calcium salt of tartaric acid is less dependent upon temperature, explaining why this salt is not eliminated by cold treatment. Indeed, cold stabilization can even favor its subsequent precipitation.

2.2. Contact Method

The contact method involves pre-chilling the wine to a temperature of 0°C and then seeding it with 4 g/L of potassium bitartrate microcrystals (0.005–0.05 mm) before incubating it at 0°C for 1 to 2 hours in a crystallizer with stirring. The presence of the microcrystals induces rapid growth of crystals of tartaric acid salts and therefore leads to rapid precipitation. The process generally takes 4 hours (from the chilling of the wine to filtration: chilling → stirring → filtration → heat exchanger). The advantage over traditional cold treatment is clear: since the wine is chilled at a higher temperature and for a shorter period of time, less energy is required. The main drawback of this method is the quantity of microcrystals required (400 g/hL) and their cost. Nevertheless, some of the crystals can be recovered after treatment. It has been suggested that microcrystals can be reused up to four times without loss of efficacy in white wines. It is more difficult to reuse them in red wines, however, since their effectiveness is reduced when they become covered with coloring matter.

2.3. Pseudo-Contact Method

Methods are now being developed in which wines are seeded with smaller amounts of microcrystals (30–40 g/hL) and exposed to longer periods of cold treatment. The wines are shaken for 36 hours and then left to repose for 24 hours prior to filtration. The process can be completed in around 62 hours (compared with a week for traditional cold treatment). In addition, the temperature used is higher than that employed in conventional cold treatment (−2°C versus −4°C), although it remains lower than the temperature used in the contact method.

2.4. Treatment with Ion-Exchange Resins

In a different stabilization technique, the wines are passed over cation- or anion-exchange resins. In the exchange process, the resins become enriched with ions from the wine and vice versa. Cation-exchange resins donate protons or sodium ions according to whether the resin has been preconditioned with an acid or a sodium salt, respectively. Anion-exchange resins are pretreated with a basic solution such as sodium hydroxide and therefore donate hydroxyl ions to the wine and deacidify it. They are therefore rarely used in the food industry. The most commonly used resins for this type of treatment are sodium cation-exchange resins, which replace the calcium and potassium ions in the wine with sodium ions. The sodium salts of tartaric acid that are formed as a result of this cation-exchange process are more soluble in the wine.

The affinity of the ions in the wine for cation-exchange resins depends on the charge and size of the cation, such that a sodium-based resin can be used to achieve a 90% reduction in the concentration of calcium ions and a 40% reduction in the concentration of potassium ions in the wine. A downside of this treatment is that it can cause a four- or five-fold increase in the concentration of sodium ions in the wine. The affinity of the cations in wine for cation-exchange resins decreases in the following order: $Ca^{2+} > Mg^{2+} > K^+ > Na^+ > H^+$.

A negative effect of the treatment is a brutal denaturation of the wine, since it has an all-or-nothing effect, and its use for the tartrate stabilization of wine is limited by the OIV.

The two types of ion-exchange resins have the following mode of action:

- *Cation-exchange resins:* These improve the tartrate stability of the wine by eliminating calcium and potassium ions. They also lessen the risk of casse by reducing the concentration of Fe^{3+} ions. Cation-exchange resins that donate protons reduce the pH of wine, whereas those based on sodium ions increase the sodium concentration.
- *Anion-exchange resins:* The most widely used anion-exchange resins contain hydroxyl anions that reduce the acidity and increase the pH of the wine. They can also reduce the concentration of certain acids such as tartaric and acetic acid. They have a profound effect on the flavor and composition of the wine.

2.5. Treatment Using Membranes: Electrodialysis

The use of appropriate semipermeable membranes allows elimination of tartrate anions and calcium and potassium cations using electric fields (electrodialysis). Osmotic pressure can also be used with appropriate membranes to separate potassium ions and tartaric acid. This technique is known as reverse osmosis. Electrodialysis has been widely used in the food industry and therefore appears to be more viable for future applications.

The principle of the technique is based on circulation of the wine at low pressure and velocity in the opposite direction to an aqueous salt solution under the same pressure and flow conditions. The two liquids remain separated by anionic and cationic membranes that are permeable to tartrate and bitartrate anions and to calcium and potassium cations. Ion transport is ensured by a continuous electrical field applied by two electrodes situated at the ends of the compartments containing the selective membranes.

The reduction in the concentration of potassium and calcium ions in the wine is directly related to a reduction in conductivity. The same is true of the reduction in bitartrate and tartrate content, although the proportionality of the relationship between tartaric acid concentration and conductivity is lower than that of potassium ions. The reduction in the concentration of other ions is undetectable when the conductivity of the wine is reduced by less than 10%.

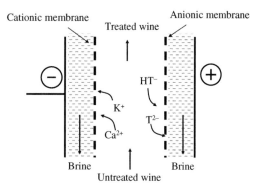

FIGURE 16.1 **Electrodialysis.**

The principal non-ionic or weakly charged constituents (polysaccharides, polyphenols, volatile compounds, amino acids, etc.) remain unaffected by electrodialysis, as does the colloidal structure of the wine. As with other treatments, it is advisable to use fining prior to stabilizing wines using this method.

The advantage of electrodialysis is that the resulting wines have a genuine tartrate stability compared with those treated using cold stabilization techniques, which do not give an absolute guarantee of stability in tests. Although the technique is not widely used, pilot-scale tests have shown that it provides good stability against tartaric acid precipitation and is less costly than traditional cold treatment.

2.6. Effectiveness of Physical Treatments for Tartrate Stabilization

Figure 16.2 shows the effects of physical treatments on total tartaric acid and potassium ion concentrations at a pH of 3.5, a temperature of $-1°C$, and an ethanol concentration of 10% (vol/vol).

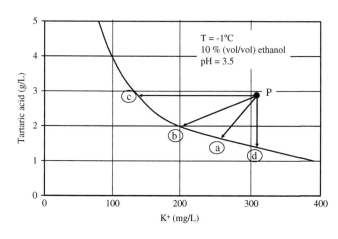

FIGURE 16.2 Solubility curve for potassium bitartrate under the conditions shown. Effect of four physical stabilization treatments on the total concentration of tartaric acid and potassium ions in the wine.

The letters a, b, c, and d represent four different treatments of a solution containing concentrations of tartaric acid and potassium ions corresponding to point P.

a. Cold treatment or treatment by osmosis: the stoichiometric quantities of HT^-, K^+, T^{2-}, and Ca^{2+} ions are eliminated
b. Electrodialysis: eliminates more K^+ ions than HT^- or T^{2-} ions
c. Cation exchange: eliminates K^+ ions and replaces them with other ions, without altering the concentration of HT^- or tartaric acid
d. Anion exchange: eliminates the HT^- anion but does not alter the K^+ concentration

According to Figure 16.2, treatment with ion-exchange resins (c and d) eliminates only one of the ions responsible for the precipitation of tartaric acid and, therefore, does not eliminate the risk of subsequent precipitation. On the other hand, cold treatments, osmosis, and electrodialysis (a and b) simultaneously eliminate the two ions involved in the precipitation equilibrium and, therefore, reduce the risk of precipitation in the short and long term.

The physical techniques mentioned are subtractive methods, since they reduce the concentration of certain ions and compounds in the wine. Techniques involving ion-exchange resins, on the other hand, are substitutive, since they replace ions. They have a powerful effect on the sensory characteristics of the wine. Electrodialysis is a controllable method that has practically no effect on sensory profile. It should nevertheless be remembered that fining and filtration are required prior to the use of these techniques. It is important to monitor the concentration of dissolved oxygen in the wine carefully when using cold treatments, since the solubility of gases increases at lower temperatures and the risk of subsequent oxidation is increased substantially.

3. CHEMICAL TREATMENTS FOR TARTRATE STABILIZATION

Chemical tartrate stabilization treatments involve the addition of compounds that form colloids in solution and inhibit the growth of salt crystals in the wine. These colloids are known generically as protective colloids, and they protect the wine against precipitation of tartaric acid salts. Their use has the following drawbacks:

1. Their efficacy is difficult to control over time.
2. All additives can, sooner or later, seriously alter the sensory properties of wine.

The main chemical agents used to protect against tartaric precipitation are metatartaric acid, carboxymethyl cellulose, and mannoproteins.

3.1. Metatartaric Acid

Metatartaric acid is a polylactide (polyester) with an undefined formula that is commercially available as a powder or colored tablets. It is obtained by heating tartaric acid or a mixture of tartaric and citric acid (4:6). The reaction through which it is formed is shown schematically in Figure 16.3.

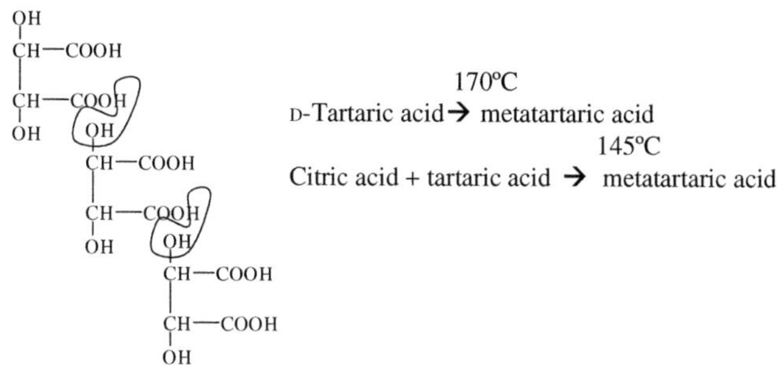

FIGURE 16.3 Formation of metatartaric acid.

$$\text{D-Tartaric acid} \xrightarrow{170°C} \text{metatartaric acid}$$

$$\text{Citric acid + tartaric acid} \xrightarrow{145°C} \text{metatartaric acid}$$

The inhibitory effect of metatartaric acid on crystallization depends on its degree of esterification, which is increased if heating is carried out under partial vacuum during synthesis. Commercially available metatartaric acid is not pure and commonly contains oxalacetic acid and, particularly, pyruvic acid as impurities.

Mode of action and dose: Metatartaric acid binds close to the active sites of the microcrystal nuclei, which form the starting point for crystal formation. In doing so, it impedes binding of bitartrate and potassium ions and inhibits growth of the crystal.

It is commonly used at a dose of 5 g/hL, although the legal limit is 10 g/hL. The effect depends on the quantity of the product added, its degree of esterification, and its hydrolysis in wine, which is dependent on temperature. It is normally effective for 6 to 9 months, but if treatment is carried out in wine bottled in autumn or spring, the wine will only remain stable for the summer (i.e., it will be unprotected the following winter).

TABLE 16.1 Inhibition of Potassium Bitartrate Precipitation by Metatartaric Acid with Varying Degrees of Esterification

		Metatartaric Acid Added (mg/L)					
		0	0.8	1.6	2.4	3.2	4.0
Tube	Esterification Index			K^+ (mg/L) After Addition			
1	40.8	17.2	15.8	17.2	17.2	17.2	17.2
2	37.3	17.2	15.3	17.2	17.2	17.2	17.2
3	31.5	17.2	11.0	15.3	15.9	16.5	17.2
4	22.9	17.2	7.6	11.2	13.6	15.6	16.8

It is advisable to carry out tests to determine the efficacy of a metatartaric acid preparation. Table 16.1 shows the results of tests performed using different doses of metatartaric acid on a saturated solution of potassium bitartrate in 10% ethanol (vol/vol) after exposure to a temperature of 0°C for 12 hours.

Solutions of metatartaric acid are very sensitive to temperature, and hydrolysis can cause an increase in acidity. It is important for winemakers to be aware of this instability and its dependence upon time and temperature. As shown in Figure 16.4, the wine remains

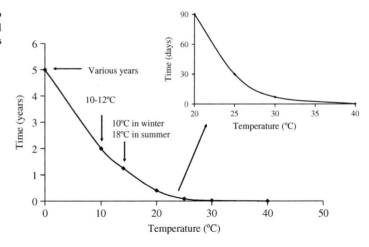

FIGURE 16.4 Relationship between storage temperature and duration of stability in wines treated with metatartaric acid.

stable for approximately 5 years at a temperature of 0°C, for 2 years or more at a temperature of between 10 and 12°C, and for 12 to 18 months at a temperature of 10°C in winter and 18°C in summer. At higher temperatures, there is a marked reduction in stability, with the wine becoming unstable after 3 months at 20°C, 1 month at 25°C, and just a few hours at 40°C.

3.2. Sodium Carboxymethyl Cellulose

Sodium carboxymethyl cellulose is an esterified cellulose containing carboxymethyl groups on carbons C6 and C2. The carboxyl groups carry negative charges at the pH found in wine, and as a consequence this compound can sequester potassium and calcium ions in the wine and thereby reduce their active concentration. The charges on the molecule also confer an affinity for the active sites of the forming crystals of potassium bitartrate and as a result impede their growth.

Carboxymethyl cellulose is not altered over time and, in theory, its effect on the precipitation of potassium bitartrate is unlimited. Since it is non-toxic, it can be consumed without adverse health effects at levels of up to 20 g per day. This compound is therefore of particular interest in the food industry and is widely used as an agglutinating agent. The dose in wine should not exceed 100 mg/L.

3.3. Mannoproteins

Wines naturally contain proteins that have a low charge density and act as protective colloids against precipitation, impeding it but not inhibiting it completely. For some time, efforts have been made to eliminate these colloids through fining and aggressive filtration, since they reduce the efficacy of cold treatments.

However, wines aged for several months on yeast lees are particularly stable against tartaric precipitation and in some cases cold temperature treatments are unnecessary.

TABLE 16.2 Reduction in the Concentration of Potassium Ions in White Wines Following Cold Treatment at $-4°C$ for 6 Days After Prior Addition of Different Doses of Mannoproteins

Wine	Mannoproteins Added (g/hL)				
	0	15	20	25	30
1	52	72	17	0	0
2	104	53	33	0	0
3	62	21	0	0	21
4	155	52	0	0	62
5	51	0	0	0	0
6	52	0	0	0	11

Given that the main macromolecules introduced into the wine during this type of aging are mannoproteins, it seems reasonable to assume that these compounds are responsible for the stability against tartaric precipitation displayed by wines aged on lees.

Mannoproteins are industrially obtained from yeast cell walls by enzymatic treatment (Glucanex™). The compound used to stabilize precipitation should not be confused, however, with that obtained using the same enzymatic treatment for protein stabilization. The first contains a high proportion of highly glycosylated mannoproteins with a molecular weight of around 40 kDa, whereas that used for protein stabilization is a 31.8 kDa mannoprotein called MP32. Both commercial preparations are purified from the same mannoprotein preparation.

The dose necessary is between 15 and 25 g/hL. However, tests on certain wines have revealed that higher doses (30 g/hL) reduce the stabilizing effect (wines 3, 4, and 6 in Table 16.2). Tests must therefore be carried out to determine optimal doses.

The use of metatartaric acid or mannoproteins is a technical choice based on the storage conditions and length of time between treatment and sale. For a wine that will be stored at a low temperature (either due to the climate in the region or the season), the use of metatartaric acid may be recommended, whereas mannoproteins may be more appropriate for a wine to be stored at a higher temperature.

4. ASSESSMENT OF TREATMENT EFFICACY

A number of tests are used to assess the efficacy of different treatments.

4.1. Cold-Storage Test

Four days at $0°C$ may be sufficient for dry wines and 6 days at $-8°C$ is required for natural sweet wines. If crystals do not appear by this time, the wine can be considered stable.

It is a simple, inexpensive test but has the drawback of not providing information on the degree of instability and is not very reliable. It is also a long test and does not indicate the efficacy of an ongoing treatment.

4.2. Tests Based on the Contact Method

The following tests are based on the seeding of crystal nuclei to induce precipitation of bitartrate crystals:

Müller-Späth

The wine sample is maintained for 2 hours at 0°C following addition of 4 g/L of potassium bitartrate microcrystals. The separation and subsequent quantification of the mass of bitartrate deposited provides an indication of the instability of the wine. The quantity of bitartrate precipitated can also be determined by comparing the acidity produced by the addition of microcrystals with that obtained after separation of the precipitate.

The drawback of this test is that it overestimates the efficacy of treatment, since it provides a measure of the stability of wine at a temperature of 0°C but after 2 hours only 70% to 90% of the tartrate in the wine (not that added) is precipitated. Although the test is relatively simple, it is not very reliable.

Martin Viallate

To increase the reliability of the previous method, a proposal was made to add 10 g/L of tartrate microcrystals and measure the reduction in conductivity at 0°C for 10 minutes. If during this time the reduction in conductivity does not exceed 5% of the initial value (measured prior to addition of tartrate), the wine can be considered stable. If, on the other hand, the reduction is greater than 5%, the wine is considered to be unstable.

Although the test is more rapid, it still fails to take into account factors such as the size of the potassium bitartrate granules added, and this has a significant influence on the formation of the precipitate and therefore the conductivity. The length of contact is also too short; consequently, although the test is relatively simple its reliability is limited if we take into account factors such as the granule size of the potassium bitartrate added.

Escudier

The Escudier test involves measuring the conductivity of a sample of wine before and after a process which consists of filtering the wine with 0.65 μm filters, adding 4 g/L potassium bitartrate with a standard granule size, and incubating the sample at −4°C with shaking for 4 hours.

If C_i = conductivity at t = 0 and C_f = conductivity at t = 4 hours, by extrapolation, the conductivity at infinity (generally 24 hours) is calculated using a previously established mathematical model.

From the values of the conductivity at infinity, the percentage drop in conductivity corresponding to a critical indicator of stability is calculated:

$$CIS\% = \frac{C_i - C_{inf}}{C_i} \times 100$$

In the case of dry wines, if the reduction in conductivity is less than 3%, the wine is considered stable. If the reduction is greater than 3%, stabilization treatment is required.

It is a simple and reliable test, although it is relatively long. It can also be used to determine the percentage of deionization in a manner that is individualized and adapted to the wine in question.

4.3. Tests Based on the Saturation Temperature

The idea for a test based on saturation temperature stems from the observation that, at low temperature, the more potassium bitartrate that is dissolved, the less saturated the wine will be and the greater stability it will exhibit against tartaric acid precipitation. The saturation temperature (T_{sat}) is defined as the lowest temperature at which a wine is able to dissolve potassium bitartrate. Experimental determination of this temperature is relatively straightforward. However, it takes a long time, and therefore does not provide rapid information on the efficacy of a treatment. Nevertheless, the following equations have been established based on statistics compiled from the analysis of hundreds of wines:

$$T_{sat} = 20 - \frac{\Delta C_{20°C}}{29.30}; \text{ (white wines)}$$

$$T_{sat} = 29.91 - \frac{\Delta C_{30°C}}{58.30}; \text{ (red and rosé wines)}$$

ΔC = the variation in conductivity at 20 or 30°C after addition of 4 g/L of potassium bitartrate to the wine.

Once the saturation temperature is known, it is of interest to determine the temperature at which crystalline precipitation does not occur, in other words the true stability temperature (T_{stab}). Calculation of this temperature is similarly laborious, but there are a series of functions relating saturation temperature and stability temperature that facilitate decision-making:

a. For a white wine with an ethanol concentration of 11% (vol/vol): $T_{stab} = T_{sat} - 15°C$
b. For a white wine with an ethanol concentration of 12.5% (vol/vol): $T_{stab} = T_{sat} - 12°C$
c. For a wine with an ethanol concentration of 12.5%, the saturation temperature should not exceed 8°C if we want the wine to be stable at −4°C.

To calculate the stability of red and rosé wines, the saturation temperature is related to stability in such a way that the wine will be stable if:

$$T_{sat} < 10.81 + 0.297 \times ITP \text{ (ITP = index of total polyphenols)}$$

An equation of this type can also be used for white wines, such that the wine will be stable if:

$$T_{sat} < 12.5°C$$

The test is relatively simple, rapid, and moderately reliable, since it is constructed from data obtained from hundreds of wines. Nevertheless, not all wines will necessarily fit the mathematical equations shown.

4.4. Tests Based on Analytical Results

Analyses are focused on determining the pH and the concentrations of tartaric acid, potassium ions, and ethanol in order to assess the stability of a wine according to chemical criteria established for the solubility and precipitation equilibria. With the concentrations obtained through analysis, the relative supersaturation (R) can be calculated:

$$R = \frac{S_A - S}{S}$$

S_A = g/L of potassium bitartrate in the wine
S = g/L of potassium bitartrate in an aqueous alcoholic solution with an ethanol concentration equal to that found in the wine

According to the relative supersaturation obtained, it is possible to determine whether or not the wine is stable against precipitation of potassium bitartrate. In general terms, if R is lower than 0.2, the wine is stable, although stricter criteria exist, as discussed in Chapter 15.

The method is simple and rapid; the fact that solubility in aqueous alcoholic solutions has been tabulated makes it much easier to calculate the value of R, and it is therefore only necessary to determine the tartaric acid, potassium, and ethanol concentrations and the pH in the treated wine.

5. SPOILAGE PROCESSES THAT AFFECT ACIDS

The acid content of wine can be affected by a range of changes that can inadvertently occur during storage and aging. These changes are caused by the development of microorganisms in wine that has not been appropriately treated and stored, and they therefore correspond to true spoilage of the wine. The following are the most important:

5.1. Pricked Wine

Small quantities of acetic acid are produced by yeasts during alcoholic fermentation and by bacteria during malolactic fermentation. The growth of aerobic bacteria belonging to the genus *Acetobacter*, however, leads to pricked wine. These bacteria convert ethanol into acetic acid and water, thereby causing an increase in volatile acidity. Esterification of acetic acid with ethanol generates ethyl acetate, which is responsible for the organoleptic properties associated with this type of spoilage (overpowering, unpleasant aroma and a hard finish in the mouth).

The legal limits for acetic acid content are currently 18 meq/L for white and rosé wines (1.04 g/L acetic acid) and 20 meq/L for red wines (1.20 g/L acetic acid). There are some exceptions, for example, for wines aged for a long period in the barrel, dessert wines, and wines produced from grapes affected by noble rot.

Concentrations of acetic acid below 720 mg/L are not easily detectable. Above these values, sharp aromatic notes and hardness in the mouth become apparent. Nevertheless, the influence of acetic acid on the organoleptic properties of the wine depends upon the particular type of wine, and generally, white wines are less tolerant of high levels of volatile acidity than are red or dessert wines. Consequently, a slightly elevated volatile acidity can be analytically correct or even good, but low values are always advisable.

The formation of acetic acid from ethanol occurs through a reaction involving acetaldehyde as an intermediate and involves two steps:

1. Generation of acetaldehyde in a reaction catalyzed by alcohol dehydrogenase (step 1)
2. Generation of acetic acid in a reaction catalyzed by aldehyde dehydrogenase (step 2)

$$CH_3-CH_2-OH \xrightarrow{\;①\;} CH_3-CH=O \xrightarrow{\;②\;} CH_3-C\overset{\displaystyle O}{\underset{\displaystyle OH}{\diagup\diagdown}}$$

Ethanol   Acetaldehyde   Acetic acid

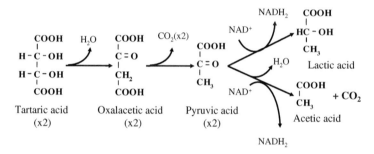

FIGURE 16.5 Degradation of tartaric acid through mechanism I.

5.2. Tourne

Tourne is a type of spoilage caused by fermentation of tartaric acid. It causes a reduction in fixed acidity and generally an increase in volatile acidity. This fermentation follows two pathways. The first produces lactic and acetic acid and releases CO_2, via mechanism I (Figure 16.5). *Lactobacillus plantarum* and *Streptobacterium* sp. mainly use this pathway. In the second mechanism (Figure 16.6), succinic acid, CO_2, and acetic acid are produced from oxalacetic acid. This mechanism (mechanism II) is used by *Lactobacillus brevis* and *Betabacterium* sp.

Tourne is characterized by silky clouds that move more slowly in the wine when it is swirled around the glass. White wines also have clear browning.

At the beginning of the process, affected wines have a flat taste and a substantial increase in volatile acidity. As spoilage advances, they develop a characteristic, unpleasant, rancid flavor of rotting that is attributed to the formation of acids such as formic, propanoic, butanoic, and valerianic acids. This flavor does not disappear even following treatment with activated charcoal.

Grapes infected with gray mold carry microorganisms responsible for tourne, although these are also present in dirty corners of the winery.

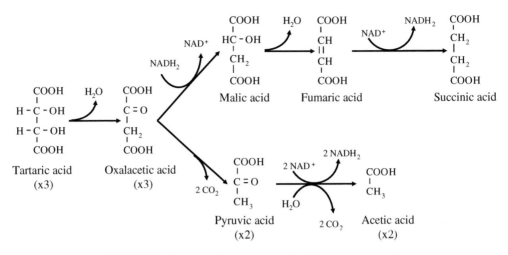

FIGURE 16.6 Degradation of tartaric acid via mechanism II.

5.3. Bitterness due to Fermentation of Glycerol

Spoilage due to fermentation of glycerol can be caused by various species of lactic acid bacteria, although other groups such as *Bacillus polymyxa* and *Bacillus macerans* can also be involved. This type of spoilage almost exclusively affects aged red wines when they are bottled under anaerobic conditions.

Lactic acid, acetic acid, and acrolein are all produced during fermentation of glycerol, but it is acrolein that is responsible for the characteristic bitter flavor — caused by binding to

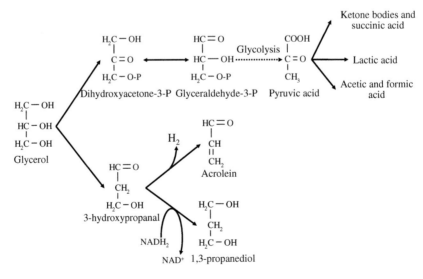

FIGURE 16.7 Degradation of glycerol.

polyphenols — associated with this type of spoilage. Affected wines have increased acidity and acquire a putrid aroma of butyric acid and a sharp, very bitter flavor.

5.4. Degradation of Citric and Sorbic Acid

Various lactic acid bacteria can cause decomposition of citric acid (Figure 16.8) to generate a range of products, principally lactic acid, acetic acid, and other products such as acetoin and 2,3-butanediol.

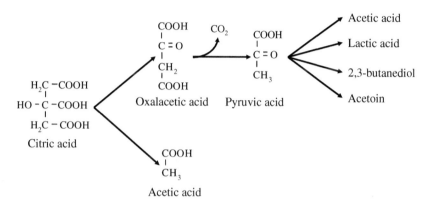

FIGURE 16.8   Degradation of citric acid.

The effect is an increase in volatile acidity and in the concentration of ketone compounds, such as acetoin, and their derivatives, such as butanediol.

Sorbic or 2,4-hexanedioc acid is a compound with antifungal but not bactericidal properties. It is commonly used in the food industry at doses of up to 200 mg/L. Degradation of sorbic acid does not cause substantial changes in the acid composition of wine but it does lead to the appearance of compounds with an aroma of geraniums.

Lactic acid bacteria such as *Leuconostoc oenos* and heterolactic acid bacteria of the genus *Lactobacillus (brevis, hilgardii)* are responsible for this transformation (Figure 16.9). One of the compounds responsible for the aroma of geraniums is 2-ethoxy-3,5-hexadiene, which is a powerful odorant.

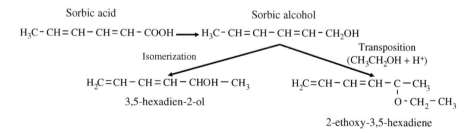

FIGURE 16.9   Degradation of sorbic acid.

TABLE 16.3 Diagnosis of the Precipitates That Can Be Found in Vessels Containing Wine

Appearance Under the Microscope	Color of the Precipitate	Solubility in	Behavior	Additional Test	Observed Property	Diagnosis	Cause
Crystalline	Whitish	Boiling water	Soluble	Examination of the crystals	Laminar	Potassium bitartrate	Cold
			Insoluble		Granular	Calcium tartrate	Deacidification
Amorphous	Whitish	NaOH 0.1N	Soluble	Microscopic examination	Amorphous	Protein casse	Excessive fertilization of the vine Overfining
			Insoluble		Microorganisms	Yeasts and/or bacteria	Spoilage
			Visible flocculation			Bentonite	Inadequate fining
		HCl 0.1N	Soluble	Treatment with HCl and potassium ferrocyanide	Blue coloration and formation of a precipitate	White phosphate-ferric casse	Excess iron in white wines
			Insoluble	Microscopic examination	Microorganisms	Yeasts and/or bacteria	Spoilage
	Reddish	H$_2$O$_2$ 3%	Soluble	Treatment with HCl and ammonia	Blue coloration	Copper casse	Excess of copper
	Blueish-black	HCl 0.1N	Soluble	Treatment with HCl and potassium ferrocyanide	Blue coloration and formation of a precipitate	Phosphate-ferric casse	Excessive iron in tannic wines
			Insoluble	Treatment with sodium hydroxide	Dissolves and generates a yellow color	Iron ferrocyanide	Inadequate treatment to remove metals

All of these spoilage processes are caused by bacteria and can be linked to poor hygiene in the winemaking facilities giving rise to excessive development of microbial populations. In addition, the development of these microorganisms is favored by low doses of sulfite and high pH. Significant attention must be paid to winery hygiene and equipment must be cleaned regularly during the harvest. Modern stainless steel equipment is easier to clean than tanks made of concrete or wood. Aging barrels should also be adequately maintained, particularly when they are empty.

17

Redox Phenomena in Must and Wine

1. INTRODUCTION

Atmospheric oxygen dissolves in liquids such as water, must, and wine at a concentration that depends on temperature and ethanol content. In water at atmospheric pressure and a temperature of $20°C$, it dissolves at a concentration of approximately $8\,mg/L$. The presence of oxidase enzymes and elemental oxygen in recently obtained musts causes the oxidation of several species during enzymatic activation. The species generated in this process oxidize other, less potent, molecules, triggering a series of reactions that can persist throughout the winemaking process and the subsequent storage of wine. Furthermore, elemental oxygen can react via electron exchange, at the pH of must and wine, giving rise to powerful oxidizing and reducing agents, which, in turn, contribute to oxidation-reduction (redox) reactions during the production and storage of wine. Indeed, the management of oxygen levels in wine has emerged in recent years as a key process in controlling and improving the aging of wines.

2. REDOX REACTIONS: BASIC CONCEPTS

Redox reactions involve the transfer of electrons between two species. One of these species is reduced (i.e., it gains electrons) while the other is oxidized (i.e., it loses electrons).

The reducing species donates (loses) electrons and therefore becomes oxidized, while the oxidizing species accepts (gains) electrons and is therefore reduced.

Examples:

$Ag^+ + e^- \rightarrow Ag$ (silver gains electrons and is reduced)
$Cu \rightarrow Cu^{2+} + 2e^-$ (copper donates electrons and is oxidized)

2.1. Oxidation Number

Oxidation numbers reflect the charge of species that participate in redox reactions. The following rules apply when assigning oxidation numbers:

1. All elements have an oxidation number of 0.
   Example: Pb^0; O_2^0
2. The oxidation number of oxygen is -2 in all cases except for peroxides, in which it is -1.
   Example: $H_2SO_4^{2-}$; $H_2O_2^{1-}$
3. The oxidation number of hydrogen is $+1$ except for hydrides, in which it is -1.
   Example: $H_2^{1+}SO_4^{2-}$; $H^{1-}K^{1+}$
4. Alkaline metals have an oxidation number of $+1$ in compounds.
   Example: $K^{1+}Mn^{7+}O_4^{2-}$
5. Alkaline earth metals have an oxidation number of $+2$ in compounds.
   Example: $Ca^{2+}O^{2-}$
6. Halogens have an oxidation number of -1 in halides.
   Example: $Na^{1+}Cl^{1-}$
7. The sum of oxidation numbers of all the elements of a chemical species is equal to the net charge of the species.
   Example: $(S^{6+}O_4^{2-})^{2-}$

Oxidation numbers are increased by oxidation and decreased by reduction.

For a reaction to be considered a redox reaction, the oxidation numbers of the participating species must change.

3. REDOX POTENTIAL

Redox potential is a measure of the oxidation or reduction state of a medium; it is measured in volts (V). Just as pH represents the acid-base state of a solution at a given moment, redox potential indicates the tendency of a species to become reduced or oxidized.

A scale of standard redox potentials ($E°$) has been established by comparing the hydrogen couple (H^+/H_2) with other redox couples in their standard states (analyte concentration of 1 M, pressure of 1 atm, and 25°C).

When two redox couples interact, the couple with the highest standard redox potential is reduced while that with the lowest potential is oxidized. Total potential is calculated by subtracting the potential of the oxidized species from that of the species that is reduced the least. In a battery, reduction occurs at the cathode, while oxidation occurs at the anode.

Example: Let us calculate $E°$ for the following reaction:

$$2\,Ag^+ + Cu \;\rightarrow\; 2\,Ag + Cu^{2+}$$

1. Silver is reduced (cathode). The $E°$ of the half-reaction $Ag^+/Ag = 0.8$ V
2. Copper is oxidized (anode). The $E°$ of the half-reaction $Cu/Cu^{2+} = 0.337$ V

$E° = E°_{cathode} - E°_{anode} = 0.8 - 0.337 = 0.463$ V

3.1. Nernst Equation

The Nernst equation is used to calculate redox potential in redox reactions that do not occur in standard conditions.

$$E = E° - \frac{0.059}{n}\log\frac{(Products)}{(Reagents)}$$

n = number of electrons exchanged.

$$E° = E°_{cathode} - E°_{anode}$$

The Nernst equation can also be used to calculate the concentrations of ionic species based on the redox potential obtained. One of the most widespread applications of the Nernst equation is the calculation of the pH of an aqueous solution.

4. MEASURING REDOX POTENTIAL IN WINE

The redox potential of a liquid solution depends on oxygen content and pH. The expected reaction is:

$$O_2 + 4\,H^+ + 4\,e^- \;\rightarrow\; 2\,H_2O \quad E° = 1.23\text{ V}$$

Because the conditions are not standard, the Nernst equation is applied, yielding the following potential:

$$E = 1.23 - \frac{0.059}{4}\log\frac{1}{(H^+)^4\,(O_2)} = 1.23 - \frac{0.059}{4}\log\frac{1}{(H^+)^4} - \frac{0.059}{4}\log\frac{1}{(O_2)}$$

$$E = 1.23 + 0.059\log(H^+) + \frac{0.059}{4}\log(O_2)$$

Expressing voltages in mV, and taking into account the definition of pH, the above expression becomes:

$$E = 1230 - 59\,pH + 14.75\log(O_2)$$

It should also be taken into account that wine is an alcoholic solution, meaning that the standard redox potential may change if the conditions of the medium change (presence of ethanol). These changes can also alter the coefficients applied to pH and oxygen content. Therefore, generally speaking, for a wine, the above equation can be expressed as:

$$E = E_{hydroalcoholic \, sol} - \alpha \, pH + \beta \log (O_2)$$

Where α and β will vary for each wine, albeit with certain limits:

- 50 to 150 for α
- 14.8 and 15.9 for β (within the pH of wine)

While the value of $E_{hydroalcoholic \, sol}$ will vary throughout fermentation, it is constant for a given wine.

Using a combined electrode, the redox potential of a wine can be calculated from the following equation:

$$E_{wine} = 1264 - 59.9 \, pH + 150.9 \log (O_2)$$

From a technological perspective, it is interesting to know the redox potential of a wine that is ready to be aged or bottled, as this provides a very useful indication of changes that are likely to occur over time and thus helps with planning appropriate interventions.

Of the two variables that have the greatest influence on the redox potential of wine, pH and oxygen, pH has only a minimal effect as it changes only slightly (by between 0.1 and 0.3 units) during the winemaking process. Oxygen, in contrast, has a major impact as its levels can vary from zero to saturation (8 mg/L at 20–25°C). Because oxygen is an active molecule in redox processes in wine, it modifies the relationship between the oxidized and reduced forms of compounds present in the wine. If the introduction of oxygen is not constant, any increase in oxygen content will change the redox potential.

Normally, when there is an increase in the initial oxygen content, the redox potential will increase, and the corresponding wine will be more prone to oxidation. This potential decreases, however, with an increase in pH. Redox potential also decreases with an increase in temperature, probably because of the reduction in oxygen solubility associated with increasing temperature. By way of example, if we were to assign a value of 100 to the contribution of oxygen, pH would have a value of 10 and temperature a value of 1.

4.1. Changes in Redox Potential During the Making of Wine

Redox reactions occur throughout the winemaking process, as is demonstrated by the changes detected in its redox potential. The prefermentation treatment of both red and white wines gives rise to extensive oxidation, which is even more pronounced with the enzymatic activity of polyphenol oxidases from healthy grapes (tyrosinase) and from grapes infected with *Botrytis cinerea* (laccase). The alcoholic fermentation of red and white grape must creates a reducing environment, in which redox potential values are at their lowest, mostly due to the absence of oxygen. Redox potential also decreases during malolactic fermentation.

After fermentation, the redox potential stabilizes at values of between 200 and 300 mV, depending on storage conditions and treatments. Generally, operations performed

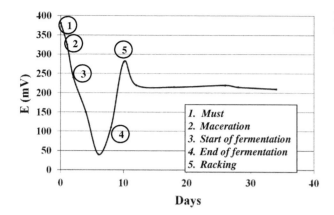

FIGURE 17.1 Changes in redox potential (E) during the production of white wine.

1. Must
2. Maceration
3. Start of fermentation
4. End of fermentation
5. Racking

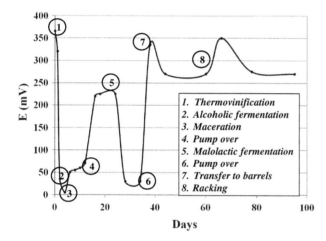

FIGURE 17.2 Changes in redox potential (E) during the production of red wine.

1. Thermovinification
2. Alcoholic fermentation
3. Maceration
4. Pump over
5. Malolactic fermentation
6. Pump over
7. Transfer to barrels
8. Racking

during storage cause an increase in redox potential. When redox reactions cause redox potentials of less than 150 mV, the wine should be monitored regularly for the appearance of sulfur flavors associated with reducing environments. Gentle racking is normally sufficient to prevent this fault. When values drop below 50 mV, more vigorous aeration is required.

There is also a relationship between microbial growth and redox potential. Aerobic microorganisms, for example, need potentials of over 180 mV to grow at a pH of 3, while facultative anaerobes can grow at potentials of below 120 mV at the same pH.

Light and temperature both reduce the redox potential of wine, meaning that problems such as copper casse arise more rapidly in wine stored in colorless bottles or exposed to light and moderate temperatures. White and blue casse, which are caused by the presence of iron(III), need high redox potentials, such as those caused by aeration, to develop.

TABLE 17.1   Relationship Between Oxygen Content and the Redox Potential of Red Wine

Oxygen (mg/L)	Potential (mV)	Increase in Potential
0.1	263.1	
0.8	280.2	17.0
2.5	339.7	59.5
4.8	424.3	84.6
5.0	433.6	9.3

Adapted from Vivas, 1999.

TABLE 17.2   Effect of Adding 8 mg/L of Fe^{2+} and 2 mg/L of $Cu°$ on Redox Potential and Oxygen Consumption in Wine

Type of Wine	Instantaneous Rate of Oxygen Consumption (mg O_2 /L/min)	$E_{(max)}$ (mV)	dE/dt	E°
Control white wine	0.20	574	−0.27	223
Control white wine + Fe + Cu	0.38	530	0.41	330
Control red wine	0.45	528	−0.70	273
Control red wine + Fe + Cu	0.90	461	−1.42	252

$E_{(max)}$ = maximum redox potential after O_2 saturation; dE /dt = slope of the curve for redox potential versus time; E° = standard redox potential of wine.

TABLE 17.3   Factors That Influence Redox Potential and Oxygen Consumption in Wine

Factor	Effects Observed
1. Ethanol	Increase in instantaneous rate of oxygen consumption and reduction in redox potential.
2. H_2T, H_2M, and HL	Few variations in instantaneous rate of oxygen consumption and redox potential.
3. pH	Increase in pH and decrease in redox potential, with changes within the pH range of wine leading to reductions of 3 to 4 mV.
4. Glycerol	No changes in either instantaneous rate of oxygen consumption or redox potential.
5. Phenols	Inhibition of changes in redox potential. Anthocyanins consume oxygen rapidly and reduce redox potential. Oligomeric catechins and procyanidins are more active than their polymers. Flavonols and condensed tannins consume more oxygen than condensed tannins alone.
6. Temperature	Changes of 100 mV seen with an increase in temperature from 0 to 30°C.
7. Containers	Effect depends on permeability of container to oxygen. New oak is more permeable than both used oak and steel.

5. OXYGEN AND WINEMAKING

As has already been seen, the redox potential of wine depends primarily on the concentration of dissolved oxygen.

The solubility of a gas in a liquid depends on the partial pressure of the gas on the surface of the liquid and on temperature. While the solubility of a gas is directly related to its pressure, in the case of temperature, solubility decreases with an increase in temperature.

Henry's Law relates the partial pressure of a dissolved gas to its concentration:

$$P_{gas} = K_H \times C_{gas}$$

C_{gas} = molar concentration of gas

P_{gas} = pressure of gas on liquid

Table 17.4 shows the influence of temperature and ethanol content on the concentration of saturated oxygen in an aqueous alcoholic solution.

TABLE 17.4 Concentration of Oxygen (mg/L) at 760 mm Hg in Aqueous Alcoholic Solutions at Different Temperatures

Ethanol (% vol/vol)	Temperature (°C)						
	0	5	10	15	20	25	30
0	14.6	12.8	11.3	10.15	9.2	8.4	7.7
10	12.6	11.3	10.1	9.2	8.3	7.7	7.2
20	11.7	10.6	9.6	8.9	8.1	7.5	7.1

TABLE 17.5 Concentration of Saturated Oxygen (mg/L) at 760 mm Hg in Aqueous Alcoholic Solutions and Wine at a Temperature of 20°C and a Pressure of 760 mm Hg

	Aqueous Alcoholic Solution					Wine	
Ethanol % (vol/vol)	0	5	10	15	20	11.2	20.5
Dissolved oxygen (mg/L)	9.2	8.5	8.3	8.2	8.1	8.25	7.8

The absolute density of oxygen under normal conditions (0°C and 1 atm) is 1.428 mg/mL. It is therefore easy to convert the unit mg O_2/L_{wine} to the unit mL O_2/L_{wine}, which is used in certain studies of wine.

5.1. Measuring Dissolved Oxygen: The Clark Electrode

The Clark electrode is an amperometric membrane electrode consisting of a gold cathode and a silver anode immersed in a solution of potassium chloride. This arrangement is separated from the medium by a Teflon membrane that is permeable to oxygen.

FIGURE 17.3   **Clark electrode.**

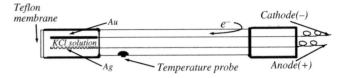

The gold cathode must be previously polarized by applying an electrical potential of 850 mV. When the oxygen in the medium containing the electrode diffuses through the Teflon membrane, the following reactions occur:

Anode half-reaction: $4\,Ag \rightarrow 4\,Ag^{+} + 4\,e^{-}$

Cathode half-reaction: $O_2 + 2\,H_2O + 4\,e^{-} \rightarrow 4\,OH^{-}$

An electric current is generated at the electrode. This current is proportional to the partial pressure of the oxygen in the medium and can be converted to units for the concentration of dissolved oxygen (generally, mg O_2/L or % saturation). It is important to remember that the Clark electrode is destructive in that it consumes oxygen, and this obviously affects the measurement procedure as the medium needs to be stirred continuously to deliver a sufficient supply of oxygen to the electrode to maintain the reaction and ensure that the electrical current measured remains constant.

5.2. Influence of Certain Operations on Oxygen Content

A small quantity of oxygen per liter of wine is generally sufficient to oxidize it or to trigger ferric casse. This is why it is important to control all possible means by which oxygen can enter wine. To do this, it is first necessary to understand the factors that influence the solubility of atmospheric oxygen in wine under normal pressures:

1. Contact surface. Solubility increases with contact surface area.
2. Stirring. Solubility increases with stirring.
3. Temperature. Solubility increases with a decrease in temperature.
4. If air or pure oxygen gas is bubbled through the solution, the size of the bubbles also has an impact on solubility as the smaller the bubbles, the more efficient the transfer of oxygen will be.

The main operations that influence oxygen content are:

1. Racking. When wines are racked using a siphon, without stirring and with the end of the tube immersed in the racked liquid, the increase in dissolved oxygen content will be minimal (at most 0.1–0.2 mL/L). When racking is performed, however, by allowing the wine to fall from the top of the tank or flow through a large funnel (on whose walls it forms a layer), oxygen content increases by several milliliters per liter. It has been reported that a thin stream of wine allowed to fall from a height of 1 m is sufficient to generate an oxygen saturation of close to 8 mg O_2/L.
2. Stirring. Vigorous stirring of wine in the presence of an equal volume of air causes oxygen saturation in less than half a minute. This saturation occurs more quickly with wine than with pure water because ethanol forms a persistent emulsion with air. Lower oxygen solubility has been observed in wine with more extract.

3. Contact with air. When oxygen-free wine is exposed to air via a contact surface area of 100 m^2, only a few milliliters of oxygen are dissolved per liter of wine in 15 minutes. The amount will increase if the surface is agitated.
4. Diffusion through wood. It has been estimated that 1 mL of oxygen diffuses through the wood of wine barrels per liter of wine and per month.
5. CO_2 content. All wines contain several tenths of a milligram of CO_2 per liter, but this concentration is insufficient to prevent the dissolution of atmospheric oxygen. The rate of dissolution is considerably decreased in wines with a CO_2 concentration of over 100 mg/L.
6. Bottling. The amount of oxygen that dissolves in wine during bottling varies (from 0.2 to 1.5 mL/L) according to the pressure of the liquid in the bottle. There is practically no difference in the content of dissolved oxygen in wine added to bottles in a solid stream (this reduces the contact surface with air but increases emulsion) or when the wine is allowed to run down the walls of the bottles (this reduces the emulsion effect but increases the contact surface area and the time during which the wine is in contact with the air). The amount of air added to the bottle is a risk factor for bottle sickness.
7. Transfer pumps. The amount of oxygen that is dissolved when wine is transferred from one container to another by gravity or pumping depends on several factors. The effect is normally negligible unless pockets of air develop in the pump or in the pipework. Oxygen saturation (at levels of 6–7 mL/L), however, can occur when the pump is placed above the level of the liquid to be pumped, as this can cause a loss of prime in the pump itself or at the intersection of two pipes, causing the penetration of persistent, fine air bubbles.

5.3. The Presence of Oxygen in Wine: Technological Implications

The solubility of oxygen in wine at ambient temperature is approximately 6 mL/L. As this oxygen is consumed, there is a gradual decrease in the redox potential of the wine. This improves the organoleptic properties of the wine, but only in the initial stages. If the wine is left to undergo successive oxygen saturation and consumption, it will gradually lose quality due to browning, precipitation, and the appearance of oxidized aromas.

The rate at which oxygen is consumed is highly variable and is notably influenced by iron and copper concentrations as these minerals catalyze the transformation of molecular oxygen into its active forms. The rate of consumption is initially slow but then increases. It has also been seen that the consumption of oxygen exceeds the theoretical rate of one atom per molecule of oxidized phenol. This can be explained by the establishment of new, complex redox mechanisms involving the molecules formed in the early stages of oxidation and by condensation reactions, radical coupling, and nucleophilic additions.

TABLE 17.6   Influence of Iron and Copper Concentrations on the Reaction of Oxygen

Type of Wine	O$_2$ (mL/L) that Reacts in 7 Days
Test wine (24 mg/L Fe + 1.4 mg/L Cu)	5.5
Treated wine (0 mg/L Fe + 0 mg/L Cu)	0
Treated wine (24 mg/L Fe + 0 mg/L Cu)	1.9
Treated wine (0 mg/L Fe + 1.4 mg/L Cu)	2.3

The enzymatic oxidation of polyphenols in must obtained from white grapes does not generally cause problems in the final wine, but a high concentration of phenolic compounds (due to overprotection against oxidation) can give rise to a higher proportion of potentially oxidizable substrates that can cause serious problems. Must hyperoxidation is designed to reduce the presence of readily oxidizable substrates in the future wine.

Enzymatic oxidation has a limited impact on red wines, unlike chemical oxidation, which plays a key role in the aging of these wines. Not only polyphenols themselves but also the oxidation products of ethanol and tartaric acid are important, as they play a key role in the transformation of polyphenols by participating in their polymerization. This polymerization leads to reduced astringency and contributes to the color change that young wines undergo as they age and change from a bright red to a brick-red color.

6. OXIDATION OF POLYPHENOLS IN MUST AND WINE

Phenolic compounds are the main components of wine affected by redox reactions. These processes influence color, and they are generally responsible for browning. They also influence flavor by increasing or reducing astringency.

The oxidation of phenols occurs via two pathways: an enzymatic pathway and a purely chemical pathway. The first pathway, or mechanism, is known as enzymatic browning and involves metalloenzymes such as polyphenol oxidases that act on phenolic substances (mainly hydroxycinnamic acids). These oxidases require the presence of iron and molecular oxygen.

The second pathway involves the oxidation of phenolic compounds by quinones formed via the enzymatic pathway.

6.1. Enzymatic Oxidation

Enzymatic browning is mediated by two distinct polyphenol oxidases, which give rise to the formation of quinones (brown compounds).

1. A catechol oxidase, derived from healthy grapes, which has both cresolase activity (hydroxylation of monophenols to ortho-diphenols) and catecholase activity (oxidation of ortho-diphenols to ortho-quinones).
2. A laccase, of fungal origin, that does not have cresolase activity but does catalyze the oxidation of several types of phenols, in particular para-diphenols.

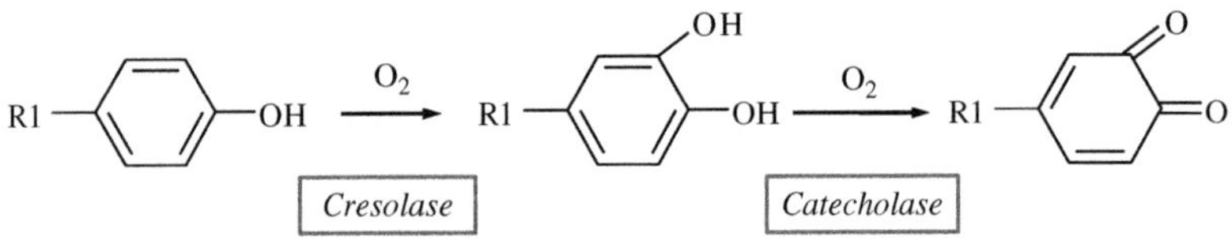

FIGURE 17.4   Mechanism of action of an enzyme with cresolase and catecholoxidase activity.

Enzymatic activity begins when the enzyme comes into contact with its substrates (oxygen and phenolic compounds); that is, when the berry is burst.

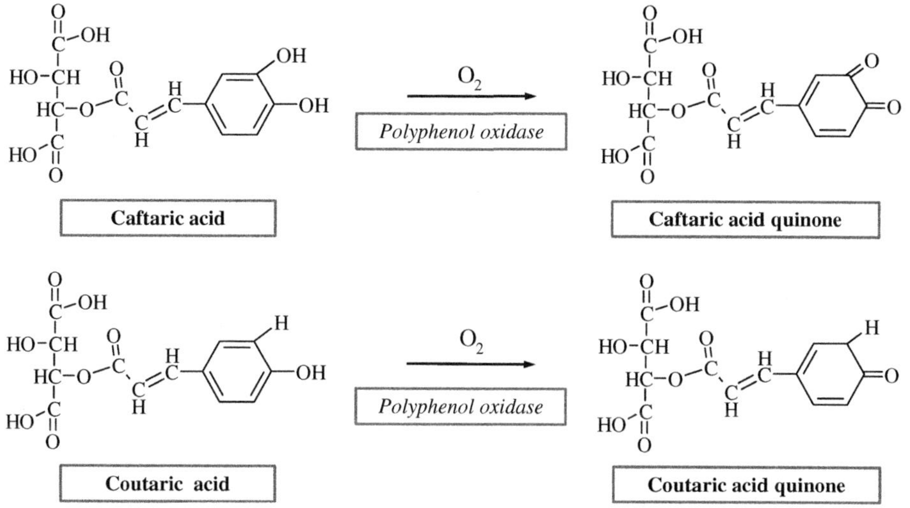

FIGURE 17.5   Formation of caftaric acid and coutaric acid quinones.

Because caffeoyl tartaric acid (caftaric acid) is the most abundant phenol in must and because polyphenol oxidases have a high affinity for this acid, the corresponding ortho-quinone is the main enzymatic oxidation product of must. In addition to the high concentration of the quinone, the caftaric acid quinone/caftaric acid redox couple has a high redox potential. This molecule is therefore highly reactive, meaning that it participates in other redox reactions such as the oxidation of ascorbic acid and sulfites, and also of ortho-diphenols that are not substrates of polyphenol oxidases. Accordingly, it is also involved in oxidation reactions. Another, less abundant — but not less important — quinone, is p-coumaryl tartaric acid, which is derived from the oxidation of coumaryl tartaric acid (coutaric acid).

6.2. Chemical Oxidation

The enzyme polyphenol oxidase catalyzes the formation of primary quinones. The phenol/quinone redox couple has a high redox potential, meaning that quinone tends to be reduced, leading to the oxidation of other phenolic compounds, notably oligomeric flavanols and anthocyans. This phenomenon is known as coupled oxidation. Quinones formed in this manner are called secondary quinones. They are highly unstable and give rise to condensation products (with their reduced form or with caftaric acid).

6.3. Browning of Must and Wine

The oxidation of must and wine involves an initial enzymatic oxidation stage that gives rise to the formation of primary quinones. These molecules are responsible for successive coupled oxidations with other polyphenols, leading to secondary quinones. These redox equilibria are responsible for the browning of all wines due to the condensation of secondary quinones. The

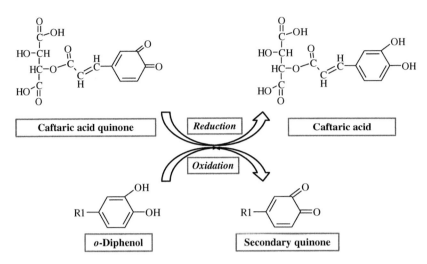

FIGURE 17.6   Formation of secondary quinones from caftaric acid quinones.

formation of a secondary quinone from anthocyans causes a considerable loss of red color intensity, explaining why wines that have been aged for a long time develop a red-orange color.

Several compounds in must and wine have natural antioxidant activity and react with primary quinones, reducing their oxidizing action. One of these compounds is glutathione (a tripeptide formed by glycine, cysteine, and aspartic acid), which is highly abundant in certain grape varieties. Glutathione reacts with the quinone of caftaric acid to form the acid 2-S-glutathionyl caftaric acid, also known as grape reaction product (GRP).

The formation of GRP, which is colorless and does not provide a substrate for polyphenol oxidase, paralyzes the formation of brown products, as it inactivates the quinone of caftaric acid and therefore does not give rise to coupled oxidation reactions. GRP, however, can be oxidized in the presence of an excessively high concentration of caftaric quinones once glutathione is depleted, giving rise to intense browning. The browning of grape must therefore depends on the relative proportions of glutathione and hydroxycinnamic acids.

Depending on the relative proportions of hydroxycinnamic acids (HCA) and glutathione (GSH), musts can be classified as having:

- Low sensitivity to browning if HCA/GSH < 1
- Sensitivity to browning if $1 <$ HCA/GSH < 3
- High sensitivity to browning if HCA/GSH > 3

Must also contains other substances, such as ascorbic acid (an antioxidant), capable of regenerating caftaric acid from its quinone, thereby delaying oxidation. SO_2, which has antioxidant properties, is also added to must to inhibit enzymatic oxidation.

The presence of large quantities of glutathione does not guarantee that browning will not occur, as the protection afforded is temporary and accumulated GRP functions as a reserve of oxidizable products in the long term. This is of particular concern in musts made from grapes infected with *B. cinerea*, as laccase is capable of oxidizing GRP to quinone (meaning that

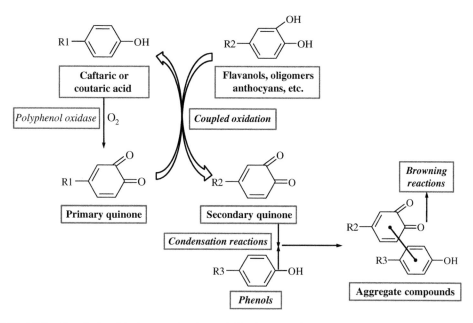

FIGURE 17.7 Coupled oxidation mechanisms following the formation of primary quinones via the enzymatic pathway.

FIGURE 17.8 Mechanisms of action of glutathione.

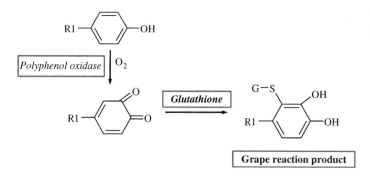

browning is not prevented). In the presence of excess glutathione, the GRP quinone may be reduced again, giving rise to 2,5-di-S-glutathionyl caffeoyl tartaric acid, but musts and wines containing laccase are highly unstable, and are therefore prone to spoilage during storage and aging.

6.4. Technical Considerations

The bursting of grape cells during pressing simultaneously triggers all the above-described reactions. The rate of these reactions depends mainly on the relative proportions of the intervening chemical species and the pH of the medium.

The formation of brown compounds has more serious consequences for must made from white grapes and the resulting wine. The addition of SO_2 is important in such cases, as is the use of prefermentation treatments such as maceration and hyperoxidation.

1. The addition of SO_2 during or after pressing reduces the formation of quinones and even when quinones are formed, SO_2 leads to the regeneration of approximately 50% of the caftaric acid initially present in the must, thereby limiting oxidation and leading to a considerable recovery of the original color of the must. When the activity of SO_2 ceases, or in musts that have not had sulfites added, coupled oxidation and condensation reactions continue to occur, leading to browning without an additional consumption of oxygen.

2. The enrichment of must with phenolic compounds, and flavonoids in particular, due to the maceration of solid parts or vigorous pressing, favors browning.

3. Hyperoxidation of white grape must essentially accelerates the development of condensation products of oxidized forms, giving rise to high-molecular-mass compounds that are easy to eliminate using classical clarification techniques. Hyperoxidation in pre-fermented musts is only efficient when these contain high quantities of flavonols (catechins and epicatechins); it has no effect in the presence of high concentrations of glutathione (HCA/GSH < 1), however.

The Colloidal State

Enological Chemistry. DOI: 10.1016/B978-0-12-388438-1.00018-2

1. INTRODUCTION TO DISPERSED SYSTEMS

Analysis of materials distinguishes between pure substances and dispersed systems, dispersions, or mixtures. The simplest dispersed systems comprise two pure substances, one of which is dispersed in the other. The substance in the lowest proportion is referred to as the dispersed phase and that found at the higher proportion is known as the dispersion phase or dispersant. Dispersed systems are classified according to the diameter of the dispersed particles (assuming them to be spherical), or when the particles are elongated, as frequently occurs with organic molecules, they are commonly classified according to the number of particles that make up the mixture. Traditionally, a distinction has been made between homogeneous and heterogeneous mixtures or dispersions. Homogeneous mixtures appear to have a single phase whereas heterogeneous mixtures contain two or more phases.

A phase is defined as a region in space that contains the same physical and chemical properties. State refers to the degree of aggregation of the material.

In general terms, the difference between a homogeneous and heterogeneous mixture is easily discernible to the naked eye. However, between these two categories lies a third: colloids or colloidal dispersions.

True Solutions

True solutions are homogeneous mixtures that are present in a single phase and are usually made up of low-molecular-weight ionic or covalently bonded compounds that disperse rapidly and homogeneously within the dispersant. The compound present at the lowest proportion is known as the solute and that present at the highest proportion is known as the solvent. For instance, when a few grams of salt (sodium chloride) or sugar (sucrose) are shaken in water, a solution containing just a single phase is obtained.

TABLE 18.1   Types of Dispersed Systems or Mixtures

Type of Dispersion	Particle Size	Number of Atoms in a Particle	Characteristics
Molecular (true solutions)	<1 nm	$<10^3$	Pass through filter paper and ultrafilters Not visible by microscopy or ultramicroscopy Diffuse and dialyze Do not form sediments
Colloidal (particles can be microcrystalline or macromolecular)	1 – 100 nm	$10^3 - 10^9$	Pass through the finest filter paper but not through ultrafiltration devices Visible by ultramicroscopy but not microscopy Do not diffuse or dialyze Sediment very slowly
Suspensions	>100 nm	$>10^9$	Do not pass through filter paper Visible under the microscope Do not diffuse or dialyze Sediment rapidly

Suspensions

Suspensions are heterogeneous mixtures that are easily recognized by eye, since they contain two clearly differentiated phases. The dispersed particles are large and tend to deposit rapidly on the bottom of the vessel to form more or less compact sediments. For instance, when sand and water are mixed together the system constitutes a suspension and its components separate rapidly. The finer the sand, in other words the smaller the dimensions of the particles, the longer they take to settle.

Colloidal Dispersions or Colloids

At first sight, colloids have a homogeneous appearance (a single phase) and are made up of a dispersed phase (corresponding to the solute) and a dispersant phase or medium (corresponding to the solvent). The particles in the dispersed phase are suspended in the dispersant phase, and are sufficiently small for precipitation to be extremely slow. However, they are also sufficiently large to scatter light, giving rise to a cloudy and often opaque appearance. They are therefore apparently homogeneous systems that in time can convert into heterogeneous mixtures. Examples include wine, milk, and some types of soup (such as Spanish gazpacho). The phases can separate as a result of mechanical or chemical manipulation. For instance, milk separates upon the addition of a few drops of lemon juice.

Many colloids have the appearance of true solutions or homogeneous mixtures, and in these cases they must be identified through the presence of characteristic properties associated with colloids. A homogeneous mixture can be distinguished from a heterogeneous mixture through its physical properties, such as the phase transition temperature of its components. In a heterogeneous mixture, the phase transition of one of its components is not affected by that of another, whereas in a solution or a homogeneous mixture, the property is affected by the presence of the other components.

2. GENERAL CHARACTERISTICS OF COLLOIDS

Colloidal systems are made up of a wide range of substances belonging to very different chemical families. However, they possess a series of common properties that distinguish them from true solutions:

- The dispersed phase is not soluble in the dispersant phase
- They do not diffuse or dialyze
- They sediment very slowly

Some dispersions can display the properties of a true solution or of a colloid depending on the conditions. For instance, sodium chloride and other salts that produce ions in solution form true solutions with water (they are electrolytes). However, the addition of alcohol induces the formation of colloidal dispersions, with all of the intermediate degrees of dispersion between a solution and a precipitate. The same occurs in wine with potassium bitartrate and phenolic compounds, which can be present as true solutions or colloidal dispersions. The appearance of ethanol in the course of fermentation induces this phenomenon.

Aggregation State of the Dispersed Phase

Individual particles that make up the dispersed phase of a colloid are given the generic name of micelles, and they can be considered as a molecular aggregate. The size of the micelles ranges from 10^{-3} to 1 μm. Although at first sight colloids appear amorphous, many have a crystalline structure derived from the spatial organization of their constituent molecules. These unstable, hydrophobic dispersions are referred to as microcrystalline colloids. Another type of dispersed colloidal particle is not made up of molecular aggregates but rather of giant molecules that give rise to stable macromolecular hydrophilic colloids. (The term micelle is reserved exclusively for microcrystalline colloids.)

3. CLASSIFICATION OF COLLOIDS

Colloids are classified according to two main criteria: the affinity of the dispersed phase for the dispersant and the state in which the material is found in the phases that make up the colloid.

The simplest colloids are made up of only two phases and, as in solutions, all combinations of the phases can occur. Given that gaseous mixtures always produce a single phase, there are only eight possible phase combinations (Table 18.2).

Colloids in which the dispersed particles display an affinity for the dispersant medium are referred to as lyophilic, whereas those that repel the dispersant are referred to as lyophobic. Lyophobic colloids are characterized by the absence of bonds between the dispersed phase and the dispersant. When the dispersant is water, these colloids are referred to as hydrophilic or hydrophobic, respectively.

Hydrophilic colloids absorb water, swell, and can give rise to a special type of colloid known as a gel. Here, the liquid appears to be completely absorbed into the dispersed phase.

TABLE 18.2 Types of Binary Colloidal Dispersions

Dispersed Phase	Dispersant Phase	Denomination	Example
Liquid	Gas	Liquid aerosol	Mist, spray
Solid	Gas	Solid aerosol	Smoke
Gas	Liquid	Foam	Foam on beer, bubbles in sparkling wine, whipped cream
Liquid	Liquid	Liquid emulsion	Mayonnaise, milk, vinaigrette
Solid	Liquid	Sol Gel	Whitewash, skimmed milk, paints, puddings, jelly, mud
Gas	Solid	Solid foam	Meringue, ice cream, bread, sponge cake
Liquid	Solid	Solid emulsion	Butter, cheese
Solid	Solid	Solid sol or solid solution	Some metal alloys such as steel and duralumin, and some gemstones

A gel has the appearance of a solid and is elastic and gelatinous, as seen in the classic example of this type of substance, namely gelatin. These colloids are generally made up of macromolecules dispersed in water.

Hydrophobic colloids are formed by the aggregation of spatially ordered particles and are referred to as microcrystals.

4. APPROACHES TO THE ANALYSIS OF SOLS AND GELS

Most colloids of interest in the food and wine industry have a liquid dispersant phase made up mainly of water. This restricts analysis to foams, sols, gels, and emulsions. Sols and gels are of particular interest in winemaking, since they explain phenomena occurring during vinification and aging and also the properties of various products used in the manufacturing process. Nevertheless, we should not underestimate the importance of liquid foams, particularly in the case of cava and other sparkling wines. It should also be remembered that winemaking involves yeasts, which, when lysed, release proteins and polysaccharides into the aqueous alcohol solution of the wine and form colloidal systems.

Sols and gels are derived from complex and unstable equilibria that develop in one way or another according to the conditions to which they are exposed. The variability of these conditions over time is a critically important factor affecting the colloids used in foods, both during production and over the course of the usable life of the foodstuff. It is therefore important to be aware of the terms used to describe the phenomena that occur in these classes of colloids.

A sol is a fluid, non-rigid colloid with a low viscosity that contains free particles. The particles exhibit spontaneous, random movement caused by contact with neighboring particles. In the case of gels, the particles join together to form larger or smaller aggregates, and as a result, both the dispersed and dispersant phases extend throughout the entire system. In a gel, the dispersed phase forms fine filaments or a thin network that traps the dispersant phase in a semi-rigid structure that can even behave as a true solid. Gels are not a clearly defined state and can be destroyed, for instance, by mechanical force.

4.1. Thixotropy

The application of force or shaking of some gels ruptures the three-dimensional structure of the dispersed phase. This causes the gel to form a sol, which is no longer viscous and can flow freely. If the force or shaking that caused the rupture of the network ceases, the gel reforms. This phenomenon is known as thixotropy, a classic example of which is non-drip paint. These paints are viscous and thick in the tin, but they become liquid when a brush is inserted and agitated, causing them to transform into a sol. When the agitation stops, the paint thickens on the brush, re-forming a gel that does not flow. When the paint brush is applied to the surface, the gel once again becomes liquid before returning to a gel state after application.

4.2. Peptization and Flocculation

Depending to the concentration of the particles, and therefore the distance between them, a gel can be transformed into a sol by reducing the concentration of the dispersed particles through addition of a dispersant. This phenomenon is known as peptization. A sol can become a gel if the quantity of dispersed particles is increased or the volume of dispersant phase decreases, and also if the size of the particles increases. This phenomenon is known as flocculation. The two phenomena are respectively comparable to the process of dissolving or precipitating a salt.

A distinction is sometimes made between flocculation and coagulation, with flocculation used to describe the separation and sedimentation at the bottom of the vessel of *floccules* or *flakes*, which are thought to be generated by the aggregation of particles that were previously free. These floccules generate crystalline or amorphous deposits with a variable volume that can remain in suspension for a variable period of time, causing the liquid to appear visibly turbid. Alternatively, they can fall to the bottom of the vessel under their own weight, leaving the liquid clear and transparent. In contrast, the term coagulation is reserved for a process through which a sol is transformed into a gel.

4.3. Syneresis

When the two phases of a gel separate, the process is referred to as syneresis. It occurs as a result of a contraction in the colloid mass and a retraction with expulsion of part of the dispersant phase (excipient) that forms the gel. A very graphic example occurs in yoghurt when it has been stored for a few days. Here, the phases separate and a liquid phase covers a denser and more viscous solid phase than that found in freshly prepared yoghurt.

5. PROPERTIES OF COLLOIDS

The physical properties of colloidal systems are very different to those of homogeneous systems such as true solutions, and these differences are further accentuated in multiphase systems, in which a simple colloid system can be the dispersed phase in a more complex dispersant.

Many foodstuffs cannot be considered as just emulsions, sols, or foams, since these three types of colloids can occur simultaneously, particularly when the dispersed phase contains lipids, proteins, and gases. Many whipped creams, condiments, and sauces (such as mayonnaise) fall into this category.

Colloids have a series of characteristic properties that fundamentally depend on the size and number of dispersed particles. These properties can be grouped into four main categories: optical, mechanical, adsorption, and electrical.

5.1. Optical Properties

When light is passed through a colloidal dispersion and observed from a position perpendicular to the pathway of the light, the suspension is illuminated. This phenomenon is known as the Tyndall effect and is due to the reflection and refraction of light by the particles in the dispersed phase of the colloid. A solute particle that is too small will not generate this phenomenon. Consequently, this effect is not observed in a true solution. Examples of the Tyndall effect include the scattering of light from a cinema projector caused by particles and dust suspended in the air of a dark room, and the scattering of light from car headlamps in fog.

5.2. Mechanical Properties

All of the particles dispersed in a medium display a random movement known as Brownian motion, which depends principally on the temperature and the concentration gradient. Brownian motion is irregular, does not follow a fixed trajectory, and is responsible for the tendency of the particles present in a liquid to occupy the maximum possible volume and become uniformly distributed after a period of time. Colloidal particles display weaker Brownian motion than solute particles in true solutions, and this reduction increases with the size of the particle. It is this movement (or restriction of movement) that explains the failure of colloidal particles to deposit immediately at the bottom of a vessel.

5.3. Adsorption Properties

One of the most important properties that defines colloidal systems is the high degree of separation present in the dispersed phase, and as a consequence their high surface area to volume ratio. This characteristic generates a large area of contact between the dispersed phase and the dispersant, which results in a high interface energy at the surface separating them. The stability of a colloid depends directly on the formation and nature of the interactions that occur at the point of contact between the two phases, and surface phenomena need to be analyzed in order to understand their behavior.

The atoms at the surface of a colloidal particle are only chemically bonded to other atoms surrounding them at the surface of the particle and to those lying inside it. These surface atoms therefore have a strong tendency to interact with any substance that comes close to them. In the process of adsorption, species other than those of the colloid particles adhere to the surface of the particle.

An example of adsorption is the formation of a brilliant red complex that is produced when hot water is mixed with a concentrated aqueous solution of iron(III) chloride. Each colloidal particle in this sol is formed from a hydrated aggregate containing many units of Fe_2O_3. Each particle attracts Fe^{3+} ions in the solution to its surface, leaving the micelle surrounded by a positive charge, and as a result, the micelles repel each other and cannot group together in sufficient quantity to cause precipitation. Figure 18.1 shows the phenomenon of adsorption of Fe^{3+} ions to micelles of iron(II) and iron(III) oxide.

$$2 \left[Fe^{3+}(aq) + 3\ Cl^- (aq) \right] + x(3+y)\ H_2O\ \rightarrow\ \left[Fe_2O_3\ yH_2O \right]_x (s) + 6x \left[H^+ + Cl^- \right]$$
(Yellow solution)   (Brilliant red dispersion)

In summary, the forces of attraction that hold atoms, molecules, or ions together in a solid are balanced within that material but not at the surface, and as a result these forces can attract other atoms, ions, or molecules, leading to the phenomenon of adsorption. The forces involved in adsorption are predominantly electrostatic, van der Waals-type forces and hydrogen bonds.

FIGURE 18.1   Adsorption of Fe^{3+} to Fe_2O_3
micelles dispersed in a solution of $FeCl_3$.

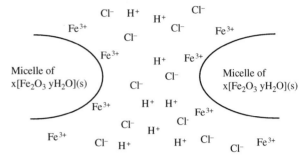

Hydrophobic sols are good adsorbents and are stabilized by electrolytes in solution through the formation of a double layer of electrical charge. Aqueous solutions of globular proteins can be considered as hydrophilic colloids, since the protein chain is folded in such a way that nonpolar groups on the amino acids are orientated towards the inside of the molecule, whereas polar groups ($-NH_2$ and $-COOH$) orientate towards the outside. These functional groups can therefore form hydrogen bonds with water molecules in the medium, and this increases the stability of the colloid formed.

Some hydrophilic colloids, particularly polysaccharides with only weakly polar functional groups, exhibit a protective behavior towards other hydrophobic or less-hydrophilic colloids with an electrical charge of the same sign, thereby impeding their flocculation. It is thought that this occurs through a mechanism in which the protective colloid surrounds the protected colloid. Typical examples of protective colloids are gum arabic and colloidal carbohydrates, which protect wine from casse and precipitation of potassium bitartrate crystals.

5.4. Electrical Properties

When a colloidal dispersion is introduced into a U-shaped glass tube containing a small quantity of water and an electrode in each of its arms, it can be seen that the dispersion moves towards one or other of the electrodes when a current is applied. This process is called electrophoresis, and occurs in dispersions or liquid solutions containing electrically charged particles. The electrical charge on the colloidal particles is determined in some cases by their character as electrolytes, in others by the adsorption of ions from the solution, and in yet others due to reactions that have occurred on the surface of particles with an acid, base, or other electrolyte present in the dispersion medium. Ions carrying an opposite charge that remain in excess in the medium surround the colloidal particle and form a double electrical layer.

5.5. Other Properties

Colloids also possess other properties:

a. Like non-colloidal particles, the composition of a micellar colloid is not perfectly defined. Thus, in a dispersion of iron oxides, micelles with varying composition and dimensions are found, unlike in a solution of sodium chloride, which only contains sodium and chloride ions, and these have an invariant composition. This is not so clear in the case of colloids formed by macromolecules, whose composition does not change through contact with the dispersant phase.

b. Ion-adsorption properties are dependent on the specific properties of the surface, and this is in turn related to the composition and size of the micelle; consequently, the quantity of ions that can be adsorbed varies from one micelle to another.

c. The freezing and boiling points of colloidal systems are very similar to those of the pure dispersant phase; in other words they do not follow the Raoult law, even in the case of concentrated colloidal dispersions. Colloidal dispersions behave as though there were two phases, which is in fact the case.

6. SOLS

Sols are formed by the dispersion of small solid particles in a liquid. Sols in which the dispersant phase is water commonly contain polysaccharide or protein particles, which can give rise to hydrophobic and hydrophilic sols. Both types of sol are usually strongly hydrated, but hydrophobic sols are less stable than their hydrophilic equivalents.

The charge on the surface of the particles that form sols depends on the pH of the dispersant medium. This charge generates a repulsive force between the particles that maintains them in motion and causes them to become uniformly distributed throughout the dispersant phase. If this repulsive force disappears, large aggregates form and flocculate.

The importance of sols in winemaking is due to their:

- Natural occurrence during aging (condensed phenolic compounds and coloring matter)
- Accidental appearance leading to problems such as iron and copper casse
- Use as treatments for stabilization (iron ferrocyanide, copper sulfate) and clarification (bentonite, silica sols, gelatin, etc.)

6.1. Hydrophobic Sols

Hydrophobic sols are made up of micelles, in other words aggregates of simple particles that bind to each other through weak electrostatic forces. The charge on these micelles depends on the pH of the aqueous phase and is responsible for their stability. A reduction in the electrical charge increases the tendency to form large aggregates and leads to the appearance of floccules that then form sediments.

6.2. Hydrophilic Sols

Hydrophilic sols are made up of macromolecules such as polysaccharides and proteins. They carry an electrical charge due to ionization of functional groups. This charge confers stability, as it does in hydrophobic colloids. Due to their hydrophilic character, they are hydrated, which further increases their stability. As a result, hydrophilic sols are commonly used as protective colloids to prevent sedimentation of hydrophobic colloids.

FIGURE 18.2   Some types of micelles.

Spherical micelle

Laminar micelle

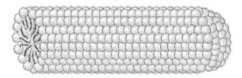

Cylindrical micelle

TABLE 18.3   Properties of Hydrophobic and Hydrophilic Sols

Hydrophobic Sols	Hydrophilic Sols
Absorb little water	Absorb large quantities of water
Sediment easily after addition of salts	Do not sediment upon addition of salts
Generate powdery floccules	Produce gelatinous deposits
Form micelles by binding of molecules through weak electrostatic forces	Do not form micelles by association of particles
Have the appearance of microcrystals	Are macromolecules
Are stable due to the presence of electrical charge or dipoles in the micelles	Stability is due to the ionization of functional groups and hydration
Examples: iron phosphate, iron ferrocyanide, copper sulfate, bentonite	**Examples:** rubbers, mucilages, proteins, condensed phenolic compounds

The electrical charge of a sol in an aqueous medium depends on the chemical nature of the functional groups that are orientated towards the outside of the micelles and in direct contact with the water. This leads us to consider two possible mechanisms for the generation of electrical charge. The first involves the direct ionization of functional groups contained within the macromolecule itself. The second is due to adsorption of ions dissolved in the dispersant phase.

6.3. Mechanisms of Electrical Charge Formation

Mechanism I: Ionization of Chemical Groups Within the Molecule Itself

This mechanism is observed particularly when the dispersed particle is made up of macromolecules. The most characteristic example is seen with proteins, since their amino and carboxyl groups ionize easily and generate a positive or negative charge according to the pH of the medium in which they are found. If the pH is higher than the isoelectric point of the protein, it acquires a negative charge due to ionization of carboxyl groups, whereas

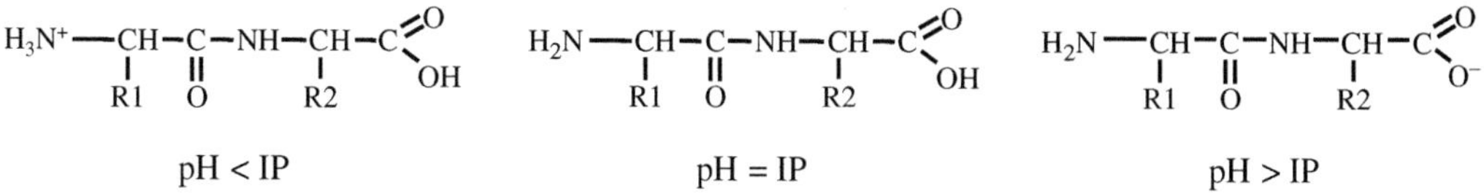

$$H_3N^+ \text{---CH---C---NH---CH---C} \overset{O}{\underset{OH}{\diagdown}}$$
$$\underset{R1}{|} \quad \overset{||}{O} \quad \underset{R2}{|}$$

pH < IP

$$H_2N \text{---CH---C---NH---CH---C} \overset{O}{\underset{OH}{\diagdown}}$$
$$\underset{R1}{|} \quad \overset{||}{O} \quad \underset{R2}{|}$$

pH = IP

$$H_2N \text{---CH---C---NH---CH---C} \overset{O}{\underset{O^-}{\diagdown}}$$
$$\underset{R1}{|} \quad \overset{||}{O} \quad \underset{R2}{|}$$

pH > IP

FIGURE 18.3   Electrical charge of a dipeptide as a function of pH. IP = isoelectric point.

TABLE 18.4   pH of the Isoelectric Point of Colloids Used in Winemaking

Fining Agent	Isoelectric Point
Casein	4.62
Gelatin	4.70
Serum albumin	4.64
Serum globulin	4.80 − 6.40
Wine proteins	≈4.0

at pHs lower than the isoelectric point, the protein becomes positively charged due to ionization of the amino groups. When the pH of the medium is equal to the isoelectric point, the net charge on the protein is zero. Proteins and other macromolecules such as phospholipids and some carbohydrates form hydrophilic colloids and can be charged.

Some hydrophobic colloids, such as ferrous or ferric iron ferrocyanide, can form electrically charged micelles. This occurs in wines treated with blue fining, which is the addition of potassium ferrocyanide, which forms a blue precipitate with iron(II) or iron(III) in the wine. Each micelle can be considered to be formed by an indeterminate number of molecules of iron(II) or iron(III) ferrocyanide, which are bound by potassium ferrocyanide and generate two ions according to the following equilibrium:

$$\{[Fe(CN)_6]_3 Fe_4\}_n \, [Fe(CN)_6]K_4 \rightleftarrows \{[Fe(CN)_6]_3 Fe_4\}_n \, [Fe(CN)_6]^{4-} + 4\,K^+$$

The formation of negatively charged micelles generates the sol, which is responsible for the blue haze (blue casse) that develops in wine following treatment with ferrocyanide. The presence of proteins, which have a positive charge at the pH of wine, facilitates the growth of these micelles and therefore their flocculation.

Mechanism II: Adsorption of Ions or Molecules from the Dispersant Phase

In the case of neutral colloidal particles, the charge originates from the binding or adsorption of ions to the surface. The temporary dipoles or residual electrical charge on the micelle attract ions of the opposite charge from the dispersant phase with a force that depends on the temperature, the pH, the ionic strength of the dispersant phase, and of course, the nature of the colloidal particle. Acidic polysaccharides are an example of the strong interaction with metal ions in the medium, leading to the acquisition of positive charge (Figure 18.4).

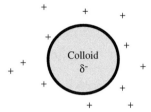

FIGURE 18.4   Adsorption of positive ions to a negatively charged colloid.

A special case occurs when polar molecules rather than ions are adsorbed to the colloid. An example is shown in Figure 18.5, where dipolar water molecules orient their negative poles towards the micelle, which carries residual positive charge. The negative pole will remain orientated towards the micelle and the positive pole towards the dispersant medium. In this case, the micelle acquires an overall positive charge.

In the particular case of aqueous dispersions, such as must or wine, a double layer forms around the colloidal particle. Water situated in the immediate proximity of the colloidal particle remains immobile and is referred to as concrete water; the more distant layer is called diffuse water.

The nature of the electrical charge that forms around a micelle can be explained in more detail through the idea of a double electrical layer. This double layer is formed by an internal, immobile layer of ions or molecules from the dispersant phase bound to the surface of the colloidal particle by electrostatic or van der Waals forces. The charge density of the ions or molecules situated in the first layer surrounding the colloidal particle accounts for between 60 and 80% of the total charge surrounding the particle. The second or external layer has a diffuse structure made up of the same ions or molecules as the internal layer. It surrounds

FIGURE 18.5   Adsorption of a dipolar water molecule to a positively charged colloid.

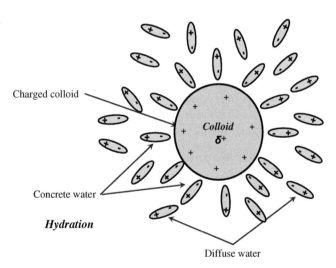

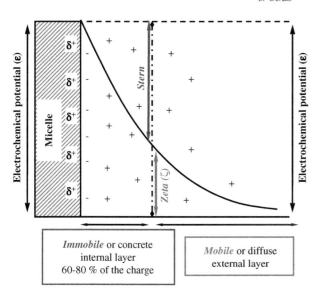

FIGURE 18.6 Double-layer theory. Adsorption of charged particles.

the micelle as a cloud with a charge density that reduces the further away it gets from the micelle until it reaches zero in the dispersant liquid.

The charges of the double electrical layer create an electrochemical potential known as the epsilon potential (ε), which is generated between the free surface of the particle and the liquid itself. This potential is in turn divided into two zones: the so-called Stern potential, which corresponds to the charge developed between the micellar surface and the internal layer of counterions, and the zeta potential (ζ) or electrokinetic potential, which is the potential difference between the immobile layer of counterions and the liquid.

Formation of a double layer can also be associated with mechanism I, since a colloidal particle that contains ionizable groups can ionize and generate a charge that induces the formation of a double electrical layer. In fact, many properties of sols are based on the value of the zeta potential. For instance, the isoelectric point of proteins is characterized by having a ζ of zero. The addition of low concentrations of salt to hydrophobic sols also reduces their ζ potential until it reaches zero, which causes precipitation. This phenomenon is known as salt precipitation and is due to a reduction in the electrokinetic potential corresponding to a shortening of the distance between the two layers of ions.

Charge in Sols

The charge of a sol can be easily determined by placing it in a U-shaped tube, connecting a platinum electrode to each arm, and applying a continuous current of 110 V for 8 hours.

Colloids formed in wine from the following materials have a negative charge: tannins, gums, proteins coagulated by tannin, copper sulfate, iron ferrocyanide, colloidal sulfur, bentonite, kaolin, colloidal silica, decolorizing carbon, and colloidal pigments. Colloids formed by non-coagulated proteins and celluloses have a positive charge.

7. STABILITY OF SOLS

The particles that form sols are small but have a large specific surface area. This in turn translates into a large surface available for contact between the dispersed phase and the dispersant, and generates a high interface energy. The stability of these colloids depends directly on the interactions between the two phases.

7.1. Stability and Flocculation of Hydrophobic Sols

The hydrophobic sols present in wine are either not derived from the grape or are not found as colloids at that point, although some compounds can develop into this type of colloid during the vinification process (condensed phenolic compounds and coloring matter). They can also form accidentally (iron phosphate and copper sulfate) or as a result of certain treatments (iron ferrocyanide, copper ferrocyanide, denatured proteins).

Hydrophobic sols are unstable and the phases gradually separate. Nevertheless, some hydrophobic sols remain stable for a long period of time thanks to two factors:

1. Brownian motion, present in all particles and responsible for the homogeneous distribution of the particles in the medium
2. Electrical charge on the surface of the micelle due to adsorption of ions or molecules from the dispersant phase

The phenomenon of flocculation occurs fundamentally because of a reduction in the electrical charge on the colloidal particles. When this occurs, the repulsive forces between the particles are weakened and they no longer compensate for the attractive forces between two molecules that approach each other through Brownian motion. At this point, rather than move away from each other, the two molecules are attracted, bind to each other, and form a larger particle. If this phenomenon is repeated by many pairs of particles, the total number of particles is reduced and their size is increased. As a consequence, Brownian motion and the surface to volume ratio of the particle is reduced and there is less charge adsorption (and consequently less repulsion) and an increasing tendency for the particles to form aggregates.

The size of the particles also influences the sedimentation rate, which according to Stokes Law is directly proportional to the square of the radius of the particle and to the difference in density between the particle and the medium, and inversely proportional to the viscosity of the medium according to the following expression:

$$v = \frac{2gr^2(d_{particle} - d_{medium})}{9\,\eta_{medium}}$$

Where:

v = sedimentation rate of the particle
g = gravity
r = radius of the particle
$d_{particle}$ = density of the particle
d_{medium} = density of the medium
η = viscosity of the medium

Since all of the parameters can be considered constants in a given medium, the sedimentation rate increases considerably when the radius of the particle increases.

Any factor that reduces the electrical charge on the particles reduces the stability of the colloidal suspension and, below a critical value, the particle begins to flocculate. The neutralization of the charge on the particles can occur through an ion with an opposite charge derived from a metal salt or through another colloid carrying the opposite charge. In the case of iron phosphate ($FePO_4$) or iron ferrocyanide {$Fe_4[Fe(CN)_6]$}, both of which carry a negative charge, neutralization of the charge occurs through cations such as K^+, Ca^{2+}, and Mg^{2+}. The coagulant activity of these ions at similar concentrations depends on their valency, and the following sequence can be established: $Al^{3+} >> Ca^{2+} > K^+$. In other words, a much lower concentration (around 1000 times less Al^{3+} than K^+ or 100 times less Ca^{2+} than K^+) is required to achieve the same flocculent effect in the same time.

7.2. Stability of Hydrophilic Sols

Hydrophilic sols are more stable than hydrophobic sols since the phases separate more slowly. The electrical charge of the particles is the main factor that determines the stability of both types of sol. The greater stability of hydrophilic sols is due to hydration, which does not occur in hydrophobic sols, and the ability of these hydrophilic colloids to remain in suspension can be explained by three factors:

1. Brownian motion
2. Presence of electrical charge
3. Hydration phenomena

Hydration involves the orientation of the dipolar water molecules around an electrically charged colloidal particle. The water molecules closest to the particle are orientated towards the center of the particle and are called concrete water. As the distance from the particle surface increases, the water dipoles become gradually less organized until they resemble the pure solvent. The region in which the dipoles are partially organized is referred to as diffuse water. Consequently, the particles that form hydrophilic sols do not have a clear interface between the dispersed phase and the dispersant.

Flocculation of Hydrophilic Colloids

Suppression of hydration and charge in hydrophilic sols causes flocculation. This can occur independently through one of the following two pathways:

1. Dehydration and subsequent discharge
2. Discharge and final dehydration

The dehydration of hydrophilic sols is a reversible phenomenon, whereas discharge is irreversible. In any case, both phenomena must occur in order for flocculation of the sol to take place.

Firstly, dehydration can be caused by a reduction in the activity of the water or of the dielectric constant of the medium caused by the presence of alcohol. If a small amount of ethanol is added to a hydrophilic colloid, the diffuse water separates, leading to the

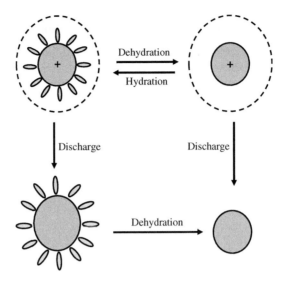

FIGURE 18.7   Suppression of stabilizing factors in hydrophilic sols.

appearance of an interface between the concrete water surrounding the particle and the dispersant phase, which is now a mixture of water and alcohol. This leads to the appearance of a surface energy that is manifested in the joining of the two layers of concrete water into one, while the particles themselves remain independent but surrounded by a single layer. This is known as coacervation and is a surface phenomenon similar to that of the joining of two or more soap bubbles to form one larger bubble.

Coacervates are concentrated colloidal dispersions that have formed through dehydration of a more diluted dispersion without any change in the state of the dispersed particle. The quantity of water found in the coacervate is the sum of that present as concrete water around each particle. The desolvating or dehydrating (as in the case of ethanol) agent transforms the system into a dispersion similar to a weakly stable hydrophobic sol.

Secondly, discharge of a hydrophilic sol particle is only produced by ions or colloidal particles with an opposite charge that are found in the dispersion or are added in order to cause this effect. For instance:

- Gelatin is a positively charged colloid at the pH of wine and flocculates in the presence of tannins that have a negative charge.
- Bentonite is a negatively charged colloid that precipitates excess positively charged protein at the pH found in must and wine.

Finally, according to other theories, hydration is not involved in determining stability. Instead, the loss of stability of a hydrophilic sol is due to the denaturation of the colloidal particle caused by the presence of alcohol or by the presence of tannins, or indeed through heating of the colloidal dispersion. The resulting colloid is hydrophobic and stable in the absence of salts but flocculates easily when salts are added.

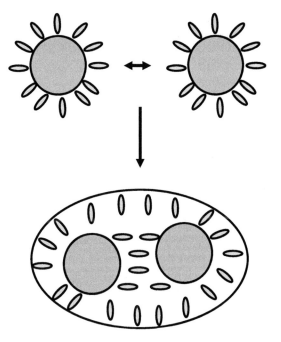

FIGURE 18.8   The phenomenon of coacervation.

8. GELS

Gels are systems formed by a continuous network of macromolecules that are interlinked to form a three-dimensional structure in which the dispersant phase is trapped within the network. The main gels found in foodstuffs are formed by polysaccharides and proteins that gel with differing degrees of elasticity and rigidity.

Gel formation is influenced by the nature and concentration of the dispersed phase, the pH, the concentration of salts in the dispersant phase, and the temperature of the system. Gels are produced more rapidly if the colloid has an affinity for the dispersant phase (e.g., a hydrophilic colloid in an aqueous dispersant).

Low temperatures accelerate gel formation, whereas high temperatures cause gels to liquefy. Gels display a phenomenon known as hysteresis, in which the gel-formation temperature differs from that at which the gel is transformed into a liquid.

Syneresis is a phenomenon exhibited by all gels and involves the exudation of the dispersant phase. The expelled liquid contains molecules that originally made up the gel but in a highly diluted form. The most common example of syneresis is seen in jelly and similar desserts when they are chilled. Syneresis involves a contraction of the gel, which leads to expulsion of water. This contraction is due to a physical redistribution of the macromolecules to achieve a more stable structure, in turn producing a readjustment of the interactions between the dispersed phase and the dispersant. Syneresis is influenced by concentration, pH, and temperature, or rather changes in those variables, and the presence of other agents can accelerate or inhibit the process.

9. FOAMS

Foams are colloids in which the dispersed phase is a gas and the dispersant is a liquid. They can be defined as a dispersion of gas bubbles suspended in a liquid. Sodas, beer, and sparkling wines are examples of these types of colloid. Sodas exhibit a tumultuous effervescence involving large bubbles and a very transient foam. In contrast, beer exhibits only a moderate effervescence with small bubbles and a very persistent foam. Sparkling wines have intermediate characteristics.

The formation of foam, its stability, and the size of the bubbles in a sparkling wine are directly related to the surface tension. This can be defined as the force per unit area that maintains the bond between the molecules at the surface of the liquid; it is a measure of the resistance of the liquid to penetration. The surface tension of water or an aqueous alcohol solution is very high, and it opposes the formation of bubbles within the liquid. The presence of surfactants such as soap reduces the surface tension of the liquid and allows the formation and persistence of bubbles.

Sparkling wines are generally obtained using the Champenois method, which essentially consists of carrying out a second fermentation in the bottle. The fermentation leads to the formation of carbon dioxide, which increases the pressure inside the bottle and causes the gas to dissolve in the liquid (around 12 g/L). Some of the carbon dioxide escapes during disgorging. Nevertheless, once dosage has been added and the bottle sealed with its final cork, part of the carbon dioxide dissolved in the liquid occupies the upper part of the bottle, leaving the final pressure on the surface of the liquid at around 6 atm. When the bottle is opened, the difference between the pressure in the bottle and that of the atmosphere causes the dissolved gas to spontaneously and very rapidly leave the liquid. Once the pressure on the surface of the liquid has been equalized with atmospheric pressure, bubbles continue to form inside the liquid.

Traditionally, the capacity to form bubbles has been attributed to microscopic imperfections in the glass, since these imperfections have a lower surface tension, which allows formation of the bubble. A high-quality sparkling wine should have a persistent effervescence that is not tumultuous, and it should contain small bubbles. The presence of microcrystals of bitartrate, however, causes a tumultuous release of foam and thus reduces the quality of the wine. The presence of surfactants in the wine is also a very important factor, having a major effect on the quality of the foam and therefore on the quality of a sparkling wine.

Currently, the quality of sparkling wine is linked to the formation and persistence of foam. Both are measured with a Mosalux, developed by the Station Oenotechnique de Champagne in collaboration with the enology faculty of the University of Reims. The device allows measurement of the maximum height reached by the foam, which generally approximates to the foaming capacity of the wine, and the stability of the foam, which generally approximates to its persistence.

All of these characteristics are related to the presence of surfactants, particularly proteins in the case of wine. The problem is that the proteins are also responsible for precipitation, which can be minimized through the use of bentonite. However, the use of bentonite causes loss of aroma due to adsorption, and in the case of sparkling wines, a significant reduction in the quality of the foam. The dilemma that this presents is whether to produce a stable but only weakly sparkling wine, or a sparkling wine with a higher degree of haze in the bottle. One of the solutions that has been proposed is to use mannoproteins derived from the cell walls of yeast, since these confer protein stability and improved foaming characteristics on the wines. Another solution is to use a combination of bentonite and alginates as tirage adjuvants.

Enological Chemistry. DOI: 10.1016/B978-0-12-388438-1.00019-4

1. INTRODUCTION

When the alcoholic fermentation of must is complete, and after leaving some time for particles still suspended in the wine to settle out, the resulting wine is drawn off. This operation, known as racking, consists of separating the wine from the solid residue, also known as sediment or fermentation lees, deposited at the bottom of the tank. Dried fermentation lees account for between 1% and 2% of the total weight of the must and have a variable biological and chemical composition. The most noteworthy biological particles are grape tissue fragments, yeasts, and certain bacteria. Chemical particles include potassium and calcium tartrates, iron phosphate, copper sulfide, protein-tannin complexes, pectic substances, and coloring matter.

Most of the material that forms the sediment in recently fermented must, or in wine that has been stored for some time, originates from previously suspended haze-forming particles. Indeed, for a particle to cause haze, it needs to be suspended, in a stable form, in the medium for some time. Particles that settle out quickly do not cause haze. Consequently, the first criterion used to classify haze-causing substances is the rate at which particles in suspension settle out, and slow-settling particles are those ones that cause haze.

Of the components of fermentation lees, chemical particles make the greatest contribution to haze as plant particles and microorganisms have a higher sedimentation rate.

There are two types of haze-forming chemical particles:

1. Particles formed by organic compounds
   a. Crystalline precipitates formed by potassium bitartrate and by tartaric acid and mucic acid calcium salts
   b. Amorphous precipitates, formed mainly by:
      Proteins associated with tannins and/or heavy metals
      Coloring matter condensates
2. Particles formed by inorganic compounds
   a. Iron combined with the phosphate ion (white casse), tannins (blue casse), or coloring matter (black casse)
   b. Copper combined with the sulfide ion (copper casse)
   c. Tin combined with proteins

All of these substances cause haze and are considered to be colloids because they form large aggregates.

2. ISOLATION AND CLASSIFICATION OF COLLOIDS IN WINE

Several methods exist for destabilizing (or separating) colloids in must or wine, but before proceeding with a complex destabilization technique, it is advisable to first obtain a mixture of purified colloids by acting on the properties that stabilize these particles. Figure 19.1 shows a simple process for destabilizing and purifying colloids in wine.

Once the purified colloidal precipitate has been obtained, pectins can be separated using the procedure shown in Figure 19.2.

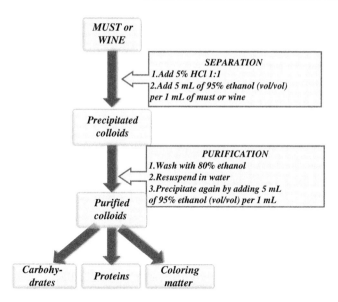

FIGURE 19.1 Destabilization of colloids in must or wine.

The colloidal particles present in wine may have been present in the must or may have been generated during the winemaking process. They can be classified as:

- Colloidal carbohydrates
- Colloidal coloring matter
- Colloidal proteins

Both colloidal carbohydrates (also known as protective colloids) and coloring matter are mainly derived from grapes, while colloidal proteins originate from yeast autolysis.

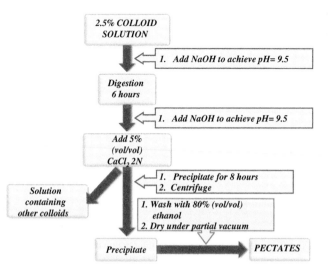

FIGURE 19.2   Destabilization of pectins in must or wine.

2.1. Colloidal Carbohydrates

The most abundant carbohydrate macromolecules in must and wine are acidic and neutral polysaccharides. Acidic polysaccharides, or pectic substances, can, in turn be divided into three categories: pectins, pectic acids, and protopectins.

Acid Polysaccharides

PECTINS

Pectins account for between 40 and 50% of all colloids found in must made from recently pressed grapes. They are heteropolysaccharides formed by galacturonic acid molecules, mostly esterified with methanol, linked by α-1,4 glycosidic bonds.

The galacturonic acid chains in pectins are covalently linked, via α-1,2 bonds, to pentose molecules such as L-arabinose and hexose molecules such as L-rhamnose (6-deoxy-L-mannose) and D-galactose.

Galactans, which form part of the backbone of pectin, are polymers formed by 120 units of D-galactose with β-1,4 bonds.

Pectins are soluble in water but not in alcohol. Chemical hydrolysis produces galacturonic acid (65–95%), methanol (3–8%), acetic acid (0–6%), and monosaccharides (8–10%) such as L-arabinose, L-rhamnose, D-galactose, and D-xylose.

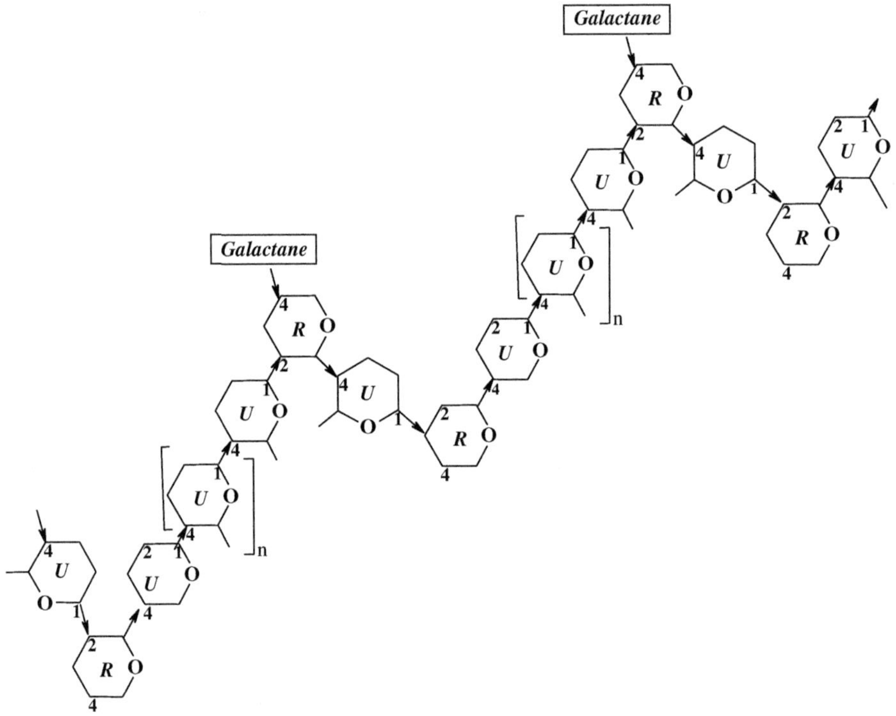

FIGURE 19.3 **Structure of a pectin**. U = galacturonic acid; R = rhamnose. For the pectin structure 4 < n < 10. Galactan = 120 units of D-galactose.

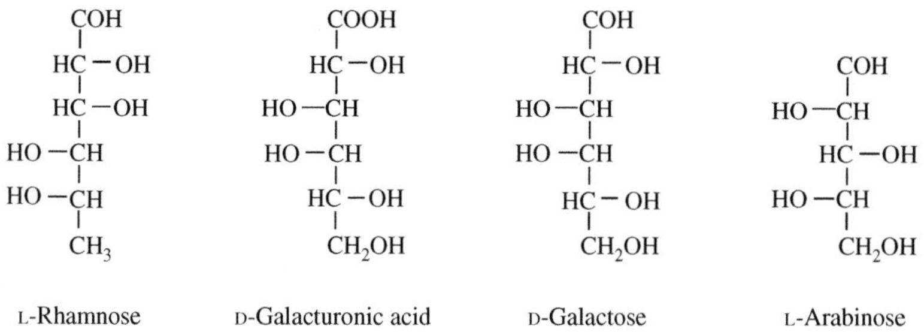

COH	COOH	COH	
HC—OH	HC—OH	HC—OH	COH
HC—OH	HO—CH	HO—CH	HO—CH
HO—CH	HO—CH	HO—CH	HC—OH
HO—CH	HC—OH	HC—OH	HO—CH
CH₃	CH₂OH	CH₂OH	CH₂OH
L-Rhamnose	D-Galacturonic acid	D-Galactose	L-Arabinose

FIGURE 19.4   Structures of colloid-forming monosaccharides.

PECTIC ACIDS

Pectic acids are chains of non-esterified galacturonic acid. They are insoluble in water.

PROTOPECTINS

Protopectins are found in grapes, but their hydrolysis in an acid medium gives rise to pectic substances. They have a highly complex molecular structure and are thought to be composed of networks of galacturonic acid units and plant components such as hemicelluloses and cellulose, linked together by bonds between carboxyl groups and calcium ions, and also hydrogen bonding between −OH groups.

> Hydrolysis
> Protopectin → Pectins + Pectic acids + Hemicelluloses + Celluloses

Protopectins are preferentially located in the cell wall and lamellae of grapes, where they contribute to the consistency of the berry. After ripening, they become soluble and the berry becomes soft.

Neutral Polysaccharides

Neutral polysaccharides are exclusively composed of simple sugar molecules. They are termed homopolysaccharides or heteropolysaccharides depending on whether they have one or several types of monosaccharides.

HOMOPOLYSACCHARIDES

Homopolysaccharides are formed by a single type of monosaccharide. Some neutral polysaccharides, such as cellulose, originate from the cell walls of the grape. Cellulose is formed by β-1,4-linked D-glucose molecules.

Other homopolysaccharides originate in infected grapes and wines. The most noteworthy polysaccharides of fungal origin are glucans, and in particular β-glucan, which is produced

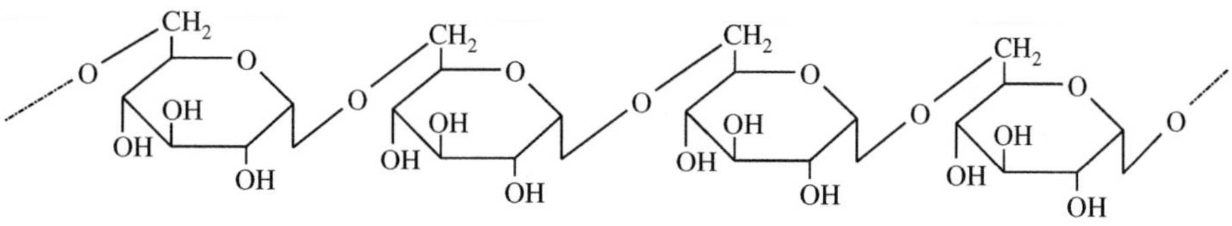

FIGURE 19.5   Structure of dextran.

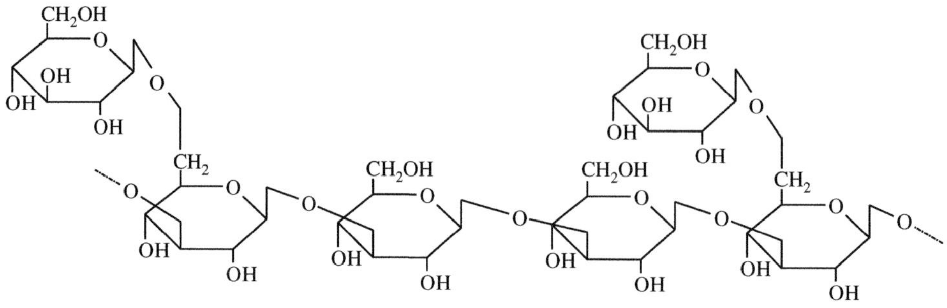

FIGURE 19.6   Structure of β-glucan produced by *Botrytis cinerea*.

by *Botrytis cinerea*. For many years, β-glucan was confused with dextran, which is formed by α-1,6-linked glucose molecules. β-glucan, in contrast, consists of a backbone of glucose molecules linked by β-1,3-bonds, with chains of β-1,6-linked glucose molecules branching off.

The presence of β-glucan causes serious filtration problems. The clogging behavior of wine containing glucan depends on numerous factors, including ethanol content, temperature, and the methods used to obtain the must, as all of these influence the glucan content of must and wine. Ethanol increases the size of the colloidal particles, explaining why glucan-containing wines cause increasing clogging as the ethanol content increases. Glucan, however, begins to precipitate at ethanol strengths of 17% (vol/vol), with complete precipitation occurring at 23% (vol/vol). Low filtration temperatures (4°C or lower) favor the growth of the polysaccharide chain and its subsequent flocculation, and therefore minimize filtration problems. At ambient temperatures and above (30–40°C), the colloidal particles decrease in size and rapidly cause clogging.

Because β-glucan is synthesized by *B. cinerea* in the grape (between the flesh and the skin), aggressive harvesting operations (such as crushing, pumping over, etc.) produce musts with high concentrations of this polysaccharide. When infected grapes are crushed gently, however, the must will have a minimum content of glucan and the resulting wine will be easy to filter.

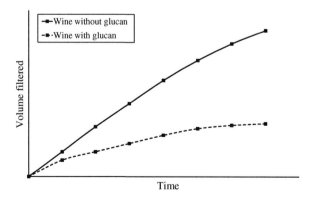

FIGURE 19.7   Filtration of wines made from grapes with and without *Botrytis cinerea* infection.

TABLE 19.1   Polysaccharides Composed of a Single Monosaccharide

	Monomers	Polysaccharide
Pentoses	Arabinose	Araban
	Xylose	Xylosan
Hexoses	Galactose	Galactan
	Fructose	Fructosan
	Glucose	Glucans, Dextrans
	Rhamnose	Rhamnan

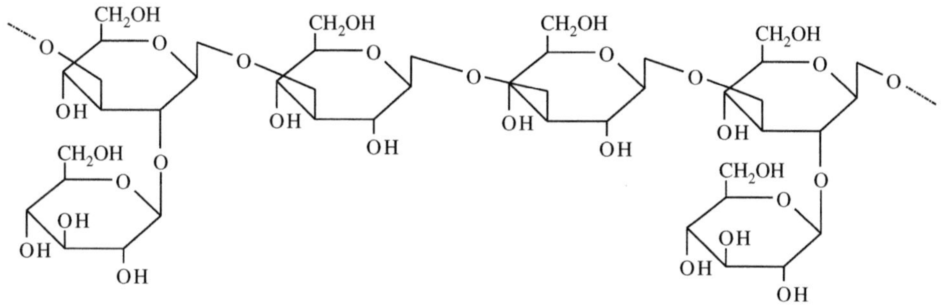

FIGURE 19.8   **Structure of the glucan responsible for ropiness.**

Filtration problems can be resolved by adding glucanase, an enzyme obtained from *Trichoderma* cultures, as certain *Trichoderma* species are natural enemies of the botrytis fungus. The enzyme starts by breaking the polysaccharide chain at its non-reducing end, producing glucose or gentiobiose, depending on whether or not the non-reducing end is linked to another glucose molecule via a 1,6 bond. Gentiobiose is a disaccharide formed by two glucose molecules linked by β-1,6 bonds. Commercial preparations contain both glucanase and glucosidase, and their use is authorized in the European Union.

Using traces of glucose, certain lactic acid bacteria can produce a polysaccharide that gives wine a viscous, oily appearance. This condition is known as ropiness. The polysaccharide in question has a backbone formed by β-1,3-glucose bonds with β-1,2-linked glucose molecules branching off. Although the polysaccharide is very similar to that produced by *B. cinerea*, glucanase enzymes are not useful in this case, possibly because their action is blocked by steric hindrance caused by the branches.

Wines affected by this bacterial disease need to be racked off and stirred vigorously, and subsequently passed through filters with an intermediate pore size to retain the colloidal particles. Finally, they should be sterilized to eliminate the causative bacteria.

HETEROPOLYSACCHARIDES

The family of neutral polysaccharides comprised of several types of monosaccharides is almost exclusively made up of gums. These gums are formed by pentoses (arabinose and rhamnose) and hexoses (galactose and mannose). Although glucose and fructose are the most abundant hexoses in must, they are not usually present in gums.

Hemicellulose is a heteropolysaccharide composed of xylose and arabinose which, together with cellulose, forms part of the cell walls of plants (and therefore of grapes).

Ropiness can also be caused by certain streptococcal fungi. They are capable of converting malic acid to lactic acid and of secreting highly viscous heteropolysaccharides formed by arabinose, mannose, galactose, and galacturonic acid units.

2.2. Colloidal Coloring Matter

Most of the coloring matter in wine is formed by phenolic compounds from grapes, in particular, anthocyans and flavanols as these are capable of forming colloids.

Flavanols undergo polymerization reactions, giving rise to what are known as condensed tannins, which are well ordered polymers that are highly reactive with proteins and increase the tannic character of the wine.

When flavanols are polymerized through acetaldehyde bridges, or condensed with anthocyans, they yield molecules that react less readily with proteins due to their bulkier structure. Furthermore, condensation with anthocyans stabilizes the color of wine.

Tannins are not present in grapes. Rather, they are formed in wine from highly labile phenolic compounds such as catechins. Flavanol polymerization typically occurs in the aging of both red and white wine. The reactions involved cause an increase in the average molecular weight of the coloring compounds. For example, the molecular weight of these compounds in wines that have undergone aging is three to four times higher than in young wines.

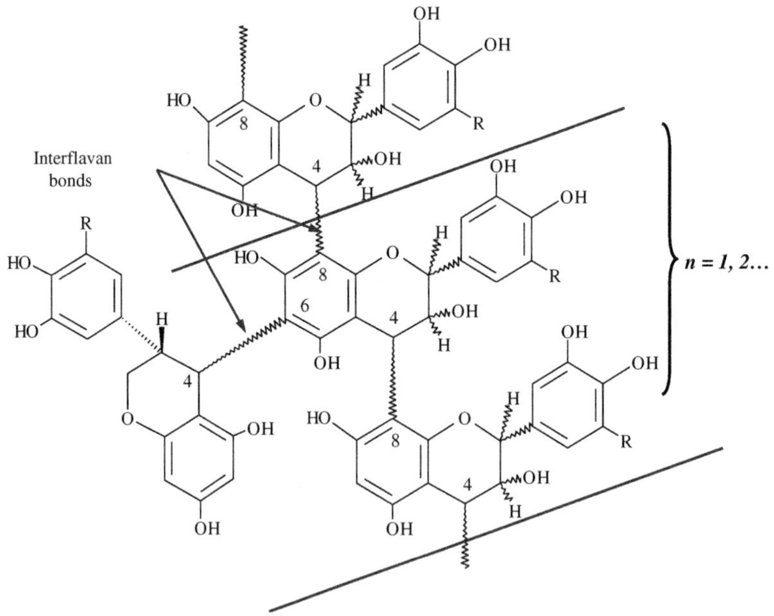

FIGURE 19.9   Cleavage of condensed tannins. (Procyanidin: R=H; Prodelphinidin R=OH).

TABLE 19.2 Average Molecular Weights and Other Properties of Coloring Matter

Fraction/Composition	Molar Mass (Da)	Color	Stability
Pure anthocyans (free anthocyans)	500	Violet red	Not very stable
Pure flavanols (catechins, leucoanthocyans)	600	Yellow	Stable
Flavanol + anthocyan	1000–2000	Red orange	Stable
Anthocyan and certain tannins. Flavanols + anthocyans + salts Tannin-anthocyan combinations	1000–2000	Dark red	Stable red color
Highly condensed tannins + anthocyans	2000–5000	Red	Not very stable
Tannin-polysaccharide combination Flavanols + anthocyans + salts	> 5000		Not very stable

Tannins have numerous properties of interest to winemakers, in particular:

1. Degree of polymerization. The degree of polymerization of tannins increases with aging. In aged wine, for example, these molecules have a molecular weight of 3000 to 5000 Da compared to just 500 to 600 Da in the case of new wines. When the polymer particles are sufficiently large, they form a negatively charged colloid, which interacts with positively charged colloids such as proteins and colloidal iron(III), causing flocculation.
2. Flocculation through combination with proteins. The most remarkable property of tannins is their ability to flocculate. However, excessively large polymers, that is, those with a molecular weight of over 3000 Da, will not interact with proteins to cause flocculation due to steric hindrance. This is because they are unable to gain sufficient access to the active sites of the proteins. This phenomenon is related to astringency and bitterness. Aged wines, for example, are less astringent than young wines because they contain tannins which are still in their polymerized form.
3. Reaction with coloring matter. Tannin condensation during the aging of red wine occurs at the same time as the wine acquires brick-red hues.
4. Redox phenomena. Condensed tannins are intermediate oxidants with antioxidant properties that protect coloring matter from the action of oxygen.
5. Blue casse. Condensed tannins combine with the iron in wine to yield a blue colloid.

2.3. Protein Colloids

Proteins are amino acids linked by peptide bonds formed between the acid group of one amino acid and the amine group of the next one. At the normal pH of wine, they exist as positively charged colloids.

They are responsible for protein casse, which is the haze that forms in white and rosé wines stored at a high temperature. Reactions between the proteins in wine and tannins from the cork can also cause this type of protein casse. It is less common in red wines because the proteins precipitate out with polyphenols during the winemaking process.

Proteins derived from must are thermally unstable and can also cause protein casse in poor storage conditions. Proteins released during the autolysis of yeast, in contrast, are thermally stable. Within this group, mannoproteins provide protein stability.

The most common method used to eliminate excess proteins responsible for protein casse is treatment with bentonite (a negatively charged colloid). Compared to heat treatment, it offers a higher degree of stabilization.

3. NATURAL CLARIFICATION AND STABILIZATION

Recently obtained must contains solid grape residues, which rapidly sink to the bottom of the fermentation tank during prefermentation treatment. Other suspended particles of chemical origin and colloidal size, such as proteins and phenolic compounds and crystalline compounds, as well as microorganisms such as yeast, settle out more slowly, both during and after alcoholic fermentation.

The natural clarification that occurs in wine after alcoholic fermentation is the result of natural sedimentation, mostly involving yeasts, protein-phenol combinations, and potassium bitartrate. This separation of the solid phase stabilizes the wine once it has been racked off the lees; although the stabilization effect is incomplete, the wine is protected against physical factors such as temperature as natural clarification occurs in the winter months following the harvest.

The rate of clarification varies depending on the size and density of the particles that make up the solid fractions that settle out of the wine. The time that a wine takes to clarify naturally, therefore, depends on its composition. It is quite short in red wine but takes longer, and is often incomplete, in white wine.

For sedimentation to occur, the density of the particles must be higher than that of the wine, which ranges from 0.992 for dry wines to 1.050 for sweet wines. It is also important that the particles are not excessively small as Brownian motion counteracts the sedimentation process. These factors are important in wines that are not stirred, that is, wines that are left to rest completely. Nonetheless, it should be noted that different temperatures throughout the mass of liquid cause different local densities, and, accordingly, convection movements which slowly agitate the wine and interfere with sedimentation. These movements are greater in containers that are good conductors of heat, such as stainless steel tanks.

It is also important to note that the presence of considerable quantities of protective colloids, which limit flocculation, reduces the rate of sedimentation, explaining why wines made from grapes containing large quantities of polysaccharides can take years to clarify naturally.

Contact with air is also considered to favor clarification. For example, the upper layers of young non-sulfite white wines clarify better than the lower layers because oxygen affects the formation of iron(III), which has flocculation ability. Tannins from wood also favor clarification.

Once the desired level of clarification has been achieved, the wine should be devatted. This must be done carefully so as not to disturb the lees at the bottom of the tank. Leaving the wine in contact with the fermentation lees for too long during natural clarification can cause serious problems. Such wines always develop strange odors and flavors as the result of anaerobic processes that give rise to hydrogen sulfide, mercaptans, etc. While these odors can be eliminated by gentle aeration, undesirable flavors are more difficult to eliminate.

Once the clean wine has been racked off, a sufficient quantity of sulfites should be added to protect the wine during storage.

It should be noted, however, that natural clarification is slow and often does not produce the desired level of stability. Accordingly, it is more common to use artificial treatments that offer guarantees of stability and take into account the target market. When the aim is to achieve microbiological stability, it is necessary to perform sterilization operations such as filtration or heat treatments, as the natural clarification of wines left to rest never completely eliminates all the bacteria and yeast present. It is generally desirable to choose a treatment, whether physical or chemical, that leads to rapid, lasting stabilization that will protect the wine from spoilage once sold.

4. CLARIFICATION WITH PROTEINS: FINING

The term *fining* refers to the process of adding substances to a wine to improve its quality. In this section, we will look at protein fining, which is the addition of proteins to trigger the flocculation of particles that cause or are capable of causing wine haze.

Most of the proteins used for this purpose carry a positive charge at the pH of wine and are therefore considered to be positively charged hydrophilic colloids. They interact with colloidal particles that carry a negative charge at the pH of wine, such as phenolic compounds and suspended particles such as copper sulfide and iron phosphate.

The coupling of two or more colloidal particles causes these particles to flocculate, for two primary reasons:

1. The formation of large particles with a zero charge that readily precipitate (e.g., coupling of protein and copper sulfide particles).
2. The formation of electrically charged particles that precipitate in the presence of salts (e.g., coupling of protein and tannin particles).

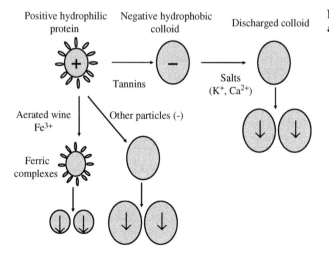

FIGURE 19.10   **Action of protein fining agents.**

The outcome of the addition of protein fining agents depends on the electrical charge of the particle formed by the interaction between the protein and its negatively charged colloidal counterpart. The stronger the discharge, the more efficient the flocculation − and clarification − will be as the aim of fining is to act on the mechanisms of attraction, repulsion, dehydration and hydration, which are responsible for the stability of colloidal systems.

Common agents used in this type of fining are gelatin, casein, egg albumin, blood albumin, and fish glue (also known as isinglass).

4.1. Gelatin

Gelatin is obtained from the acidic hydrolysis of animal products such as cartilage and collagen from skin and bones. The main components of these products are the amino acids glycine, proline, hydroproxyline, and glutaminic acid. The use of gelatin as a wine fining agent is very recent, and the wine industry accounts for between 1% and 5% of the total food market for gelatin.

Gelatin can be classified according to several criteria, including Bloom strength (gelling ability), solubility, and preparation (enzymatic or non-enzymatic hydrolysis), but the most useful classification system for winemaking is that described in the International Oenological Codex, which consists of three categories.

1. Heat-soluble gelatins: molecular weight of over 10^5 Da, protein content of 30 to 50%, high charge (0.5–1.2 meq (+)/g).
2. Liquid gelatins: extensive chemical hydrolysis, relatively high molecular mass ($<10^5$ Da), low charge, and abundance of strongly charged peptides.
3. Cold-soluble gelatins: enzyme hydrolysis, low charge, and few peptides and proteins with a molecular weight of $<10^5$ Da.

Commercially, gelatin is often characterized by a tannin precipitation index, which reflects the grams of tannin that will precipitate for every gram of gelatin. This method, however, is not very useful for measuring the effectiveness of gelatin as it does not take into account the structural diversity of tannins, or the fact that the tannin-protein reaction is not stoichiometric.

To predict and interpret the effect of a gelatin, it is more accurate to use a system based on the molecular mass and the charge of the proteins that form it. It is known that extensively hydrolyzed gelatins bind more readily to polyphenols and are therefore more suited to wines with a high polyphenol content. Less extensively hydrolyzed gelatins should be used in wines with a lower polyphenol content as they will only bond with the more reactive tannins. The use of a highly reactive (slightly polymerized and strongly charged) gelatin will give rise to a wine with a poor mouthfeel and also possibly cause overfining.

Gelatin used in winemaking is soluble in water, but it needs to be swelled before use. The dose for white wines ranges from 3 to 5 g/hL while that for red wines ranges from 8 to 15 g/hL. Gelatin is particularly recommended for the clarification of red wines because it contains a higher concentration of tannins, precipitation is more complete and the risk of overfining is reduced. Casein is more commonly used in white wines, but gelatin combined with bentonite or colloidal silica is also used to prevent overfining.

4.2. Casein

Casein is a phosphorus-containing heteroprotein. It is mainly found as a colloidal dispersion in milk in the form of caseinate-calcium phosphate. Casein is extracted from milk by precipitation in an acid medium (HCl, H_2SO_4, or CH_3COOH) or by enzyme precipitation (rennet casein). It is insoluble in water but soluble in a basic medium. Its potassium salt, however, potassium caseinate, is more widely used as it has the advantage that it is soluble in water.

Because casein precipitates instantaneously at acidic pHs, its use is recommended when the aim is to prevent overfining. The treatment, however, is a delicate procedure as casein coagulates rapidly and can therefore flocculate before the desired effect has been achieved throughout the liquid mass. It is therefore best to add casein to the medium using a dosing pump or a Venturi system.

Commercially available casein comes in the form of an amorphous, white-yellow powder. It is normally odorless although it can smell of cheese. It is particularly suited for the clarification of white wines. Recommended doses range from 50 to 100 g/hL for musts and 20 to 100 g/hL for wines.

Casein acts on polyphenols, preferentially eliminating flavonoid polyphenols such as catechins and procyanidins, which are the main components responsible for woody flavors and the browning of white wine. It also acts by adsorbing the oxidized, unstable polyphenol fraction. Casein also partly eliminates iron in white wines which are prone to iron casse, as it reacts with iron(III). It also reacts with iron in the ferrous state, but to a lesser degree.

The use of whole milk for fining purposes is not permitted in the European Union, although it can be highly effective in terms of improving the color of white wines and eliminating the sulfur flavors associated with reduction and the taste of mold. Skimmed milk is less efficient as an adsorbent but it is a superior clarifying agent.

4.3. Albumin

The albumin used in wine is derived from egg white or blood.

Egg-Derived Albumin

The use of egg white as a clarifying agent is undoubtedly the oldest known stabilizing agent used in wines and it has been associated with the production of great wines.

Egg albumin contains a range of proteins but the main ones are ovalbumin and ovoglobulin. It accounts for 12.5% of the fresh weight of egg white. Ovalbumin is soluble in cold water, unlike ovoglobulin, which will, however, dissolve in a diluted saline solution. This explains why winemakers often add a small quantity of salt (sodium chloride) to the water before adding egg white (to benefit from the properties of globulin).

Albumin can be used in a variety of forms (fresh, frozen flakes, or white or yellow powder). The recommended dose is 5 to 15 g/hL. When fresh egg white is used, three to eight egg whites are recommended for each 225-liter barrel. Fresh egg white has the advantage that is does not alter the flavor of wine.

The flocculation and precipitation of albumin forms a compact deposit. Its use is recommended in red wines with a high tannin concentration as it softens the characteristic astringent taste of tannins. It should, however, be used with caution in smooth red wines. Its use in white wines is not recommended as albumin requires large quantities of tannin to precipitate. It has, however, been traditionally used to clarify sherry and similar wines.

Theoretically, albumin has bactericidal properties, as egg white contains lysozyme, which triggers the lysis of lactic acid bacteria but has no effect on acetic acid bacteria or yeast. Nonetheless, studies have not succeeded in demonstrating the bactericidal effects of egg white, probably because lysozyme flocculates in the presence of albumin.

Blood Albumin

Fresh blood contains, on average, 76.8 to 83.9% of water, 2.6 to 8.1% of albumin, 0.2 to 0.6% of fibrin, 0.1 to 0.3% of fat, and 0.7 to 1.3% of mineral salts. It has a density of between 1.045 and 1.075. Fresh blood, generally from cattle, was used for a long time to eliminate excess tannins from wine, but its use is now prohibited.

Nonetheless, in certain wine-producing regions, the use of certain blood derivatives, namely defibrinated blood and powdered blood, is permitted. The dose used in the case of defibrinated blood ranges from 150 to 200 mL/hL, which corresponds to 10 to 15 g of active protein. Its use, however, is not widely recommended due to storage problems. Powdered blood is soluble in water, as is blood albumin, but the addition of a small quantity of sodium bicarbonate is recommended in both cases. The dose should be adjusted to between 10 and 20 g/hL depending on the tannin levels.

Blood derivatives are mainly used in young red wines and white wines that are wooded or have excessive vegetative aromas as they react rapidly with tannins, forming large flocculates that precipitate readily and are not very sensitive to the action of protective colloids. Their use in young red wines with a firm tannic structure softens the bitter, vegetative aromas associated with these wines. In such cases, the recommended dose is between 10 and 20 g/hL. The effect in white wines is also a softening of vegetative aromas, and the recommended dose is 5 to 10 g/hL.

Liquid blood, unlike powdered blood, is a powerful decolorizing agent. pH does not influence flocculation and when used within the recommended doses, liquid blood does not cause overfining. It is also a powerful clarifying agent and its action does not depend on tannin content, although the addition of tannins is recommended to accelerate flocculation. The use of blood or blood derivatives is recommended in wines that are difficult to clarify using other methods. The main disadvantage of these materials is that they can impart flavor to the wine and increase the concentration of soluble proteins. They are therefore normally used in association with another colloid, such as bentonite, to aid flocculation.

4.4. Fish Glue

Fish glue, together with milk and egg white, has been used as a clarifying agent since the 18th century. It is obtained from the swim bladder of fish (mostly sturgeon and similar species). The process by which the glue is obtained is very laborious, but commercial preparations (which need to be previously hydrated) are now available.

The recommended dose is 1 to 2 g/hL for white wines and up to 5 g/hL for red wines. The main advantages of fish glue are that its action is only marginally affected by protective colloids and that it flocculates slowly (in 2 to 3 days). It has a smaller influence on astringency than either gelatin or casein as it is less effective against condensed tannins. Several authors have indicated that fish glue is the best clarifying agent for white wine as it achieves unequalled clarity and brilliance in wines with few suspended particles. A disadvantage, however, is that careful racking off is required following its use as it produces a considerable amount of lees.

4.5. Factors Which Affect Fining

Cations

Iron(III) is the most active cation in wine and has the greatest impact on clarification. It combines with tannins to form a negatively charged tannin-iron complex that interacts rapidly with gelatin and causes its flocculation. A dose of just 2 mg/L is required for this to occur.

pH

The pH of must or wine is primarily determined by the concentration of tartaric acid as this is generally the strongest acid found in these media. There are, however, other substances with acidic properties, whose degree of dissociation is determined by pH. One of these substances is tannic acid, whose dissociation is shown in the following equilibrium:

$$\text{Tannic acid} \leftrightarrow \text{Tannate}^- + H^+$$

In wine, the active tannin against flocculation is that which carries a negative charge. Because its proportion decreases with a decrease in pH, it may not achieve complete fining. This effect is clearly seen at pH values of less than 3.2 and it contrasts with what occurs with protein fining agents, whose activity increases as the pH moves further away from their isoelectric point.

Protective Colloids

The presence of certain polysaccharides (dextran, glucan, etc.) complicates both filtration and protein fining. The influence of these protective polysaccharides depends on the clarifying agent used, and is greatest for gelatin, followed by albumin. Casein and fish glue, in contrast, are only minimally affected by the presence of protective colloids.

Temperature

Temperature influences the flocculation efficiency because it has a direct impact on the Brownian motion of colloid particles. Accordingly, low temperatures favor the precipitation of colloids. Gelatin is the protein that is most affected by temperature. At temperatures of between 25 and 30°C, flocculation is difficult and the clumps formed are not very compact.

In summary, before choosing the most suitable clarification method for a given wine, the following factors should be considered:

1. Clarification is complicated in wines with a low pH as they contain a lower proportion of active tannins.
2. The aeration of wine facilitates the oxidation of ferrous iron to ferric iron, thereby facilitating flocculation and clarification. It also eliminates part of the dissolved CO_2, which complicates flocculation.

3. Low temperatures favor flocculation because they reduce Brownian motion and favor the dissolution of oxygen (greater conversion of ferrous iron to ferric iron). Caution should, however, be exercised, because in such cases there will also be a greater oxidation of phenolic compounds, which can cause considerable color loss in the case of red wines or unwanted browning in the case of white wines.
4. Protective colloids complicate clarification, and on occasions, it is desirable to use pectolytic enzymes.
5. The methods used to mix and add the fining agent to the wine are important as, for clarification to be effective, the agent must be evenly distributed throughout the liquid mass. Casein is particularly problematic because it coagulates rapidly. It is therefore necessary to use a dispensing system that ensures its rapid dispersion through the medium.
6. Laboratory tests are necessary before using any protein fining agent as there is no such thing as a universal agent.

4.6. Overfining

Overfining occurs when a considerable proportion of the protein fining agent remains in the wine due to incomplete flocculation. It should not be confused with inadequate fining (as may occur in the presence of protective colloids), which results in permanent haze. Overfined wines, in contrast, are clear, but they develop haze if the tannin content of the wine increases. This can occur if:

1. The wine is blended with another wine with a higher tannin content.
2. Wine tannin is added.
3. The wine is aged in barrels (transfer of tannins from wood to wine).
4. The wine comes into contact with cork (transfer of tannins from cork to wine).

Overfining mostly affects white wines as these contain few tannins. This is why white wine should not be clarified with fining agents such as gelatin or egg albumin, as these need large quantities of tannin to flocculate. When there are no alternatives, it is important to conduct prior tests to determine the lowest possible dose. One solution proposed to prevent overfining is to add wine tannin before the protein fining agent. This, however, can increase the harshness of white wine and reduce its quality. It is therefore preferable to use casein, blood albumin, or fish glue.

Possible treatments to counter overfining include:

1. Addition of tannin to trigger the flocculation of the excess protein particles. This method has a less negative effect on red wine than on white wine.
2. Addition of a negative colloid such as bentonite or colloidal silica.
3. Filtration at a low temperature through a diatomaceous earth filter.

5. THE POLYPHENOL-PROTEIN COMPLEX

Protein-polyphenol complexes are mainly the result of the formation of hydrogen bonds and van der Waals forces between polyphenols and nonpolar regions of a protein molecule.

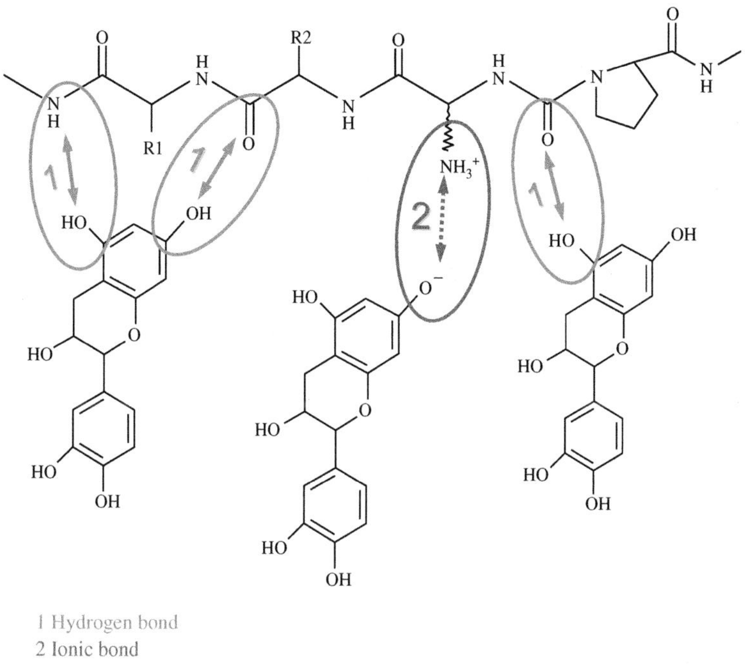

FIGURE 19.11 Most common protein-phenol interactions.

1 Hydrogen bond
2 Ionic bond

While ionic or covalent interactions can occur, they are not common. The true cause of the protein-polyphenol complex is the formation of van der Waals interactions, strengthened by hydrogen bonds between the carbonyl group of proline monomers and the hydroxyl groups of phenols. The interactions are surface interactions, meaning that complex formation is reversible, provided no covalent bonds have been formed.

When protein levels are low compared to the polyphenols, the latter form bonds at the active sites on the surface of the protein. This forms a simple layer, which primarily causes a reduction in the electrical charge of the protein and an increase in its molecular mass. These particles then flocculate as they join together. When proteins outnumber polyphenols, the same phenomenon occurs, but in this case, there is cross-bonding between different protein molecules, with phenol serving as the bridging molecule. This decreases the hydrophilic character of the protein and increases the mass of the resulting complex, which eventually flocculates.

5.1. Characteristics of Fining Proteins

Wines are treated with proteins to eliminate a proportion of the phenolic compounds present or unwanted coloring matter. Not all proteins, however, form complexes capable of flocculating with polyphenols. Amino acid composition, spatial configuration, size, and charge all have a considerable influence on the interaction between proteins and polyphenols.

Only proline-containing peptides and proteins form stable complexes with phenolic compounds. Synthetic peptides and proteins without proline do not. This suggests that

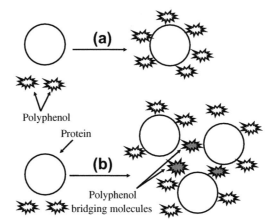

FIGURE 19.12   Formation of a protein-polyphenol complex with a high (a) and low (b) concentration of protein.

proline is a necessary component at binding sites of polyphenols and proteins, probably because proteins cannot adopt the α-helix configuration at the proline-containing site. In other words, the protein at this site would remain exposed and therefore accessible to polyphenols.

Gelatin is by far the most widely used protein in winemaking and it contains a high proportion of proline (12–14%), hydroxyproline (12–14%), and glycine (32%). Casein and potassium caseinate, in turn, contain approximately 12% proline, 2% glycine, and negligible quantities of hydroxyproline.

Small, compact proteins have little affinity for wine polyphenols and are therefore not very useful as clarifying agents.

The charge on a protein is related to its isoelectric point. At the pH of wine, proteins have a positive charge density, which increases as the difference between the isoelectric point and the pH of the medium increases.

5.2. Characteristics of Polyphenols

The strength of the interaction between polyphenols and proteins depends on their degree of polymerization, structure, and charge. Interaction increases with the degree of polymerization and the concentration of molecules esterified with gallic acid. At a given pH, the charge on the molecules also depends on the degree of polymerization, which increases with increasing size. pH also has a considerable influence on the charge of polyphenols, such that the higher the pH, the higher the negative charge.

5.3. Factors That Influence the Phenol-Protein Interaction

1. Presence of cations. While sodium, potassium, calcium, and magnesium cations all favor flocculation, the iron(III) cation exerts the greatest effect as it forms a complex with tannins and flocculates with proteins.

2. Temperature. Temperature plays a very important role in fining. The higher the temperature, the more difficult it is to achieve flocculation and clarification. Gelatin is more sensitive to temperature than either casein or albumin.

3. Acidity. pH has a considerable influence on interaction and has an opposite effect on proteins and tannins. Proteins have a higher charge as pH decreases, unlike tannins, which have a lower charge. Nevertheless, within the pH range of wine, the phenol-protein interaction increases with increasing pH. For example, for the same amount of protein, twice as much tannin is clarified at a pH of 3.9 as at a pH of 3.4.

4. Ethanol. The haze caused by contact between a protein and a polyphenol such as a tannin or a procyanidin is inhibited in a nonpolar solvent, or in the presence of molecules with hydrogen acceptor groups. Because ethanol is less polar than water, it causes less haze, although the difference is negligible for ethanol strengths of between 11 and 13% (vol/vol).

5. Time. All proteins need time to form complexes with polyphenols and to flocculate. In the case of tannins, at least 8 hours is necessary.

6. Influence of polysaccharides. The addition of gum arabic to wine hinders protein flocculation, and consequently, clarification. Gum arabic has a greater effect on gelatin than on either casein or albumin. Wine also contains grape-derived polysaccharides (such as gums and mucilages) that act as protective colloids. Other types of polysaccharides, which are derived from grape skin walls (pectins, arabinogalactans, and polygalacturonic acid), however, react strongly with tannins and favor precipitation.

7. Concentration of proteins and tannins. It is generally accepted that, for the same quantity of protein, the quantity of tannin that forms complexes increases with tannin concentration. Also, as more protein is added to the medium, more protein-tannin complexes will be formed. The reaction, however, is not stoichiometric as tannins can act as bridging molecules between proteins in the presence of large quantities of protein. It is important to perform tests before adding a protein fining agent to determine optimal doses and prevent overfining.

6. PROTEIN CASSE

Protein instability in white wines causes clarity problems known as protein casse. This consists of the formation of haze in bottled wine. It generally occurs when the wine is stored at a high temperature or when it is enriched with tannins from cork or wood. Protein casse is not common in red wines as the proteins flocculate with tannins during alcoholic fermentation.

The protein content of wine is the same as that of must as yeasts do not assimilate proteins during fermentation. Must proteins are thermally unstable, but there is no direct relationship between the concentration of proteins in a must and the haze caused by heat. This indicates that not all proteins are equally affected by heat, although, to a greater or lesser degree, they all cause haze.

6.1. Clarity and Protein Stability

When a wine is affected by significant haze, this will be visible to the naked eye. In other cases, it is necessary to use special equipment to obtain an objective measurement. An

TABLE 19.3 NTU Values for Wine

Wine	Clarity	Turbidity
White	< 1.1	> 4.4
Rosé	< 1.4	> 5.8
Red	< 2.0	> 8.0

example is a turbidimeter, which measures the intensity of light dispersed at a given angle. Another device is a nephelometer, which uses a measurement angle that is perpendicular to the incident beam of light, and the results are expressed as nephelometric turbidity units (NTU). Nephelometers are useful for measuring the efficacy of clarification or filtration treatments, or for evaluating protein stability. There is also a relationship between measured turbidity and visual appearance.

Several tests are available to determine the protein stability of wine, although the most widely used and reliable one is based on the sensitivity of proteins to heat treatment. The test consists of heating 100 mL of wine in a water bath at 80°C for 30 minutes. The wine is considered to be stable if the resulting turbidity, or haze, is less than 2 NTU. Higher values mean that the wine must be treated to protect against protein casse.

Bentonite is typically used to eliminate excess protein. In recent years, the recommended doses for stabilizing wine have increased considerably (from 40 to 120 g/hL) due to the increased use of mechanical harvesting methods and the introduction of new winemaking methods that cause a greater transfer of proteins from the must to the wine. The problem with this is that high doses of bentonite have a negative effect on aroma.

6.2. Influence of Aging on Lees on Protein Stability

Several alternatives have been proposed as a replacement for bentonite as a clarifying agent. None, however, have yielded the desired results. Tangential ultrafiltration, for example, alters the aromatic profile of wine; wine tannin only partly eliminates certain proteins and increases the wine's sensitivity to heat; and proteolytic enzymes have not produced the expected results. Nonetheless, it is known that white wines aged on fine fermentation lees (mostly formed by fermentation yeasts) acquire some stability towards protein casse. In theory, one might think that the proteins are partially adsorbed or degraded during this type of aging, but electrophoresis has shown no difference between the total concentration of proteins in a recently fermented wine and the same wine after aging. What has been observed, however, is a greater concentration of mannoproteins released by yeast.

Mannoproteins can be extracted from yeast walls using an enzyme preparation containing a protease and β-glucanase, whose use is authorized for the degradation of the glucan secreted by B. cinerea. β-Glucanase releases a mixture of mannoproteins which halve the dose of bentonite required to achieve protein stability. The purified compound is a mannoprotein with a molecular weight of 31.8 kDa (called MP32), containing 62% mannose and 27% protein. Another, more heavily glycosylated mannoprotein fraction (MP40) obtained from the same preparation, is used to stabilize wines towards tartrate precipitation.

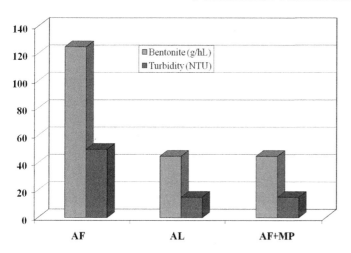

FIGURE 19.13 Turbidity test and doses of bentonite required to stabilize a wine towards protein casse. AF = after alcoholic fermentation; AL = after aging on lees; MP = exogenous mannoprotein:

Greater or lesser protein stability during aging on lees depends on several factors, including length of aging, quantity of lees, and frequency of stirring. Generally speaking, the longer the aging, the more mannoproteins that will be released into the medium, and consequently the less bentonite that will be needed to stabilize the wine. Likewise, wines in which part of the lees are eliminated before aging have a lower concentration of mannoproteins, and therefore require more bentonite for stabilization.

Wines in which the lees are stirred more frequently exhibit better protein stability as they contain more mannoproteins. Stability is also achieved more rapidly in used barrels than in new barrels.

7. CLARIFICATION WITH BENTONITE

Clays such as Lebrija clay and kaolin have been used in winemaking since ancient times. Both of these minerals have adsorbent properties due to their characteristic clay structure.

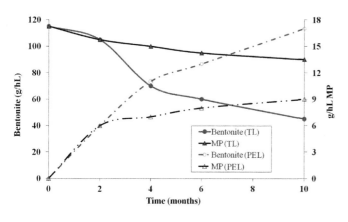

FIGURE 19.14 Concentration of mannoproteins (MP) released during aging on total lees (TL) or on partially eliminated lees (PEL) and dose of bentonite required to achieve protein stability.

The most widely used clay nowadays is bentonite, which is formed from the decomposition of volcanic ash. It was discovered in 1888, in Fort Benton (Wyoming, USA), hence its name. The most important bentonite deposits are found in Italy, Germany, northern Africa, and France.

The use of bentonite as a protein stabilizer is a universally accepted practice, but only 1% of the bentonite produced worldwide is used in winemaking as most of it is coarse and imparts unwanted flavors to the wine. Bentonite is used in other industries such as ceramics, paint, paper, cosmetics, and pharmaceutics.

7.1. Structure

Bentonite is a hydrated aluminum silicate that is a member of the clay family. It has the following formula:

$$Si_4(Al_{(2-x)} R_x)O_{10}(OH)_2(CE)_x \cdot nH_2O$$

where CE = exchangeable cations Ca^{2+}, Na^+, Mg^{2+}, and

$$R_x = Mg, Fe, Mn, Zn, Ni$$

Bentonite is a phyllosilicate that belongs to the montmorillonite group. It is composed of stacks of sheets, each of which has three layers: an octahedral layer (Oc), consisting of hydrated aluminum, sandwiched between two tetrahedral (Te) layers, formed by SiO_2. Its symbol is Te-Oc-Te (2:1). In the tetrahedral layer, several Si^{4+} ions are substituted by Al^{3+} ions, and in the octahedral layer, the Al^{3+} ions are substituted by divalent ions such as Fe^{2+} and Mg^{2+}. These substitutions impart a negative charge to the three-dimensional sheet which is neutralized by cations such as Na^+, Ca^{2+}, and Mg^{2+} in the interlayer space. This space is also occupied by a large quantity of water molecules. All clays have a characteristic interlayer space, or distance. In the case of bentonite, this distance is 10 Å, but this can increase to high as 14 Å with hydration. This swelling property is largely responsible for the enological properties of clay, and also explains why clays such as kaolin which have

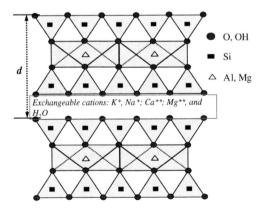

FIGURE 19.15   Structure of bentonite.

TABLE 19.4   Mineral Composition (%) of Betonite, Lebrija Clay,
and Kaolin

Composition	Bentonite	Lebrija Clay	Kaolin
SiO_2	69	63	44
Al_2O_3	15	5	40
Fe_2O_3	0.5	8	1
MgO	1	2	–
CaO	2.5	1	1.5
$K_2O + Na_2O$	3	2	0.5
Hydration water	5.5	18	12.5
Other	3.5	1	0.5

an invariable interlayer space are no longer used. The cations retained in the interlayer space of clay are exchangeable with other free ions in solution such as NH_4^+. The properties – and names – of the different types of bentonite are linked to these cations. Calcium bentonites, for example, are only barely swellable, unlike sodium bentonites, which are both swellable and highly reactive.

7.2. Physical and Chemical Characteristics

When solid bentonite is dispersed in a liquid, it forms 2- to 200-nm particles that form a negatively charged, sol-like, hydrophobic colloid. When present in high concentrations, however (approximately 10–15%), bentonite is capable of forming a gel with a very high specific surface area. In wine, bentonite particles are mainly neutralized by divalent metal cations (Mg^{2+}, Ca^{2+}) and they also adsorb other colloidal particles thanks to their large contact surface area. This area ranges from 335 to 691 m^2/g, and there is a direct relationship between adsorption surface area and the protein-removing power of bentonites.

An important property of bentonites is their ability to swell in water, which is achieved via the adsorption of molecules into the interlayer space. The degree of swelling is expressed by a swelling index, which reflects the number of cubic centimeters occupied by 5 g of dry bentonite placed in 100 cm^3 of deionized water for 24 hours. This property is reversible, as bentonite can be dried and then rehydrated, unless drying is performed at high temperatures. Above 200°C, for example, the water present in the structure of bentonite is lost, and the clay loses its swelling ability. The swelling power of a given bentonite depends on the quantity of exchangeable sodium present and the charge density. It does not depend on specific surface area or on the quantity of exchangeable calcium. The activity of bentonite, and its flocculation ability in particular, is proportional to its swelling power. It should be noted, however, that flocculation ability is also influenced by other chemical properties, such as the ability to establish long-range interactions (hydrogen bonds, van der Waals forces) with other colloids.

7.3. Types of Bentonite

Bentonites are classified into three main categories according to the proportion of exchangeable sodium and calcium cations they contain, their swelling index, and their pH. These categories are natural sodium bentonite, natural calcium bentonite, and activated calcium (sodium-calcium) bentonite.

- Natural sodium bentonite has moderate — and similar — proportions of exchangeable sodium and calcium, a moderate swelling index, and a basic pH (close to 9). It contains quite a lot of impurities but they have beneficial effects on wine.
- Natural calcium bentonite has a very low proportion of exchangeable sodium and a very high proportion of exchangeable calcium, although the proportions vary considerably from one type of bentonite to the next. It has a very low swelling index (normally $\leq 0.5\%$), and an almost neutral pH. Thanks to its low swelling index, it produces small volumes of lees, but unfortunately, it is not very efficient at eliminating proteins.
- Activated calcium bentonite is the most widely used bentonite in winemaking as it has a high swelling index and a pH of between 9 and 10. It also has a high proportion of exchangeable sodium and calcium. It is obtained by activating natural calcium bentonite with sodium carbonate, and then drying and grinding the resulting mixture. Its proportion of exchangeable sodium depends on the level of activation.

7.4. Use in Must

Bentonite can also be employed in musts used to produce white wine. In this case, it remains in the wine until fermentation is complete. It has numerous and wide-ranging advantages.

1. It produces more compact lees because the bentonite settles out with the yeast. This means that less liquid is lost.
2. It partly eliminates tyrosinase and therefore exerts a certain protective action against oxidation.
3. It stimulates fermentation as it provides a support for the yeast, which would otherwise sink to the bottom of the tank.
4. It adsorbs traces of fungicide that may have been transferred from the vineyard to the must.
5. It contributes to the production of finer, clearer wines, without interfering with aroma. Indeed it can even enhance aroma.
6. It minimizes the number of operations required during the winemaking process.

Bentonite should ideally be added to previously clarified wines, as suspended particles can reduce its efficacy. It is also advisable to add fermentation activators to the must as they can adsorb amino acids and vitamins. Bentonite does not need prior activation. It simply needs to be sprinkled over the must and stirred. In order to prevent unwanted oxidation, it is preferable to add the bentonite to must that is already fermenting, as CO_2 protects against oxidation and the stirring stimulates fermentation. The resulting wine should be racked off quickly, and aging on fermentation lees is not recommended as the periodic stirring required would disturb the settled bentonite and reduce the quality of the wine. The addition of bentonite to must should be considered when the aim is to obtain a wine of moderate quality, or when a must is made from unhealthy grapes.

7.5. Use in Wine

Bentonite can be added to wine in powder or granular form. In the first case, the wine should be stirred vigorously and in the second case, the granules need to be hydrated in water (10–20%) in a separate vessel and then added to the wine. For optimal results, the wine should have a low pH and a temperature of close to 20°C. Bentonite is normally added to wine in association with a protein fining agent such as casein or a combination of colloidal silica and gelatin.

Some authors recommend filtering the treated wine rather than waiting for the bentonite to settle out slowly, because they consider that 75% of all proteins are adsorbed rapidly and become resuspended during the sedimentation period. Others recommend activating the bentonite in wine instead of water. While this reduces its effectiveness, it reduces the quantity of lees produced. Some authors have even suggested filtering the wine during treatment to eliminate tartrate precipitates as the formation of crystals results in compact lees.

Use in White Wine

Bentonite clarification has traditionally been linked to the treatment of protein casse, which is more common in white wines because red wines contain high concentrations of polyphenols that favor the precipitation of proteins.

The protein content of a wine depends on a range of factors, including grape variety, the use of mechanical harvesting methods (destemming, for example, favors the extraction of proteins) or certain prefermentation treatments such as maceration on skins (which also increase the protein content of the must).

Regardless of quantity, the ease with which proteins are eliminated from a wine largely depends on its pH. The greater the difference between the pH of the wine and the isoelectric point of a protein, the greater the positive charge density of the protein and, consequently, the greater its reactivity with oppositely charged colloids. This is why wines with a low pH need a lower dose of bentonite compared to those with a high pH. It also explains why wines with a pH close to the isoelectric point of the proteins are more difficult to clarify.

One of the advantages of using bentonite is that it shortens the time between the end of fermentation and bottling, as natural clarification and stabilization occur sooner. At a dose of between 40 and 45 g/hL, bentonite has no significant impact on aroma, with white wines retaining their fruity character. Nonetheless, it can retain and reduce the concentration of certain aroma compounds such as γ-decalactone and β-ionone. Above this dose, there is a risk of the loss of aroma compounds.

Use in Red Wine

Treatment with bentonite has also proven to be effective in red wines with a high content of colloidal coloring matter, such as those made from grapes that have undergone thermovinification or intense mechanical processing. These wines have a high concentration of anthocyans in the form of flavylium cations. Flavylium cations can bind to poorly polymerized tannins, polysaccharides, and possibly proteins which precipitate due to the action of bentonite. Nevertheless, the effect on tannins is limited.

The use of bentonite in red wines causes color loss, which is much more pronounced in young wines as these have a higher concentration of anthocyanins that do not combine with other phenolic compounds. Wines treated with bentonite at a dose of between 6 and 12 g/hL have also been seen to be easier to filter as the bentonite eliminates other particles in colloidal suspension.

7.6. Action on Iron and Copper

Treatment with bentonite is not an efficient means of preventing ferric casse caused by the colloid formed by iron(III) and the phosphate anion as this has an overall negative charge.

Proteins favor the flocculation of the copper colloid responsible for copper casse in certain white wines. Treatment with bentonite prevents this type of casse as long as the quantity of copper does not exceed 1 mg/L. When it does, chemical treatment is required.

7.7. Alternatives to the Use of Bentonite

Some white wines have considerable protein stability problems. If treated with bentonite, their aroma and quality would be adversely affected as very high doses would be required. The alternatives in such cases are:

1. Tangential ultrafiltration. This method has the disadvantage that membranes that successfully adsorb proteins also alter the wine's aroma.
2. Wine tannin. The addition of wine tannin renders the wine harsh and does not guarantee stability.
3. Hyperoxygenation of must. This treatment is not effective as it affects the aroma of both must and wine.
4. Proteolytic enzymes. These are useful in musts but are inhibited during fermentation. Several studies have also shown that wine proteins are resistant to degradation by these enzymes.
5. Aging on lees. This is undoubtedly the most effective treatment. It does not alter the protein profile of the wine, except for the appearance of a mannoprotein after aging that acts as a protective colloid. Aging on lees considerably reduces the dose of bentonite subsequently required to stabilize the wine.

8. CLARIFICATION WITH COLLOIDAL SILICA

The colloidal silica used in winemaking is an aqueous dispersion of inorganic colloids formed by polymerized amorphous silica (SiO_2). It is highly stable and has a considerable specific surface area. The isomorphic substitution of Si with Al causes the appearance of charges in the crystalline network that are responsible for the hydroxylation of its surface, and the negative charge of the micelles formed. Thanks to this negative charge, silica can form complexes with proteins at the pH of wine.

As silica is poorly soluble in water, commercially available colloidal silica preparations contain alkaline metal silicates in 30% aqueous solution.

Compared to other colloids, it is very pure and therefore does not release any unwanted substances into the wine. It is normally used to remove the remains of other clarifying agents such as protein fining agents. Colloidal silica offers numerous advantages:

1. It accelerates the clarification process.
2. It compacts the lees produced during fining, thereby facilitating racking off and minimizing fluid loss.
3. It fixes to excess fining agents and therefore prevents overfining.
4. It eliminates other colloids such as ferric ferrocyanide, which is formed during the treatment of white wine.
5. It improves the bottling process as it facilitates filtration.
6. It partly eliminates certain proteins and the enzymes responsible for the oxidation of musts.
7. When used in association with gelatin, it contributes to the elimination of more proanthocyanidins, which are responsible for the browning of musts.
8. It is not very sensitive to the action of protective colloids, and therefore contributes to the successful treatment of wines that are difficult to clarify, such as sweet wines and wines made from grapes with noble rot.
9. It does not produce a hard wine, unlike other agents used for similar purposes, such as wine tannin.

Although silica can theoretically be combined with any protein fining agent, it should ideally be used with gelatin and fish glue. It is also very useful when associated with bentonite as this clay acts on proteins with a low molecular weight while silica acts on those with a high molecular weight.

When used in combination with gelatin, colloidal silica should be added first. However, when used with other treatments such as deferrization or addition of bentonite, these should be performed first, followed by addition of colloidal silica and, finally, the protein fining agent.

The dose depends on the ease with which the must or wine being treated is clarified. In a 30% solution, between 25 mL/hL and 100 mL/hL can be added.

9. OTHER CLARIFYING AGENTS

9.1. Vinylpyrrolidone Polymers

Polyvinylpyrrolidone (PVP) and polyvinylpolypyrrolidone (PVPP) are polymers that are synthetically obtained from vinylpyrrolidone. In an aqueous medium, polymerization yields PVP, which is soluble in water, whereas in a basic medium, the heterocyclic pyrrolidone splits and yields PVPP, which is insoluble in alcoholic aqueous solutions. These stabilizers are marketed under the Polyclar tradename.

Adsorption Mechanisms

Chemically, PVP and PVPP are polyamides that exhibit highly selective adsorption of polyphenols, giving rise to a PVP-polyphenol complex or a PVPP-polyphenol complex that is analogous to the protein-polyphenol combination. While both polymers can be used as clarifying agents, they are more effective as filtration coadjuvants.

Two mechanisms have been proposed to explain the adsorption of polyphenols to PVP and PVPP.

1. The first mechanism consists of the formation of hydrogen bonds between a hydrogen of the phenol group and the oxygen of the carbonyl group of pyrrolidone.
2. In the second mechanism, the formation of hydrogen bonds is attributed to an increase in the electrophilic character of the carbonyl group, which consequently becomes more susceptible to forming relatively high-energy bonds with neutrophilic reagents.

Not all phenolic compounds behave identically. While they are all able to form hydrogen bonds, only some, such as flavans, are capable of acting as nucleophilic reagents due to the high electron density at carbons 6 and 8. Steric hindrance also complicates this interaction and explains why phenolic compounds and polymerized pigments are not readily adsorbed.

Most polyamides of the type Nylon 11, Nylon 66, or Nylon 6, can form intramolecular hydrogen bonds between the amide hydrogen and the oxygen of the carbonyl group. The absence of an amide nitrogen in PVP and PVPP prevents the formation of these bonds, explaining why they have a greater adsorption capacity than other polyamides.

Properties in Winemaking

PVP and PVPP are white powders. They have a molecular weight of close to 1000 kDa, and variable particle size, as almost 40% of the particles are under 60 microns, 50% are between 60 and 600 microns, and the remaining 10% are larger than 600 microns. They

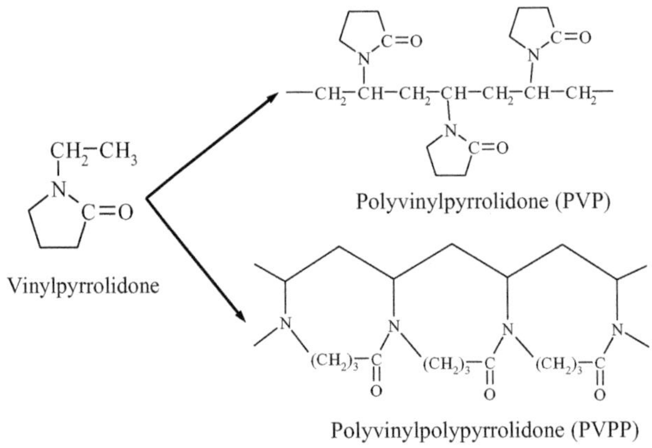

FIGURE 19.16 Vinylpyrrolidone polymers.

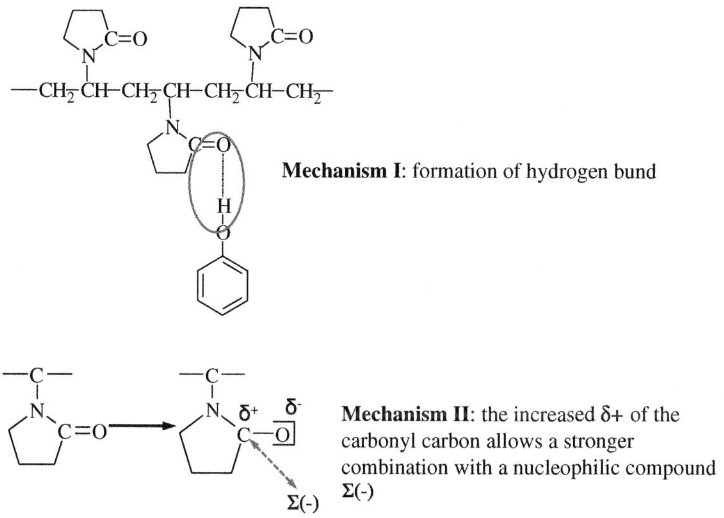

FIGURE 19.17 Phenol adsorption mechanisms in PVP and PVPP.

Mechanism I: formation of hydrogen bund

Mechanism II: the increased δ+ of the carbonyl carbon allows a stronger combination with a nucleophilic compound

are insipid and odorless, but because they are good adsorbents, they can influence certain aspects of aroma and flavor. When used at a dose of between 20 and 30 g/hL, they do not have a negative effect on the organoleptic properties of wine, and in some cases, they may even reduce bitterness.

The flocculation of PVP depends on its degree of polymerization, and when flocculation is incomplete, there is a risk of overfining. PVP is not very suitable for the treatment of wine as better clarifying agents such as gelatin are available. PVPP, however, is more suitable as it is insoluble in aqueous alcoholic solutions.

As a general rule, when PVP and PVPP are added to a white wine, they adsorb phenols and particles responsible for imparting a brown-gray color, according to the following order of affinity.

Leucoanthocyans < catechins < flavonols < phenolic acids.

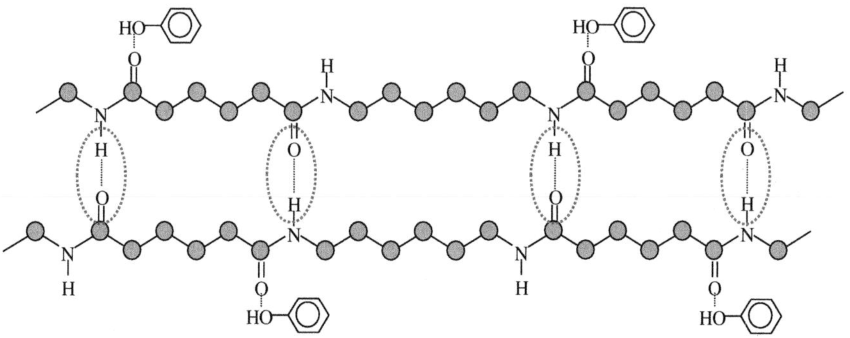

FIGURE 19.18 Formation of intermolecular hydrogen bonds in polyamides (not formed in polyvinylpyrrolidone or polyvinylpolypyrrolidone).

9.2. Alginates

Alginates are structural polysaccharides derived from certain algae. Alginic acid is a polymer formed by two mannuronic acid units linked by a β-1,4 bond. It is acidic (pK = 3.7) and only precipitates at a pH of below 3.5, which limits its use to wines with a low pH (3.1—3.2). Flocculation is more rapid when alginate is used in association with gelatin or blood albumin. Because there are clarifying agents that are much better than alginate, its use is only justified when a wine needs to be filtered a few hours after treatment. The recommended dose ranges between 5 and 15 g/hL and filtration can be performed 5 hours after the product is added.

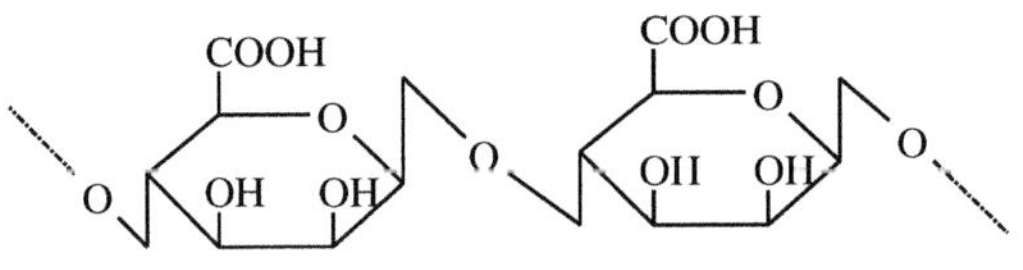

FIGURE 19.19   Structure of alginic acid.

9.3. Tannins

The term *tannin* was formerly used to refer to substances of plant origin used to harden leather and prevent it from drying out and cracking. Tannins have astringent properties and are soluble in water and partly soluble in ethanol. They are mainly obtained from chestnut and oak galls, and from grape marc. They form stable complexes with proteins, and, in the presence of ferric salts, give rise to a blue-black precipitate.

Chemically, they are polyphenols, and their reactivity with proteins depends on their size and spatial configuration. They should have a sufficient volume to yield stable combinations with proteins, but if their volume is excessive, they will not react due to steric hindrance. They give a mildly acidic reaction in water and have a negative charge at the pH of wine.

Traditionally, tannins have been classified into two groups, depending on their origin:

1. Condensed tannins, which are derived from the condensation of grape procyanidins.
2. Hydrolyzable tannins, which are mainly extracted from chestnut and oak galls. They are known as gallotannins and ellagitannins, and when hydrolyzed, they give rise to gallic acid and ellagic acid, respectively.

Hydrolyzable tannins derived from oak and chestnut are the most common commercially available tannins, and from an organoleptic perspective, they impart greater astringency, bitterness, and greenness than grape-derived tannins.

Tannin should only be added when the aim is to eliminate excess proteins following the fining of white wines. They can, therefore, be considered a clarification coadjuvant. Even in this case, however, they should be added with great care as they can increase the harshness (tannic character) of the wine. It is always preferable to use bentonite or a fining agent that does not cause overfining.

Condensed tannins eliminate more proteins than hydrolyzable tannins, explaining why they are preferred for the elimination of excess proteins. Finally, white wines accept a lower concentration of tannins (5 g/hL) than red wines (5—10 g/hL).

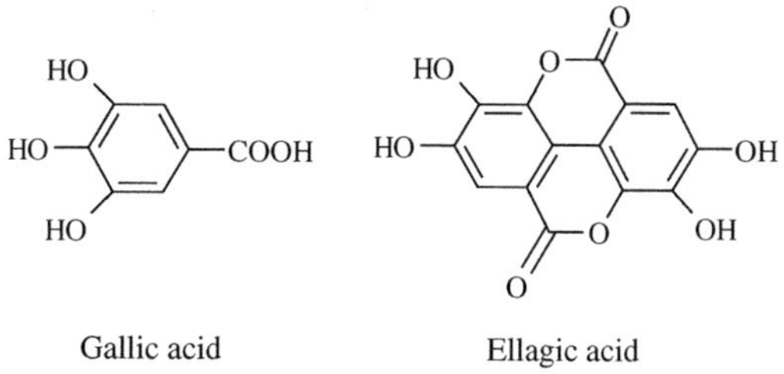

Gallic acid   Ellagic acid

FIGURE 19.20 Structure of gallic acid and ellagic acid.

10. PROTECTIVE COLLOIDS

The flocculation of colloids can be inhibited in the presence of protective colloids. The protective action occurs when several molecules of the protective colloid surround the unstable colloid, preventing its aggregation. The concentration of protective colloids should be sufficiently high to cover the surface of the unstable colloid as otherwise they can act as a bridge between two unstable particles, and therefore lose their protective effect (Figure 19.21). Very high concentrations of protective colloids can cause flocculation due to a phenomenon known as the depletion effect. This occurs when the concentration of protective colloids is so high that the osmotic pressure tends to bring the particles closer, causing them to aggregate and subsequently flocculate. This explains why coloring matter flocculates in red wines with a high polysaccharide content. While protective colloids prevent haze, they can also impede the clarification of wines already affected by haze as they slow down sedimentation.

Most wines contain mucilaginous substances that act as protective colloids. A wine that undergoes ultrafiltration or dialysis is more prone to haze, demonstrating the existence of these colloids and their protective effect. In the case of red wines, colloidal coloring matter and tannins interfere with tartrate precipitation.

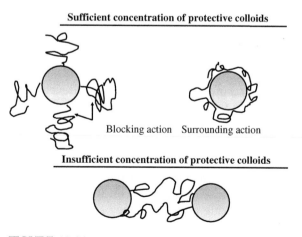

Sufficient concentration of protective colloids

Blocking action   Surrounding action

Insufficient concentration of protective colloids

FIGURE 19.21 Modes of action of a protective colloid.

Many polymers, and polysaccharides in particular, have protective properties, but little is known about them. Furthermore, they have a positive impact on the organoleptic properties of wine, demonstrated by the fact that wines that undergo intense filtration are of an inferior quality.

One much studied protective colloid is glucan, which is a polysaccharide secreted by *B. cinerea* in rotten grapes. Glucan does not interfere with the filtration of must, but when this is converted to ethanol, serious problems arise. Ethanol acts by aggregating particles and increasing their size. Glucan starts to precipitate at an ethanol content of 17% (vol/vol) and continues up to a strength of 23% (vol/vol).

10.1. Gum Arabic

Gum arabic is the most widely used protective colloid in winemaking. It is a polysaccharide with a molecular mass of 10^6 Da that is derived from the sap or bark of certain acacia trees. It provides stability in terms of clarity and has no adverse effects on the organoleptic properties of wine. Rather, it improves viscosity. Its use is authorized within the European Union.

Gum arabic is composed of D-galactose, L-arabinose, L-rhamnose, and D-glucuronic acid. This last acid forms the backbone of gum arabic and is its most abundant monosaccharide. The polysaccharide is associated with a certain protein fraction in which proline and hydroxyproline are the main amino acids.

It is used at a dose of 10 to 15 g/hL to prevent copper casse, but is only effective when copper levels do not exceed 1 mg/L. It acts by preventing the growth of colloidal particles rather than the formation of copper sulfide. It is less effective in the treatment of ferric casse as the iron complex has a greater mass. Furthermore, higher doses (in the range of 20 to 25 g/hL) are needed. Nonetheless, gum arabic does exert a protective effect, explaining why it is used to complement conventional treatments, which do not offer full protection. A similar phenomenon is observed in red wines as gum arabic prevents, at least partly, the precipitation of the colloidal ferric tannate but it does not impede the appearance of the dark color of this complex.

Treatment with gum arabic has the following characteristics:

1. It is quick and suitable for wines that need to be bottled rapidly. It is generally added before pre-bottling filtration.
2. It does not affect color.
3. Its effects are permanent presuming that more colloids develop, as occurs in aging.
4. It inhibits normal, desirable transformations in wines undergoing aging. Accordingly, its use is not recommended in wines that will be aged for a long time in the bottle. Rather, it should be reserved for young wines that are to be consumed soon after bottling.

Currently the preferred use of gum arabic is to prevent the precipitation of coloring matter in red wine. The recommended dose ranges between 10 and 20 g/hL. Lower doses may not prevent the precipitation of coloring matter but they can prevent the formation of coloring matter deposits on the bottle, which is a serious visual flaw. High doses (100 g/hL) can have a favorable effect on precipitation due to the depletion effect.

Inorganic Material and Metal Casse

1. INTRODUCTION

Must and wine contain mineral elements derived from grapes. These can be classified on the basis of their concentration into macroelements such as potassium, calcium, magnesium, and sodium — found at concentrations of between 100 and 1000 mg/L — and microelements such as the metals iron, copper, selenium, and zinc, among others, which are present at concentrations of less than 10 mg/L. These concentrations are only modified by the presence of contaminants or as a result of failing to maintain the condition of the vessels used. All of the metals present in must and wine occur as cations and their charge is neutralized by anions derived from organic acids and, to a lesser extent, inorganic compounds such as phosphates and sulfates present in the grape or introduced by prefermentation treatments.

One measure of the quantity of inorganic material in wine is the ash content, which is obtained by calcination of the dry extract at 400°C in an oven. This treatment converts all of the cations that form salts with organic anions into their corresponding carbonates; those that form salts with strong inorganic acids, mainly sulfates and phosphates, remain unaltered and are found in the ash as nonvolatile mineral anhydrides. The ammonium cation sublimes at the calcination temperature and is therefore quantified by other methods. The ash content in common wines varies between 1.5 and 3 g/L and the ratio of ash to reduced dry extract is between 1 and 10.

The dry extract of wine is a measure of the quantity of organic and inorganic material that does not evaporate at 105°C. It is determined by the evaporation of a precisely defined volume of wine in an oven until a constant mass is obtained. This extract contains a larger amount of organic than inorganic material. Another measure used on certain occasions is the reduced dry extract, which is the difference between the total dry extract and the sugar content.

The inorganic or mineral material dissolved in the must is derived from the cells that form the pulp of the grape, as well as from the skin and seeds, and its content is highly dependent upon the handling and treatments used to obtain the must. Other minerals are found at concentrations that are closely related to the contact between the must or wine and materials used to make the tanks and other winemaking apparatus. Some mineral compounds precipitate during fermentation of the must due to a reduction in solubility caused by the appearance of ethanol.

2. ANIONS

Must and wine contain inorganic anions derived from dissociation of the soluble salts that form inorganic acids and the metal cations absorbed from the soil by the roots of the vine.

The mineral or inorganic anions represent a small fraction of the total anion content of the must and wine, in which most anions are organic. Individual quantification of mineral anions tends to be of little value, except when the aim is to reveal fraudulent practices, or to assess the likelihood of casse through the analysis of anions such as sulfate or phosphate. An indirect method based on determining the alkalinity of ash is usually used. This is carried out under conditions of electroneutrality met by all solutions including, of course, must and wine.

The alkalinity of ash is expressed as mmol(+)/L and is determined by analysis of ash dissolved in a defined quantity of hydrochloric acid with sodium hydroxide. Since salts of strong acids such as sulfates, nitrates, or chlorides cannot by analyzed using this method and the amount of inorganic anions is negligible compared with organic anions, the alkalinity of ash can be considered as a measure of organic anions neutralized by metal cations. Total cation content is determined using H^+ cation-exchange resins, and according to the condition of electroneutrality, the value obtained will be equal to the sum of all anions present in the must or wine. Consequently, the difference between the total cation content and the alkalinity of ash provides a measure of the quantity of mineral anions present in the must or wine.

Generally, the most abundant inorganic anion in wine is phosphate, followed by sulfate, chloride, nitrate, and other anions present in trace quantities. Phosphate content is highly variable and is linked to the practice of adding ammonium phosphate as a fermentation activator. Red wines usually contain concentrations of between 150 and 1000 mg/L, approximately twice those found in white wines. The natural concentration of sulfate ions in must, expressed as K_2SO_4, is between 100 and 400 mg/L. These concentrations increase during aging and storage of wine, as a consequence of the repeated addition of sulfites and their oxidation to sulfates. In the case of natural sweet wines, concentrations of close to 2 g/L are commonly found after a few years of aging in wooden barrels, mainly as a consequence of the high doses of sulfite added to stop fermentation and preserve the wine. Chlorides, expressed as sodium chloride, are generally present at concentrations below 50 mg/L, although they can exceed 1 g/L in wines obtained from vines grown close to the sea and in white wines that have been fined with egg white.

Finally, although all nitrates are soluble, they are found at trace concentrations in wine. Other inorganic anions present at trace levels include bromide, iodide, fluoride, some silicates, and certain borates.

3. CATIONS

Some metal cations present in wine are important because they form relatively insoluble salts with certain anions and are responsible for haze or casse. The alkaline cations K^+ and Na^+, the alkaline earth metals Ca^{2+} and Mg^{2+}, and the metal ions Fe^{3+} and Cu^+ are particularly important. Other cations, such as those of the heavy metals Pb, Hg, Cd, As, Mn, and Zn, are important due to their effects on human health.

Potassium is the main cation present in wine and its concentration ranges from 0.5 to 2 g/L (mean of 1 g/L). The highest concentrations are found in wines made from grapes with noble rot or raisined grapes. Red wines also contain more potassium than white wines due to the capacity of phenols to inhibit the precipitation of potassium bitartrate.

Calcium forms highly insoluble salts with oxalate. It also forms poorly soluble salts with anions of tartaric, gluconic, and mucic acids, particularly in aqueous alcohol solutions. Calcium gluconate and mucate in particular form crystalline precipitates in wines produced from grapes with noble rot. The calcium content of white wines is between 80 and 140 mg/L and is much higher than that of red wine. Calcium is more active than potassium in causing flocculation of colloids such as iron phosphate and tannin-gelatin complexes.

Sodium is present in wine at low concentrations (10−40 mg/L) and the highest levels are found in wines treated with sodium bisulfite or insufficiently purified bentonite. As in the case of chloride, wines from vineyards close to the sea and those treated with egg white have higher sodium concentrations.

Magnesium salts are soluble and the cation remains unaffected by the conditions found during fermentation or aging. It is found at concentrations of between 60 and 150 mg/L.

Manganese is found in all wines in small quantities (1−3 mg/L) determined by the concentration in the grape. Vinification techniques used to produce red wines influence the concentrations found in the wine, since grape seeds contain three times more manganese than the skins and 30 times more than the pulp.

Iron and copper are also found at very low concentrations. These are responsible for casse and merit the more detailed consideration given in the following sections.

In addition to cations that occur naturally in grapes, wine contains other cations such as manganese and zinc that are mainly derived from dithiocarbamate fungicides used to treat the vines.

Finally, since wine has an acid pH, it can corrode winemaking equipment made of iron and metal alloys, particularly bronze, and as a consequence, heavy elements such as copper, nickel, and lead can dissolve in the wine.

4. HEAVY METALS

Heavy metals are generally considered extremely dangerous for living organisms as a result of their strong toxicity, which is due in part to their tendency to accumulate in tissues.

Most heavy metals form insoluble sulfates and treatment with potassium ferrocyanide partially eliminates copper, zinc, lead, and barium. Consequently, these metals are not very abundant in wines. Nevertheless, arsenic, lead, and zinc merit further consideration here — iron and copper will be discussed in detail in the sections on casse.

Arsenic can appear in wine when the vines have been treated with arsenic salts. Wines containing concentrations exceeding 1 mg/L are not fit for consumption and the International Organization of Wine and Vine (OIV) has set a limit of 0.2 mg/L for wine.

The permissible levels of lead in wine have been gradually reduced by the OIV in accordance with the results of studies addressing sources of lead in the winemaking process. The current limit is 200 μg/L. The lead content of wine is determined by the initial concentration of PbS in the must and its precipitation after alcoholic and malolactic fermentation. The paint and epoxy resins used to coat machinery are important sources of lead, as are connections and pumps made of brass and bronze alloys and tanks in which wines are stored for long periods, particularly those lined with ceramic tiles. In the past, the main source of lead in wine was tanks made of lead and covered with a thin layer of tin. Thankfully, these are now prohibited.

Wine naturally contains traces of zinc derived from the grape. Antifungal treatments with dithiocarbamate-based products and also fragments of galvanized wire damaged during mechanical harvesting can increase the zinc concentrations found in must and, therefore, wine. Another source of enrichment in zinc is equipment made from bronze or brass alloys. The concentrations of zinc in wines usually range from 0.14 to 4 mg/L and increase in direct relation to the length of maceration with the solid parts of the grapes. Ferrocyanide treatment is the most effective way to reduce the concentration of zinc and other heavy metals.

5. METAL CASSE

Most of the mineral elements described have no influence on the organoleptic characteristics or stability of wine, although it is important to control the concentrations of potassium and calcium to prevent precipitation of bitartrate and tartrate, respectively. However, other metals — particularly iron, copper, and tin — can affect the clarity of wines and cause the appearance of a type of haze known as casse.

The concentration of iron in must does not normally exceed 5 to 6 mg/L, although concentrations of 25 to 30 mg/L have been reported. Iron concentrations depend on the following factors:

- Grape variety
- Type of soil
- Materials in contact with the grape during transfer to the press
- The presence of soil in contact with the grapes
- Condition of the crusher and press, and the pressing process used: the must fractions obtained by aggressive pressing contain higher iron concentrations than those obtained by gentle pressing (15–25 and 2–5 mg/L, respectively)
- The condition of the tanks, pipes, and winery equipment, particularly if it is rusted
- Condition of concrete tanks, especially if they are poorly cast

The use of winemaking equipment made from stainless steel and plastic materials to transfer the grapes to the press has led to a reduction in the iron concentrations found in must, and has thereby reduced the risk of precipitation due to excess iron in the wine.

The copper present in wines is mainly derived from contact with copper-containing materials. Must can contain substantial quantities of copper derived from antifungal treatment (copper sulfate) of the grapes. However, the precipitation reactions that occur during alcoholic fermentation eliminate almost all of the copper present, mainly as copper sulfate. After fermentation, wines usually do not have copper concentrations of more than 0.4 mg/L, which are insufficient to provoke undesirable precipitations. One of the main sources of copper in wine, apart from pipes and taps made of bronze or copper, is the copper sulfate used to remove the sulfur derivatives that are responsible for the unpleasant aromas produced when wine is stored under reducing conditions.

Iron, copper, and tin can combine with other compounds present in wine to produce particles the size of colloids that disperse light and have a low sedimentation rate in the dispersant fluid, therefore leading to haze. When iron is combined with phosphate ions, it leads to white casse, and when combined with condensed tannins, it leads to blue casse. Finally, when it combines with coloring matter, it is responsible for black casse. Copper and tin induce casse by combining with sulfate ions and proteins, respectively.

After fermentation, wine contains almost no oxygen and is therefore, in terms of redox phenomena, a reducing medium. Consequently, most of the iron and copper dissolved in the wine has a low oxidation number. However, slight aeration of the wine causes a shift in the equilibrium between the oxidized and reduced forms of these metals. When iron is present as Fe^{2+} and copper as Cu^{2+}, they do not form insoluble compounds in wine and therefore do not affect clarity. In contrast, above a certain threshold concentration, Fe^{3+} and Cu^+ form poorly soluble colloidal compounds that generate haze and flocculate slowly.

5.1. Ferric Casse

The appearance of haze due to the formation of insoluble iron compounds is known as ferric casse. In the case of white wines, stability is linked to the formation of a complex between iron in its ferric state and the phosphate anion, which gives rise to white casse.

This colloid carries a negative charge and flocculates upon binding with proteins in the wine. In red wines, ferric iron can bind to phenolic compounds and produce a soluble complex that is responsible for the increased color intensity of the wine. This complex subsequently flocculates and causes a blackish-blue haze known as blue casse. White casse does not occur in red wines because the acidity and phosphate content are not sufficiently high and also because the iron preferentially binds to phenolic compounds.

The iron is generally present in two oxidation states (3+ and 2+), and both ions can be found in free and bound forms. The concentration of free Fe^{3+} is generally lower than that of total Fe^{3+} in wine, indicating that the complexes formed by the ion are highly stable and have a low dissociation constant. An example is the complex formed between tartaric acid and Fe^{3+}, which generates the potassium salt potassium ferritartrate $[K(FeC_4O_6H_2)_2]$. The first dissociation of this salt leads to release of potassium and ferritartrate ions. A second, more difficult, dissociation causes release of the Fe^{3+} ion:

$$K(FeC_4O_6H_2)_2 \leftrightarrows (FeC_4O_6H_2)_2^- + K^+$$

and to a lesser extent:

$$(FeC_4O_6H_2)_2^- \leftrightarrows 2Fe^{3+} + 2\,C_4O_6H_2^{2-}$$

Given that iron binds in part to alcohol functional groups and in part to acid functional groups, it is clear that iron in both the 2+ and 3+ oxidation states can form a large number of complexes with compounds containing these functional groups, mainly hydroxy acids. This explains why most of the iron in solution is sequestered and therefore does not participate in other reactions.

White wine stored under a cushion of air does not undergo the characteristic reactions of free Fe^{3+} at appreciable levels, although brief contact with oxygen leads to the appearance of this cation in increasing quantities that depend on the level and duration of oxygenation. If the wine is not aerated, almost all of the iron is found in the ferrous form (Fe^{2+}), forming complexes whose degree of dissociation increases as Fe^{2+} ions become oxidized to Fe^{3+} in the presence of oxygen, such that the equilibrium between complexed and dissociated iron is dependent upon the redox potential of the wine. Free Fe^{3+} ions can combine with various constituents of the wine to produce ferric complexes with different dissociation constants. If the combinations that are formed are soluble in wine, the wine will remain clear and transparent, but if an insoluble complex is formed, this will precipitate when its concentration exceeds its solubility in wine. As a result, the wine becomes turbid, generating what is known as casse. An example of a soluble complex is that formed between Fe^{3+} and hydroxy acids such as tartaric, malic, or citric acid. An example of an insoluble complex is that formed between Fe^{3+} and phosphoric acid, which is responsible for white casse in white wine.

Figure 20.1 shows the influence of aeration on the oxidation state of iron and on the formation and dissociation equilibria of the complexes that are formed. The capacity of Fe^{3+} to form complexes with components of the wine favors oxidation of Fe^{2+} to Fe^{3+}, since Fe^{3+} complexes are more stable, as shown by their lower dissociation constants (Table 20.1). Similarly, the addition of organic acids favors the oxidation of Fe^{2+} and, in parallel, reduces the likelihood of ferric casse, since it reduces the concentration of free Fe^{3+} available to complex with phosphate ions.

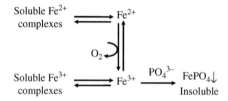

FIGURE 20.1 Equilibrium of iron in wines.

TABLE 20.1 Dissociation Constants of Some Complexes Formed Between Iron and Hydroxy Acids

Ion	Tartrate	Citrate
Fe^{2+}	$K_d = 1.26 \times 10^{-5}$	$K_d = 3.16 \times 10^{-16}$
Fe^{3+}	$K_d = 3.16 \times 10^{-8}$	$K_d = 10^{-25}$

Ferric precipitation may not occur if the $FePO_4$ ion formed does not precipitate, which can be ensured by the addition of gum arabic to the wine. The gum acts as a protective colloid and prevents $FePO_4$ from flocculating, even at concentrations that exceed its solubility. In contrast, addition of a protein causes flocculation of $FePO_4$.

In summary, white casse involves three phenomena that can occur in different degrees depending on the wine:

- The solubility of $FePO_4$ in an aqueous alcohol solution such as wine and in the presence of common and uncommon ions
- The formation of complexes with iron that "sequester" the free iron and make the formation of insoluble salts more difficult
- Colloidal protection of some compounds present in wine

The total iron content alone is therefore insufficient to determine the likelihood of ferric casse. Some wines containing 5 mg/L of iron are susceptible to ferric casse whereas others with a concentration of 35 mg/L are not. In practical terms, a total iron concentration of less than 3 mg/L is considered to offer acceptable guarantees of protection against white casse and color alterations.

Factors that favor white casse include addition of high concentrations of ammonium phosphate to must in order to activate yeast. Doses of 100 to 150 mg/L of ammonium phosphate are sufficient to activate fermentation without having a significant effect on casse. Nevertheless, the addition of ammonium phosphate only serves to increase the risk of white casse as the natural concentrations of ammonium in the must are usually sufficient for fermentation to proceed smoothly.

Enrichment of wine in copper also causes ferric casse, since copper favors the oxidation of Fe^{2+}. In fact, the greatest likelihood of finding ferric casse is observed in wines rich in both iron and copper. In general, for an iron-based precipitate to appear, the wine must contain between 12 and 25 mg/L of total iron; this precipitate will be essentially formed by iron phosphate in the case of white wines (white casse) or by a complex of iron and polyphenols in the case of red wines (blue or black casse). A copper ion concentration of around 0.5 mg/L is sufficient to cause problems that affect the stability of Fe ions and the formation of colloidal copper sulfate precipitates.

It is understood, then, that the iron content is not the only factor to be considered in white casse and that the use of complexing Fe^{3+} anions provides better protection against casse in wines with more stable complexes. Citric acid is the only chelating agent currently permitted, although several have been tested, including oxalic acid, pyrophosphoric acid, and EDTA.

pH is another factor that affects the precipitation of $FePO_4$. Specifically, an increase in pH due to partial deacidification of the wine increases the concentration of complexed Fe^{3+} and reduces the concentration of free ferric ions available to form phosphate. The solubility of the iron phosphate formed also decreases with increasing pH, possibly due to an increase in the concentration of the common PO_4^{3-} ion, leading to an increase in pH. The following equilibria are involved in this process:

$$H_2PO_4^- \leftrightarrows HPO_4^{2-} + H^+; \quad pK_2 = 6.70$$

$$HPO_4^{2-} \leftrightarrows PO_4^{3-} + H^+; \quad pK_3 = 12.4$$

$$Fe(PO_4) \leftrightarrows Fe^{3+} + PO_4^{3-}; \quad K_{sp} = 9.91 \times 10^{-16} \, [Fe(PO_4).2\,H_2O]$$

Figure 20.2 shows the total and complexed Fe^{3+} and the Fe^{3+} combined with phosphate at different pHs and after 8 days of aeration. Under these conditions, not only total Fe^{3+} but also complexed Fe^{3+} increases. Fe^{3+} from phosphate increases up to a pH of 3.3 and then decreases from that point on.

Ferric casse is only possible, therefore, over a narrow pH range, and as a consequence, addition of small amounts of acid or base within this interval can favor or inhibit this type of casse depending on the pH of the wine. Unfortunately, the optimal pH for casse coincides with that of many wines. Prevention or correction based on modifying the pH is not very effective and, therefore, casse must be eliminated using delicate procedures such as treatment with ferrocyanide.

FIGURE 20.2   Concentration of Fe^{3+} in wines.

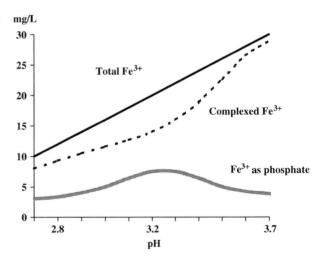

Other Types of Ferric Casse

Red wines can be affected by blue or black casse, which are caused by the formation of poorly soluble iron complexes with anthocyanins and tannins. Precipitation of these complexes reduces the quantity of free Fe^{3+} that can react with the phosphate ion and, therefore, reduces the likelihood of white casse.

In white wines, Fe^{3+} also reacts with coloring matter to form polyphenol-iron complexes that tint the wine to a greater or lesser degree and also the iron-phosphate precipitates.

Precipitation of phenolic compounds with iron can be prevented through the use of protective colloids, chelating agents, and antioxidants. Examples include the following:

- Addition of gum arabic, which discourages the flocculation of colloids but does not prevent an increase in color.
- Addition of citric acid, which forms a very stable complex with Fe^{3+}. The drawback of this treatment is the degradation of citric acid by bacteria, which is greater in wines without added sulfites.
- Addition of tartaric acid, which disassociates the iron from its complexes with polyphenols but can also promote white casse.
- Addition of sulfuric anhydride as an antioxidant.

5.2. Copper Casse

White wines that contain free SO_2 and that have been protected from the air during storage sometimes develop a reddish-brown precipitate known as copper casse at the bottom of the vessel. This precipitate contains a high proportion of copper and dissolves when the wine is stirred in the presence of air.

Copper casse is exclusive to white wines, occurs after prolonged storage in the bottle, and requires the presence of proteins, which act as flocculants for the negatively charged colloids formed by the copper. Red wines are not rich in proteins, since these have already flocculated during fermentation due to the presence of phenolic compounds. In addition, polyphenols act as buffering agents against the reduction reactions necessary for copper casse to occur.

Formation of the precipitate, which has a complex chemical composition, occurs in wines with a total concentration of copper ions that exceeds 0.5 mg/L, and the risk increases at higher concentrations of copper, higher temperatures, and in the presence of light. The major components of the precipitate are Cu^{2+} and S^{2-} ions, followed by organic substances, which usually account for half of the weight, and iron and calcium, which are adsorbed onto the precipitate in small quantities. Colloidal metallic copper is also a component of the precipitate, since copper can assume this oxidation state at acidic pH.

Mechanism of Copper Casse Formation

Step 1: Reduction of Cu^{2+} by any reducing agent

$$Cu^{2+} + \text{Reducing agent 1} \rightarrow Cu^{+} + \text{Oxidizing agent 1}$$

Step 2: Reoxidation of Cu^+ by sulfuric anhydride or sulfated amino acids

$$6\,Cu^+ + 6\,H^+ + SO_2 \rightarrow 6\,Cu^{2+} + H_2S + 2\,H_2O$$

The sulfuric acid formed reacts with Cu^+ and Cu^{2+} to generate a compound mainly made up of Cu_2S that also contains variable quantities of CuS:

$$2\,Cu^+ + Cu^{2+} + 2\,H_2S \rightarrow Cu_2S + CuS + 4\,H^+$$

Step 3: Flocculation of $Cu_2S.CuS$

$Cu_2S.CuS$ (negatively charged colloid)

+ proteins (positively charged colloid) $\rightarrow$ copper casse

Copper casse formation is favored by the presence of proteins containing sulfated amino acids, such as cysteine, due to their capacity to form complexes with copper ions (Figure 20.3). The compounds formed should not be confused with those obtained through the interaction between proteins and CuS colloids, which cause mutual flocculation. Other molecules such as gum arabic, which acts as a protective colloid, or bentonite, which causes flocculation of proteins, can prevent copper casse.

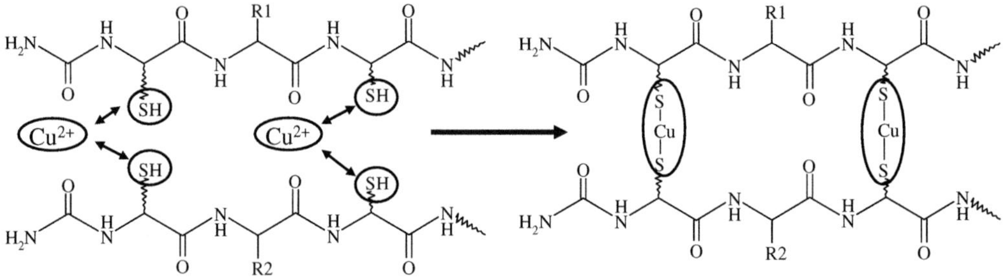

FIGURE 20.3   Formation of bonds between Cu^{2+} and $-SH$ groups on the proteins.

6. TREATMENTS TO PREVENT FERRIC CASSE

The appearance of white casse in wines is favored by aeration and the presence of iron and phosphate concentrations that exceed 10 and 100 mg/L, respectively. Blue and black casse also require aeration and are more likely in wines with low acidity. The stability of a wine towards ferric casse is therefore determined qualitatively by saturating the wine with oxygen and reducing the temperature. Unstable wines become turbid after 48 hours, whereas stable wines remain clear for at least a week. Quantification of iron can also be used to determine the likelihood of casse, but since the iron content is not the only determining factor, quantification does not provide information on the true risk of ferric casse.

Treatments used to prevent ferric casse include the addition of potassium ferrocyanide, calcium phytate, citric acid, and other iron chelating agents such as polyphosphates, EDTA, and ascorbic acid.

6.1. Treatment with Potassium Ferrocyanide

The potassium ferrocyanide salt dissociates in wine and produces ferrocyanide and potassium ions. The ferrocyanide ion reacts with iron and copper and other heavy metals (lead and zinc) and produces high-molecular-weight compounds with colloidal properties that can easily flocculate with other colloids in the wine. As a general rule, potassium ferrocyanide should be added at a lower quantity than that calculated stoichiometrically to eliminate iron and copper. This is a precaution to prevent traces of residual potassium ferrocyanide being left in the wine, as ferrocyanide can decompose to cyanide in the acidic medium of the wine.

In principle, the ferrocyanide ion reacts with Fe^{3+} to generate iron ferrocyanide with a Prussian blue color.

$$3 \ Fe(CN)_6{}^{4-} + 4 \ Fe^{3+} \leftrightarrows Fe_4[Fe(CN)_6]_3$$

However, in practice it is not so simple, since ferrocyanide can reduce Fe^{3+} to Fe^{2+} through oxidation to ferrocyanate and formation of iron ferrocyanide, which generates a mixture of iron(III) ferrocyanide and iron(II) ferrocyanide. The proportion of each depends on the conditions in the medium. Consequently, simple analysis of iron in wine is insufficient to calculate the quantity of ferrocyanide required.

$$Fe(CN)_6{}^{4-} + Fe^{3+} \leftrightarrows Fe(CN)_6{}^{3-} + Fe^{2+}$$
$$2 \ Fe(CN)_6{}^{3-} + 3 \ Fe^{2+} \leftrightarrows Fe_3[Fe(CN)_6]_2$$

The quantity of free Fe^{3+} in the wine is very small, since most of it forms poorly dissociated complexes with hydroxy acids and polyphenols also present. The addition of ferrocyanide causes free Fe^{3+} to become insoluble and shifts the dissociation equilibrium of its complexes towards the free form. Ferrous ferrocyanide is also formed, and as a result the iron in the wine precipitates slowly and progressively, in a manner dependent upon the dissociation of iron complexes.

$$Fe(CN)_6{}^{4-} + 2 \ Fe^{2+} \leftrightarrows Fe_2[Fe(CN)_6]$$

Given that Fe^{2+} complexes are more dissociated, the precipitation of ferrous ferrocyanide is rapid, although various white compounds are formed, such as $FeK_2[Fe(CN)_6]$ and $Fe_2[Fe(CN)_6]$. As a result, the proportion of added ferrocyanide and precipitated iron is not stoichiometrically constant.

Finally, the ferrocyanide ion reacts with other metal ions present in the wine, such as copper and zinc, to form $Cu_2[Fe(CN)_6]$ and $Zn_2[Fe(CN)_6]$, respectively. Both compounds, along with $Fe_2[Fe(CN)_6]$ (formed with Fe^{2+}), form more rapidly than ferric ferrocyanide ($Fe_4[Fe(CN)_6]_3$), since the divalent ions of copper, zinc, and iron are found in their free form at a greater proportion than Fe^{3+}. This reveals once again the complexity of the exact calculation of the quantity of ferrocyanide that should be added to prevent ferric casse.

Once ferric ferrocyanide is formed, and bearing in mind that it is a negatively charged colloid, it flocculates naturally as a result of the cations and proteins present in the wine. Nevertheless, this flocculation may be slow and incomplete, particularly in the presence of protective colloids. Fining is therefore carried out at the same time as treatment with ferrocyanide so that flocculation of both colloids occurs.

The reaction through which ferrocyanide compounds are formed with Fe ions is slow, and poorly soluble compounds are also formed with other metal ions present in the wine. This would seem to suggest that addition of excess ferrocyanide would result in a more efficient elimination of iron. This is true from a chemical perspective, but it should be remembered

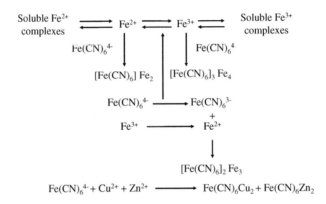

FIGURE 20.4   Reactions of ferrocyanide in wine.

TABLE 20.2   Solubility Products of Some Iron and Copper Compounds

Compound	K_{Ps} in Water
$Fe(PO_4) \cdot 2H_2O$	9.91×10^{-16}
$Fe_4[Fe(CN)_6]_3$	3.0×10^{-41}
Fe_2S_3	1.4×10^{-88}
FeS	4.9×10^{-18}
$Fe(OH)_3$	6.3×10^{-38}
$Fe(OH)_2$	7.9×10^{-15}
CuS	8.7×10^{-36}
Cu_2S	1.6×10^{-48}
$Cu(OH)_2$	1.6×10^{-19}
$Cu_2[Fe(CN)_6]$	1.3×10^{-16}

that excess ferrocyanide decomposes in acidic media such as wine to produce highly toxic hydrogen cyanide.

Calculation of the Dose of Potassium Ferrocyanide (K$_4$[Fe(CN)$_6$])

The first problem when trying to calculate the dose of ferrocyanide is the slow rate of reaction with iron and subsequent flocculation, leading to errors arising from inadequate or excessive doses. In the case of excessive doses, apart from the risk of decomposition, blue deposits can appear after bottling of the wine.

The proportion of Fe^{2+}/Fe^{3+} is also important for a higher rate of reaction during formation of ferrous ferrocyanide. As a result, tests used to calculate the dose of ferrocyanide should be carried out in the same sample of wine that is to be treated immediately prior to treatment. In this way, errors caused by differences in aeration can be minimized. The use of ascorbic acid, which reduces Fe^{3+} to Fe^{2+}, also prevents this type of error.

The dose of ferrocyanide is usually calculated through a sampling process that involves testing the wine to be treated twice. Although repeating the test may be a disadvantage, sophisticated equipment is not needed and the second test uses not only the same wine but also the same treatment products.

PRE-TEST

An initial test is used to establish the approximate range for the amount of ferrocyanide required to produce immediate precipitation of metal ions in the wine. This involves preparation of a 10 g/L solution of potassium ferrocyanide. Then, 100 mL of wine is placed in each of four tubes measuring 40 cm long and 2.3 cm in diameter and an appropriate volume of potassium ferrocyanide is added to obtain final concentrations of 50, 150, 250, and 350 mg/L.

After waiting a few minutes, a 1 mL suspension of fish meal (2.5 g/L) is added and carefully mixed to avoid formation of foam. After leaving the tubes for 10 minutes, an aliquot of the mixture from each tube is centrifuged and the supernatants are transferred to test tubes. Two drops of a saturated solution of iron alum are added to each test tube and the solution is acidified with a few drops of hydrochloric acid. The appearance of a blue color indicates an excess of ferrocyanide whereas its absence indicates an inadequate dose.

MAIN TEST

The main test consists of repeating the process used in the pre-test using ferrocyanide doses between the lowest dose that yielded a blue color and the dose immediately before this. After centrifugation and addition of iron alum and hydrochloric acid, the dose is determined based on the presence or absence of blue color. From this dose, 30 mg/L is subtracted as this corresponds to the technical safety limit. This limit is extended to 50 mg/L if filtration was used rather than centrifugation in the tests, since filter paper absorbs ferric ferrocyanide.

Bearing in mind that 1 mg of iron is precipitated per 6 mg of ferrocyanide (Figure 20.5), the previous test can be omitted if the iron content of the wine is known.

Factor 6, obtained based on the stoichiometry of the reaction of potassium ferrocyanide 3 hydrate with the Fe^{3+} ion, is also approximate because it does not take into account possible reactions of the anion with other metal cations (Figure 20.5). Consequently, it is subject to the same errors discussed earlier. Nevertheless, this is corrected by carrying out the main test with the calculated dose along with two other doses 20 mg/L higher and lower.

FIGURE 20.5 Stoichiometric calculation of the dose of potassium ferrocyanide.

According to the stoichiometry of the reaction:

$$3 [Fe(CN)_6]K_4 \cdot 3H_2O + 4 Fe^{3+} \rightarrow [Fe(CN)_6]_3Fe_4$$

$$3 \times 422.2 \text{ g/mol} \qquad 4 \times 55.8 \text{ g/mol}$$

$$\downarrow$$

5.7 mg of ferrocyanide / 1 mg of Fe

Precautions

Less ferrocyanide than that calculated based on the stoichiometry and the total iron concentration should be added.

Calculation of safe dose

Quantity of $[Fe(CN)_6]K_4 \cdot 3H_2O$ in mg/L = $6 \times [(Fe) \text{ mg/L} -3]$

Treatment of the wine is carried out by preparing a solution of ferrocyanide containing 50 or 100 g/L in water, using a commercial product that meets the specifications of the International Oenological Codex. This solution is added homogeneously to the wine followed by addition of the fining agent. The treated wine is filtered after 4 to 6 days, ensuring that all blue deposits at the bottom of the vessel are removed.

The formation of ferric ferrocyanide is very slow in some wines, particularly those with high levels of Fe^{3+} or a pH above 3.4. In these cases, it is advisable to add 50 to 60 mg/L of ascorbic acid prior to addition of ferrocyanide to reduce Fe^{3+} to Fe^{2+}.

The ferrocyanide also forms copper ferrocyanide, which has a red color, thereby partly eliminating the metal and reducing the risk of copper casse. This only occurs in the presence of a sufficient quantity of iron (5–10 mg/L), since iron and copper precipitate together and inadequate elimination of copper has been observed in wines containing low concentrations of iron. If the quantity of iron is sufficient, Cu^{2+} ferrocyanide will precipitate prior to Fe^{3+} ferrocyanide, not because it has a lower solubility but because the reaction is more rapid as a consequence of copper being complexed to a lesser extent than Fe^{3+}.

The formation of blue deposits after treatment to remove iron from wine occurs as a consequence of the flocculation of ferric ferrocyanide and proteins together. Some countries have proposed the addition of $ZnCl_2$ to wine prior to addition of ferrocyanide in order to form zinc ferrocyanide, which promotes flocculation.

Chemical Changes Following Addition of Potassium Ferrocyanide to Wine in Excess

Treatment with ferrocyanide should always be carried out under the supervision of an enologist and in accordance with established legal requirements, since there is a risk that an excess of the added compound will form hydrogen cyanide or cyanide salts in the acidic medium of the wine. This decomposition of the ferrocyanide ion complex (or hexacyanoferrate) can also occur if the formation of ferrocyanide compounds with iron is very slow or if the presence of protective colloids delays their precipitation.

The formation of hydrogen cyanide occurs in a series of steps (Figure 20.6). Firstly, complete dissociation of the added potassium ferrocyanide occurs to generate potassium ions and the ferrocyanide complex ion. This complex ion partially dissociates according to

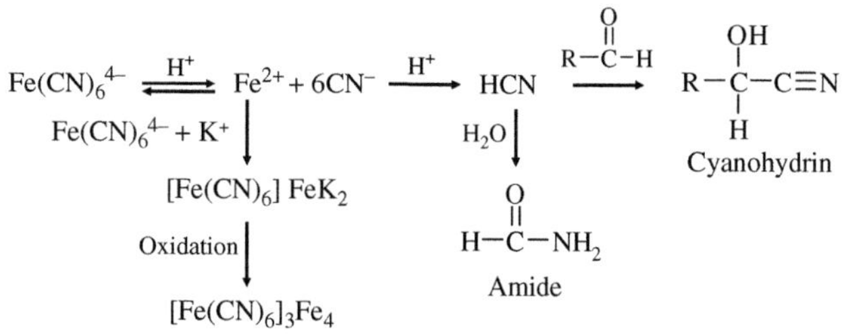

FIGURE 20.6 Chemical changes following addition of excessive quantities of potassium ferrocyanide to wine.

its dissociation constant ($K_d = 10^{-24}$) to produce the Fe^{2+} ion and the cyanide ion, which at the pH of wine is found in high proportion as hydrogen cyanide.

The cyanide so formed reacts rapidly with carbonyl compounds (aldehydes or ketones) in the wine and gives rise to the respective cyanohydrins. It also reacts with water and forms an amide. As a result, the cyanide ion that has formed disappears within a few days.

In parallel, the small quantity of Fe^{2+} that forms as a consequence of the dissociation of the complex ion reacts with excess ferrocyanide and produces ferrous potassium ferrocyanide or ferrous potassium hexacyanoferrate, which are white. This compound is easily oxidized to ferric ferrocyanide, such that in white wine, first green tones and then a blue precipitate appear even when iron has been previously eliminated. The reduction in the Fe^{2+} content caused by the formation of potassium and iron ferrocyanide shifts the dissociation equilibrium of the ferrocyanide ion towards formation of the cyanide ion.

Factors Affecting the Formation of Cyanide in Wines Treated with Potassium Ferrocyanide

The toxic nature of hydrogen cyanide is well known and it is therefore extremely important that the dose of potassium ferrocyanide added to the wine is always lower than that required. Furthermore, it is essential to establish safety limits and always take into account the following factors:

1. Decomposition of ferrocyanide is very slow at the pH of wine and decreases as the value of the dissociation constant ($K_d = 10^{-24}$) increases.
2. The risk increases with the dose of ferrocyanide used.
3. Higher temperatures favor the hydrolysis of ferrocyanide, although they also accelerate the elimination of hydrogen cyanide due to the formation of cyanohydrins and amides.
4. Long periods of time in contact with the blue precipitate and the presence of protective colloids that block flocculation favor the formation of hydrogen cyanide.
5. Significant amounts of hydrogen cyanide are only found when the dose of ferrocyanide used is much higher than that recommended and the concentrations remain far below those considered to be dangerous.
6. Yeasts themselves produce hydrogen cyanide.

Some authors have suggested that ascorbic acid should be used first to reduce Fe^{3+} to Fe^{2+} and favor the formation of ferrous ferrocyanide complexes that precipitate more completely and immediately. This would permit simultaneous fining, thereby reducing the period of contact with ferrocyanide and minimizing the formation of hydrogen cyanide.

6.2. Treatment with Calcium Phytate

Phytates are salts of the hexaphosphoric ester of inositol and phytic acid. They can be used to treat white and red wines, although they are usually reserved for red wines. (White and rosé wines are treated with ferrocyanide.)

The double salt formed by phytic acid ($C_6H_{18}O_{24}P_6$) and calcium and magnesium corresponds to phytin, which is used as a phosphorus reserve in plants. Industrially, it is prepared as a poorly soluble calcium salt from husks of wheat, rice, maize, etc.

Calcium phytate is a white powder with an acid flavor and is poorly soluble in water (0.4 g/L), almost insoluble in wine, and soluble in dilute strong acids. Aqueous solutions of this compound have an acidic pH.

At the pH values of wine, calcium phytate dissociates and the phytate anion combines with metal cations to produce poorly soluble salts. The compound is authorized for the elimination of excess iron in wines but its use should be very carefully monitored, since an excess of added phytate in relation to Fe^{3+} content leads to haze and the formation of a precipitate in treated wines.

According to the stoichiometry of the reaction, 4.22 mg of calcium phytate are needed to precipitate 1 mg of iron. Nevertheless, in practice 5 to 5.5 mg are added since phytate forms a mixed iron and calcium salt containing approximately 20% calcium.

$$Ca_6(C_6H_6O_{24}P_6)\, 3\, H_2O + 4\, Fe^{3+} \rightarrow C_6H_6O_{24}P_6Fe_4 + 3\, H_2O + 6\, Ca^{2+}$$
$$942.41 \text{ g/mol} \qquad\qquad 4 \times 55.8 \text{ g/mol}$$

When treatment is carried out, it should be remembered that only Fe^{3+} reacts and that the precipitation of ferric phytate is slow, since the rate of the reaction is influenced by the dissociation of the Fe^{3+} complexes and by the oxidation of Fe^{2+} to Fe^{3+} when oxygen is dissolved in the wine. Consequently, the wine should be aerated prior to addition of calcium phytate after adding 30 to 50 mg/L of sulfite to facilitate the formation of Fe^{3+}. The sulfite used at this dose does not inhibit the oxidation of ferrous to ferric iron.

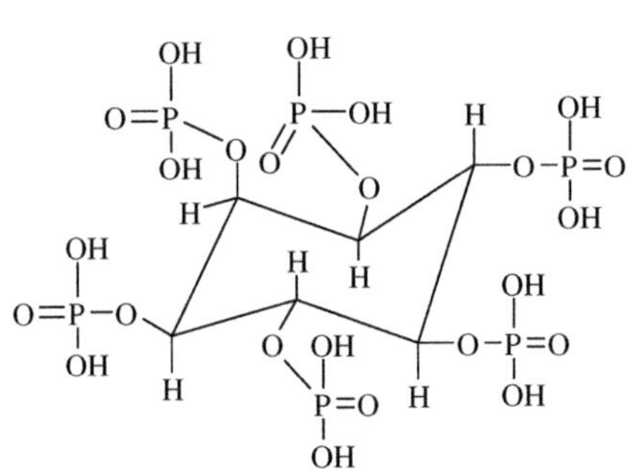

FIGURE 20.7   Molecular structure of phytic acid.

In practice, the quantity of calcium phytate that should be added is calculated by multiplying the quantity of iron to be eliminated by five and then subtracting 1 g/hL from the dose expressed in g/hL. This is because complete precipitation of iron does not occur, meaning that precipitation of iron phytate can take place. The treatment can be completed by adding citric acid and gum arabic.

Treatment with calcium phytate does not cause hygiene-related or microbiological problems but extensive aeration of the wine is always necessary and this can have a negative effect on the organoleptic quality of the wine. Finally, it should be remembered that the addition of this compound to the wine causes an increase in the calcium concentration that has been estimated at 1 mg of calcium for each milligram of iron eliminated.

6.3. Treatment with Citric Acid

Citric acid is a natural component of wine. It is normally found at higher concentrations in white wines, because in red wines it is transformed by malolactic bacteria. The addition of citric acid is not recommended without a sufficient concentration of sulfite to guarantee microbiological stability. In practice, this treatment is only used in white wines.

The protective effect of citric acid against ferric casse is based on the formation of complexes with Fe^{3+}; these complexes are yellow and have a composition that has not been fully elucidated. Nevertheless, the preference of Fe^{3+} for the citrate ion is only slightly greater than that for phosphate, and if the conditions change, casse can still occur. This treatment is therefore not recommended for wines with an iron concentration that exceeds 15 mg/L, as these have a strong tendency to undergo ferric casse. Potassium citrate, which does not have the same effects on flavor as citric acid, can be used since the process depends on the citrate anion.

The dose of citric acid ranges from 20 to 30 g/hL and treatment to prevent ferric casse is complemented by addition of gum arabic, which acts as a protective colloid to inhibit the flocculation of colloidal ferric phosphate.

6.4. Treatment with Other Chelating Agents

The efficacy of treatments with chelating agents to prevent ferric casse is based on the stability of the complexes formed. Tests have therefore been carried out with other ions that have a lower dissociation constant than that for the complex formed between Fe^{3+} and citric acid. One of these is oxalic acid, which forms complexes that have a low dissociation constant with Fe^{3+}. Nevertheless, oxalic acid is toxic and its use in the food industry is not permitted.

$$x\, C_2O_4{}^{2-} + Fe^{3+} \leftrightharpoons [Fe\,(C_2O_4)_x]^{(2x-3)-}$$

Another chelating agent is pyrophosphoric acid ($H_4P_2O_7$), which has a chelating capacity for iron that is comparable to that of oxalic acid, as long as concentrations of 200 to 300 mg/L are not exceeded, since under those conditions iron pyrophosphate would precipitate. Nevertheless, within the pH range of wine, this acid hydrolyzes more rapidly at low pH and high temperature.

TABLE 20.3   Dissociation Constants of Some Complexes Formed
With Fe^{3+}

Ion	K_d	Metal:ligand
Tartrate	3.16×10^{-8}	1:1
Citrate	6.31×10^{-12}	1:1
EDTA-Fe	1.0×10^{-25}	1:1
Oxalate	6.25×10^{-21}	1:1
Polyphosphate (n $=5$)	10^{-10}	1:6

Polyphosphates are obtained by fusion of a mixture of orthophosphates. The most important are sodium hexametaphosphate $(NaPO_3)_6$ and sodium tripolyphosphate $(Na_5P_3O_{10})$. The tripolyphosphate has a powerful chelating capacity for various metals, particularly calcium and iron. Whereas the greatest stability of the calcium complexes is found at high pH, the iron complexes are most stable at a pH approaching 2. Hexametaphosphate also forms complexes, though part of its action is achieved by precipitation of iron as insoluble complexes. It is slightly less effective than polyphosphates.

EDTA, as a disodium salt or a calcium salt, has a high chelating capacity. Both salts are widely used as additives to chelate metals in various food products.

Currently, treatment of wine with chelating ions other than citric acid is not permitted in the European Union.

6.5. Treatment with Ascorbic Acid

Ascorbic acid is an excellent reducing agent that inhibits the oxidation of ferrous iron during any process that involves oxygenation of the wine (racking, bottling, etc.). Its effect, however, is not permanent, and it is therefore only used in very specific cases, such as in wines with a slight tendency towards ferric casse that cannot be treated using other methods and that are to be consumed within a short period of time.

Finally, it should be mentioned that exposure to light for 24 to 48 hours in wine during the initial stages of ferric casse causes reduction of Fe^{3+} to Fe^{2+} and the casse disappears. Nevertheless, the problem returns after a short period of time. This solution can be useful in some cases when the wine is already in the bottle but $FePO_4$ deposits have not yet formed.

7. TREATMENTS TO PREVENT COPPER CASSE

Copper casse mainly occurs in white wines. The tendency for this type of casse to occur can be tested by exposing a transparent bottle of wine to sunlight or ultraviolet light. If no reddish precipitate forms after a week, the wine can be considered stable.

The use of bentonite indirectly results in protection against precipitation of copper since it causes the flocculation of the proteins necessary for the casse to appear. Gum arabic also has a preventive effect as it protects against flocculation of the copper colloid, although it does not prevent its formation. Nevertheless, these treatments are ineffective if the concentration of copper in the wine exceeds 1 mg/L.

The most effective treatment when copper is found at higher concentrations is the use of potassium ferrocyanide, which forms a highly insoluble compound:

$$2 \, Cu^{2+} + Fe(CN)_6{}^{4-} \leftrightarrows Cu_2[Fe(CN)_6] \downarrow . \, K_{sp} = 1.3 \times 10^{-16}$$

Copper ferrocyanide is more soluble than ferric ferrocyanide but its formation is much more rapid, since there is a higher concentration of free Cu^{2+} than free Fe^{3+}. The use of ferrocyanide is only effective, however, in the presence of iron concentrations between 5 and 10 mg/L. After treatment, the concentration of copper does not exceed 0.2 mg/L.

The use of ferric ferrocyanide instead of potassium ferrocyanide results in a greater reduction in Cu^{2+}, but it is not permitted.

Other treatments are available for the elimination of copper, although they are no longer used. One of these involves heating the wine in the absence of air to favor the formation of a copper colloid that flocculates when a fining agent is added and the wine is cooled. Another treatment involves adding $Na_2S.9 \, H_2O$ to the wine at a dose of 25 mg/L to form copper sulfate, a negatively charged colloid that slowly flocculates. The reasoning is that the addition of a positively charged colloid will favor its flocculation and increase the efficiency of fining. Given that the solubility of copper sulfate is 0.2 mg/L, 0.13 mg/L of copper will remain, although this concentration does not cause haze. A problem with this treatment is created by the presence of oxygen, which oxidizes the copper sulfate to sulfate, which then redissolves. Nevertheless, an increase in the dose of Na_2S will cause the copper to precipitate once again.

The greatest problem, however, occurs when excess Na_2S is added, since this leads to the production of H_2S and causes an unpleasant aroma. In the presence of SO_2, colloidal sulfur forms and gives a bitter flavor to the wine. Both disappear over time if the excess is not too great. If the wine does not contain SO_2, H_2S persists and needs to be eliminated by addition of $CuSO_4$.

$$2 \, SO_2 + H_2S + H_2O \rightarrow S + H_2S_2O_5 + H_2$$

The reaction of SO_2 and H_2S leads to the formation of pentathionic acid ($H_2S_5O_2$), which is unstable in acid solution and leads to the formation of pyrosulfurous acid ($H_2S_2O_5$). The method is also less efficient than precipitation by potassium ferrocyanide.

Since exposing wines with a tendency to undergo copper casse to light leads to the appearance of a copper precipitate, treatment with sunlight or ultraviolet light can be considered as an option to eliminate copper. However, this can cause photo-oxidative degradation of cysteine and methionine, which forms volatile thiols with an unpleasant aroma which are responsible for the defect known as light strike that makes many wines unsuitable for consumption.

Finally, aging on lees causes a reduction in the redox potential of the wine that makes it more sensitive to copper precipitation. Nevertheless, the lees themselves fix copper and reduce the likelihood of precipitation. As a result, aging on lees not only offers organoleptic benefits but also functions as a means of stabilizing wine against precipitation of proteins, tartaric acid, and copper.

1. INTRODUCTION: AGING AND STORAGE

Once must has been fermented, the resulting wine has various potential destinations. In the case of young wines, it is hoped that they will retain their organoleptic properties for the longest possible time — in other words, the aim is to conserve the wine in its current condition. Sometimes, however, the aim is to generate a specific change in the composition of the wine in order to obtain other target organoleptic properties. In some cases, the wines are deposited in inert vessels that contribute little or nothing to the composition of the wine and serve only as containers; they therefore provide an environment in which the chemical reactions can occur but they do not participate in them. In other cases, the wines are deposited in vessels that contribute certain components that can interact chemically and physically with the other components of the wine. This is the case for barrel aging, in which the physical and chemical properties of the wine are influenced by components derived from the container. Bottle aging, on the other hand, can be defined as a process in which the natural physical and chemical properties of wine change over time in inert containers. Storage refers to the process by which a finished wine ready for sale can be kept for the longest possible period of time without changes occurring in its composition, organoleptic properties, or hygiene.

Oxidation occurs in porous containers such as wood, whereas reduction occurs in containers made of glass or stainless steel. Between these two ends of the oxidation-reduction spectrum, a multitude of different reactions can occur according to the type of container, the period of contact with the wine, the composition of the wine, and the external conditions under which the process occurs.

Barrel-aged white wines are commonly subjected to oxidizing conditions in wooden barrels over long periods of time. The most characteristic examples are Port and Madeira wines in Portugal, Oloroso wines from Jerez and Montilla-Moriles, and *rancio* or *wooded* wines such as rancio Muscat de Rivesaltes. Great barrel-aged wines combine a period of time (6 months to several years) of oxidative aging in wooden barrels with a number of years of aging in glass bottles.

These two types of aging are known as chemical aging, since they only involve chemical reactions and are not dependent on microorganisms. Biological aging processes are also used and involve specific types of yeast, such as the *flor* yeast that is used to produce biologically aged wines in regions such as Montilla-Moriles and Jerez in Spain, Sardinia and Sicily in Italy, Jura in France, some regions of California in the USA, and parts of South Africa and Australia.

BOX 21.1

SUMMARY OF THE CHARACTERISTICS OF AGING PROCESSES

Barrel aging: Occurs in active containers (wood) or using yeast

- **Chemical:** Does not involve microorganisms
  - Compounds are contributed to wine by the wood
  - The presence of oxygen creates oxidizing conditions
  - Redox reactions, esterification, hydrolysis, addition, and condensation occur

- **Biological:** Involves microorganisms
  - *Flor* yeasts
  - Specific compounds are contributed to the wine by yeasts

Bottle aging: Occurs in inert containers (glass)

- The absence of oxygen creates reducing conditions
- Esterification, hydrolysis, addition, and condensation occur

---

2. AGING IN OAK BARRELS

Oxidative chemical aging is generally carried out in oak barrels. Wood is a porous material that allows exchange of gases, mainly oxygen, with the atmosphere and also contributes characteristic chemical components to the wine. Consequently, the processes that differentiate chemical aging in wooden barrels from the simpler aging processes in inert materials are oxidation and extraction of components from the wood.

In the barrel-aging process, it is essential for the wine to be in contact with the wood in order to effectively and selectively extract the substances that enrich its organoleptic profile and facilitate appropriate changes in the wine.

The most common method used for aging has been employed since antiquity and involves maintaining the wine in contact with the wood of the barrels. Under these conditions, slow oxidation takes place and at the same time, tannins from the wood combine with anthocyans from the wine. Although it is known that other types of wood, such as chestnut, cherry, and acacia, have traditionally been used for the construction of barrels, oak is now by far the most widely used material worldwide. Of the 250 species of oak that are found in the wild, however, only a few are used to make barrels.

2.1. Types of Oak

Oak belongs to the genus *Quercus*, and the two most widely used species in Europe belong to the subgenus *Euquercus*. *Quercus petraea* or *sessilis* grow in relatively poor soils and require little sunlight. They are tall and thin, can be planted at a high density, and have a high polyphenol content. In contrast, *Quercus robur* or *pedunculata* grows in fertile soils and requires many hours of sunlight; the methods used in its cultivation generate trees with a large diameter and a high tannin content. In Europe, the largest forests producing high-quality oaks are found in France. Forests cover 27% of the total area of France, and approximately 9% of these are oak forests. The regions of Le Fôret du Centre, Nevers, Tronçais, Allier, and Limousin in the Massif Central and Vosges in the northeast of the country are particularly important producers. Although French oak is the most highly valued in Europe, other producing regions include Hungary, Poland, Russia, Italy, and, on the Iberian Peninsula, the Basque Country. Until the 1930s, the oak that was most widely used in the châteaux of Bordeaux came from Russia rather than France.

American oak is the other major source of wood for barrels, particularly white oak (*Quercus alba*), which is the most widely cultivated, along with the species *Q. macrocarpa*, *Q. muehlenbergii*, and *Q. garryana*. Finally, *Quercus oocarpa*, which grows in Costa Rica, is also used. The growing area in the USA essentially covers the states of Ohio, Missouri, Virginia, Wisconsin, Illinois, and Iowa.

3. COMPOUNDS CONTRIBUTED TO WINE BY WOOD

Wood is the mixture of xylem tissue that makes up the roots, trunk, and branches of woody plants, excluding the bark.

The main components of wood are cellulose (35–50%), hemicelluloses (22–30%), and lignins (22–32%). Of relevance to winemaking is also a high concentration of ellagitannins (around 10%). The drying and toasting of the staves used to make the barrels transforms the natural components of the wood into compounds that will later be donated to the wine during aging, although compounds that are not transformed during drying and toasting are also released. All of the components extracted from the wood by the wine have considerable influence on its organoleptic characteristics and an initial classification groups them into volatile and nonvolatile compounds.

3.1. Volatile Compounds

Furan Aldehydes

Furan aldehydes are formed from cellulose and hemicellulose during toasting via a Maillard reaction. Notable among these aldehydes are furfural, 5-methylfurfural, and 5-hydroxymethylfurfural. They contribute an aroma of almonds and toasted almonds. Their concentration in wine depends on the amount of toasting to which the barrel has been subjected and to the length of contact with the wine.

FIGURE 21.1   Principal furan aldehydes.

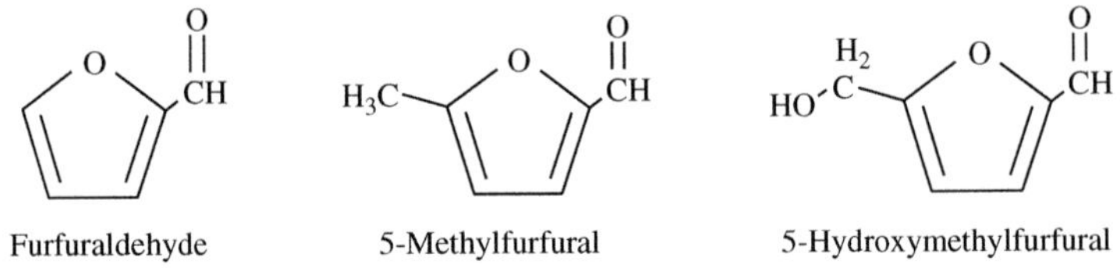

Furfuraldehyde   5-Methylfurfural   5-Hydroxymethylfurfural

Volatile Heterocyclic Compounds

Toasting also generates volatile heterocyclic compounds through a Maillard reaction. These are of two types: oxygenated and nitrogenated. The oxygenated compounds include maltol and dihydromaltol, which contribute toasted and caramel aromas. Their nitrogenated counterparts include dimethylpyrazines, which contribute aromas of cocoa, hazelnut, and toasted bread.

FIGURE 21.2   Volatile heterocyclic compounds formed during the toasting of wood.

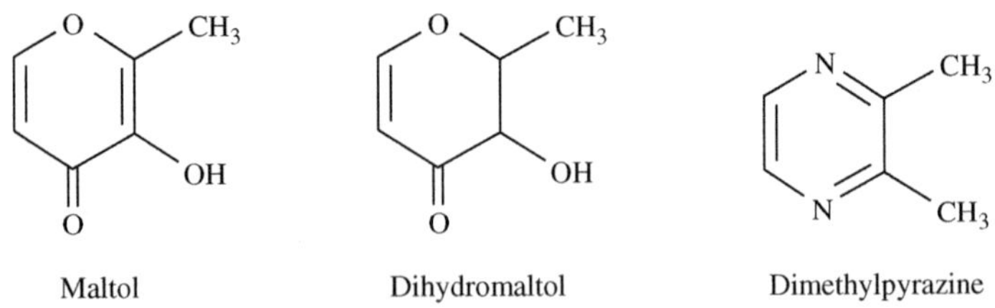

Maltol   Dihydromaltol   Dimethylpyrazine

Phenolic Aldehydes

Degradation of lignin due to heating or alcoholic hydrolysis generates aldehydes, which produce phenolic acids associated with lignin following oxidation:

Syringaldehyde → syringic acid
Sinapaldehyde → sinapic acid
Vanillin → vanillic acid
Coniferaldehyde → ferulic acid

Vanillin is without doubt the most important phenolic aldehyde and contributes a characteristic vanilla aroma to wines aged in wooden barrels. Other cyclic compounds containing a benzene ring are found in wine. These include phenyl ketones (e.g., acetovanillone), which have vanilla aromas and are also produced through the breakdown of lignin.

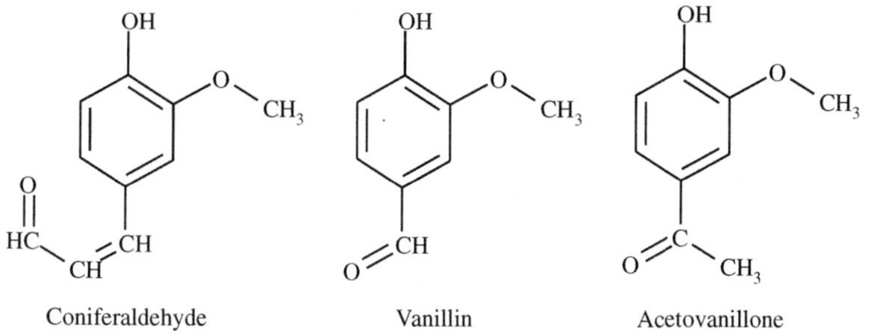

FIGURE 21.3 Compounds obtained from the breakdown of lignin.

Coniferaldehyde   Vanillin   Acetovanillone

Volatile Phenols

The volatile phenols present in wine are classified according to whether they are extracted from wood (derived from the degradation of lignin during toasting) or derived from microorganisms. Wood-derived volatile phenols include guaiacol, 4-methylguaiacol, eugenol, syringol, methyl-4-syringol, and allyl-4-syringol.

The concentration of these phenols in wine is related to the degree of toasting, with concentrations generally increasing with greater degrees of toasting.

Lactones

A number of lactones have been described in barrel-aged wines (β-octalactones, β-nonalactones, and β-decalactones). However, β-methyl-γ-octalactone is the most common. Because of its aroma of coconut and wood, it is commonly referred to as *wood lactone* or *whiskey lactone*. It contains two asymmetric carbons, and there are therefore four possible isomers, of which the most abundant are *cis* and *trans* β-methyl-γ-octalactone.

These lactones appear to be thermal degradation products of the lipids present in the wood. When present in excess, they are linked to resin notes that reduce the organoleptic quality of barrel-aged wine.

Acetic Acid

Large quantities of acetic acid that are not derived from microbial activity have been detected in wines aged in barrels, particularly new barrels. Acetic acid is a product of hydrolysis of the acetyl groups present in hemicellulose that occurs during toasting. Wood-derived

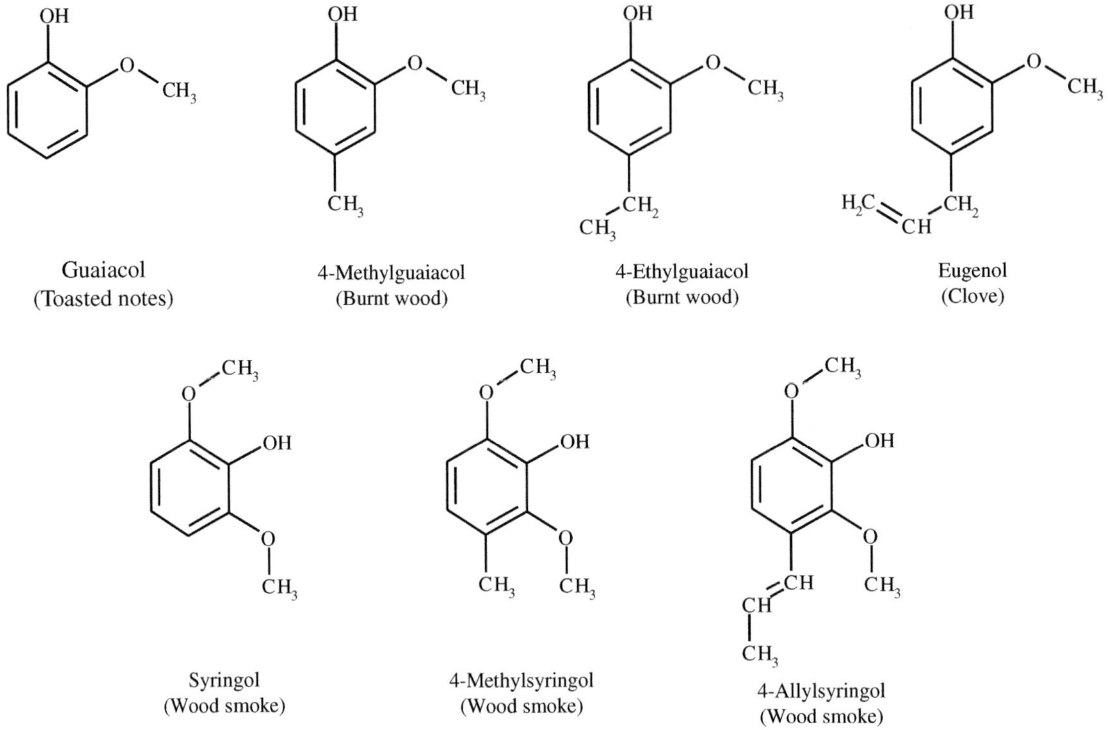

Guaiacol (Toasted notes)	4-Methylguaiacol (Burnt wood)	4-Ethylguaiacol (Burnt wood)	Eugenol (Clove)

Syringol (Wood smoke)

4-Methylsyringol (Wood smoke)

4-Allylsyringol (Wood smoke)

FIGURE 21.4　Volatile phenols extracted from wood and their characteristic aroma.

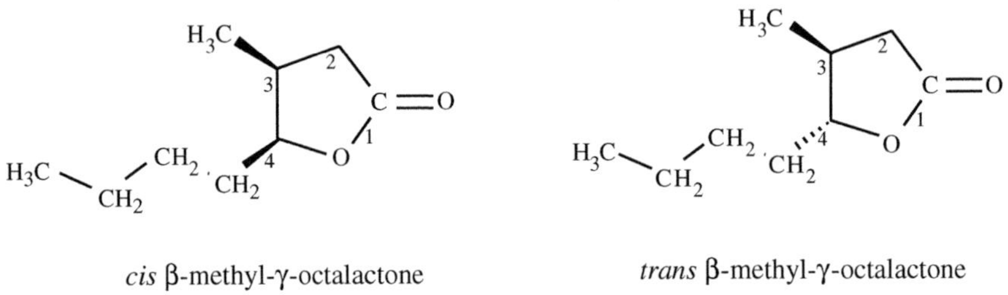

cis β-methyl-γ-octalactone

trans β-methyl-γ-octalactone

FIGURE 21.5　Structure of whiskey lactones.

acetic acid can reach levels of 0.15 g/L. Prior washing of the barrels with water and sulfuric anhydride is recommended to minimize the amount of acetic acid contributed by the wood when there is a large proportion of new wood and whenever the wine destined for aging contains high levels of acetic acid. Washing in this way, however, has the drawback of also removing other substances that are desirable for aging.

Terpene Compounds

Certain terpene compounds and C_{13} derivatives such as linalool and its derivatives, β-terpineol, β-ionone, and β-damascenone have been identified in samples of oak. Their concentration is generally higher in American than in French oak.

3.2. Nonvolatile Compounds

During aging, compounds that contribute to astringency, bitterness, and color stabilization are extracted from the wood.

Phenolic Acids

Notable among the phenolic acids are gallic and ellagic acid. Their importance centers on the stabilization of color, since they can act as co-pigments and protective agents against the oxidation of anthocyans.

Flavanols

Wood can contribute a certain amount of flavanol monomers (catechins) or oligomers (procyanidins). Their influence on astringency and bitterness is minimal in the case of red wines, since these already contain a high concentration of these compounds.

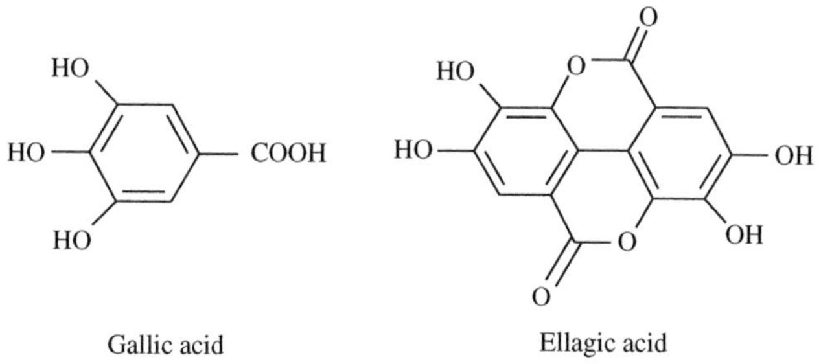

Gallic acid

Ellagic acid

FIGURE 21.6 Structure of gallic and ellagic acid.

Ellagitannins

Ellagitannins are also known as hydrolyzable tannins since they are hydrolyzed to generate gallic acid in acidic media. They are made up of a linear chain of gallic acid molecules esterified with glucose. They are very abundant in wine and contribute to consolidating its structure. The addition of ellagitannins is increasingly common, although when present in excess they give rise to the so-called "plank flavor". These tannins contribute to the stabilization of color because they have a protective effect on the oxidation of anthocyans.

Other Compounds

Coumarins and gallotannins are two groups of phenolic compounds that are characterized by their strong bitterness. The first group, found in untreated wood as a heteroside, is hydrolyzed during the drying process and transformed into the corresponding aglycone, which is much less bitter than the heteroside.

The structure of the gallotannins corresponds to a glucose molecule in which the hydroxyl groups are partially or totally glycosylated with gallic acid molecules. Fortunately, they are not especially abundant in wood and their contribution to bitterness and astringency is therefore minimal.

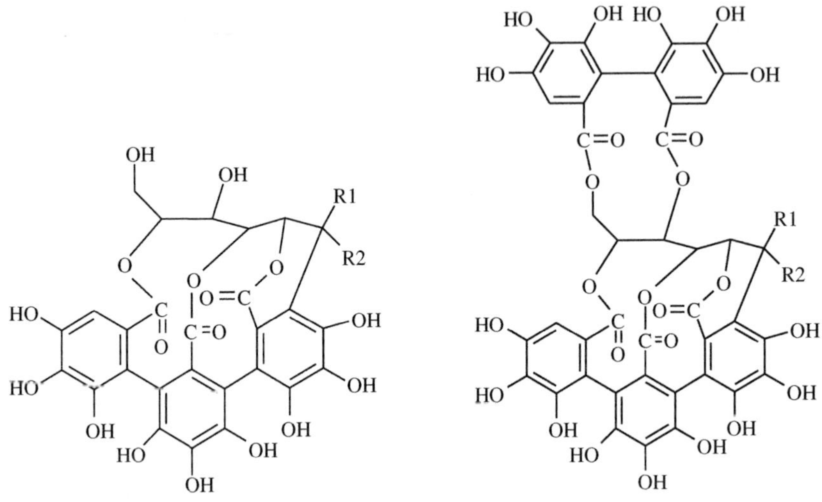

R1= H; R2=OH: Vescalin
R1= OH; R2=H: Castalin

R1= H; R2=OH: Vescalagin
R1= OH; R2=H: Castalagin

FIGURE 21.7   Structure of ellagitannins.

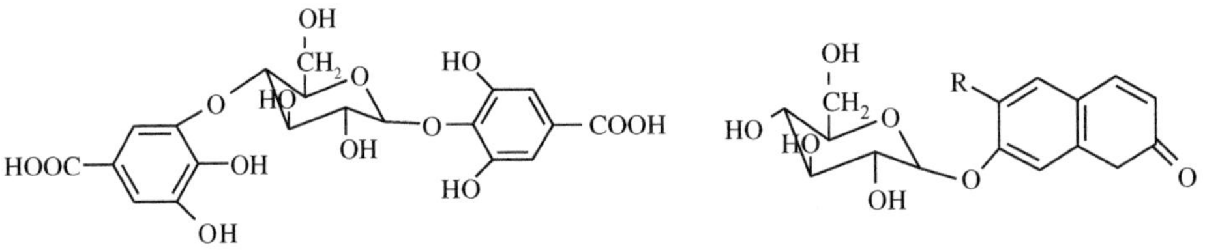

Digalloyl glucoside

Coumarin glucoside

FIGURE 21.8   Structure of a gallotannin (digalloyl glucoside) and a glycosylated coumarin (coumarin glucoside).

4. INFLUENCE OF OXYGEN ON AGING

The presence of oxygen in wine causes oxidation of the alcohols to their corresponding aldehydes and ketones. Large quantities of isobutyraldehyde and other carbonyl compounds such as benzaldehyde, 2-nonanone, and 2-undecanone are produced. However, acetaldehyde predominates over all of them.

The presence of acetaldehyde in wine is due to yeast metabolism during fermentation and the oxidation of ethanol through a mechanism of linked oxidation involving *ortho*-dihydroxyphenols and oxygen. Acetaldehyde functions as a molecular bridge (ethyl bridge) favoring the union of anthocyans and flavanols and the formation of tannins; anthocyan dimers bound to a molecule of acetaldehyde have even been described.

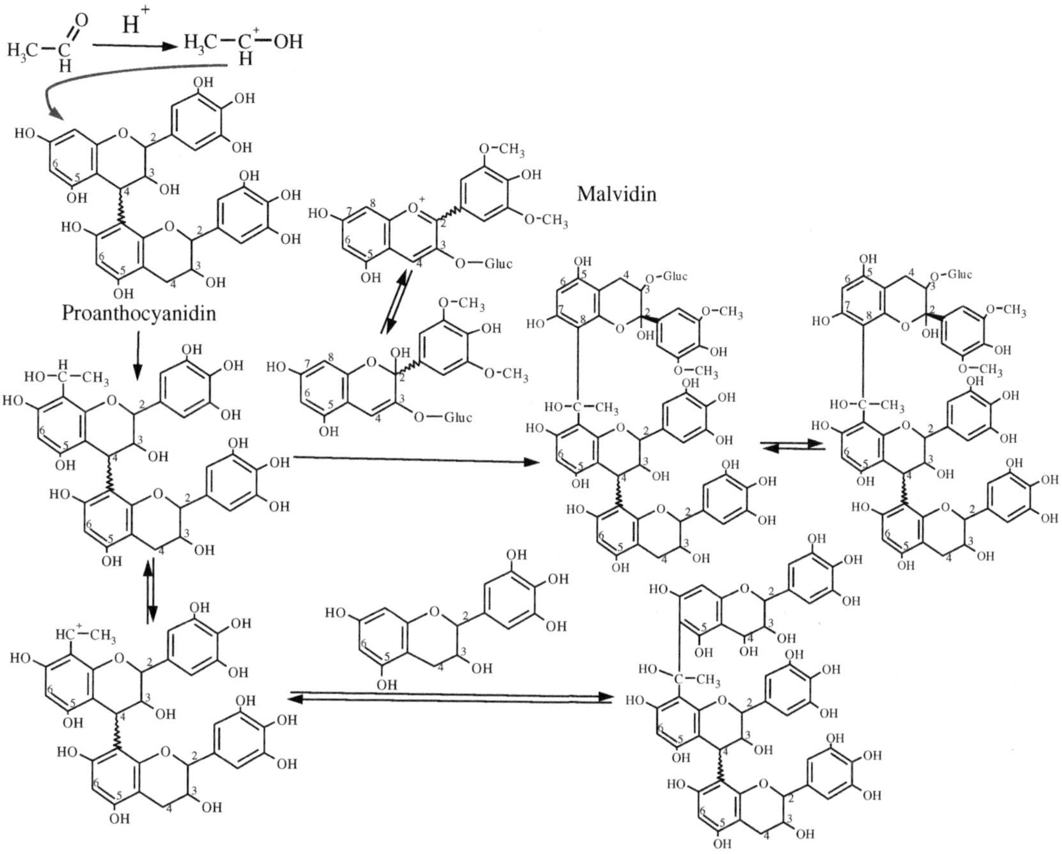

FIGURE 21.9   Reactions mediated by acetaldehyde.

Flavanol-anthocyan combinations make tannins less hydrophobic and gradually reduce the astringency of wine, thereby exerting a favorable effect on the properties of the aging wine.

Condensation reactions between anthocyans and flavanols are generally favored by low pH (<3.5), since this promotes the formation of the activated acetaldehyde cation, which is necessary for the formation of an ethyl bridge. However, advances in the understanding of the chemical mechanisms underlying the aging process have placed in doubt the true contribution of this type of pigment to the color of aged red wines, since:

- Flavanol-anthocyan condensations give rise to large oligomers with a tendency to continue to polymerize, and these would be expected to precipitate after a period of time.
- They are only partially resistant to decoloring caused by sulfites and are poorly resistant to increases in pH.
- They have bluish-red tones that are absent in wines that have undergone a moderate period of aging.

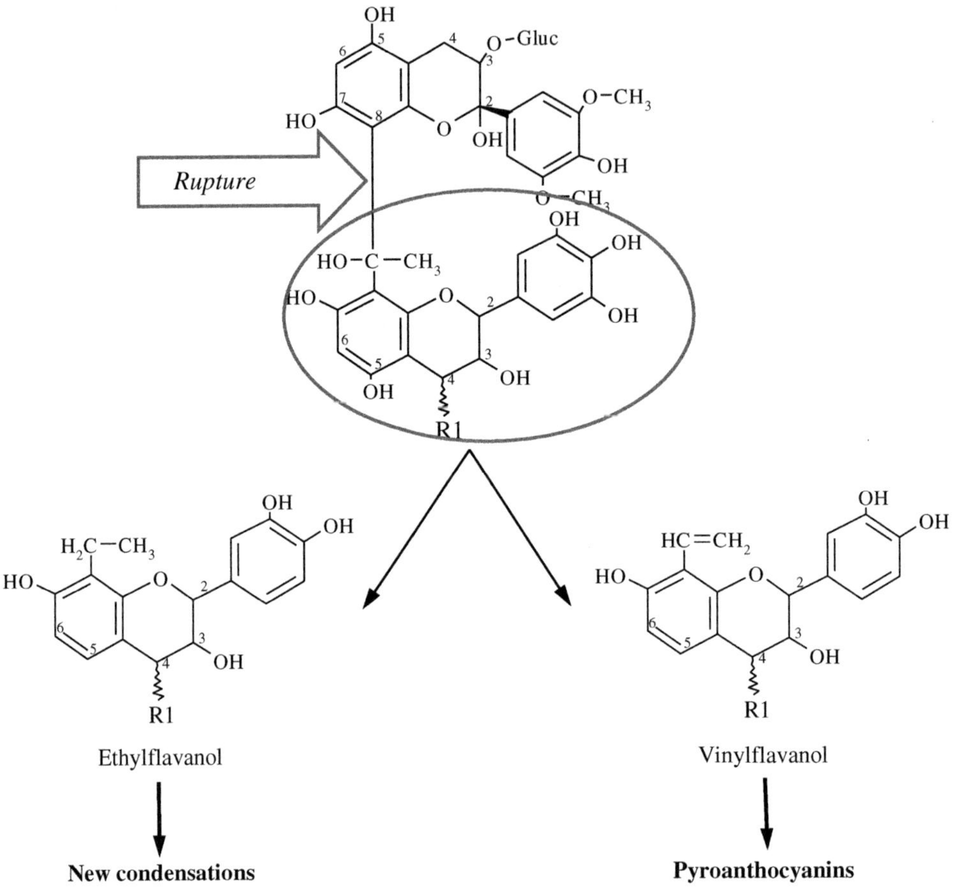

FIGURE 21.10   Rupture of an ethyl bridge and formation of ethyl- and vinyl-flavanols.

Anthocyan-flavanol combinations mediated by acetaldehyde are not as stable as was once thought, and the rupture of these ethyl bridges leads to the formation of ethyl-flavanols and vinyl-flavanols. The ethylflavanols can repolymerize and the size of the polymers is readjusted. Vinyl-flavanols can react with anthocyans to give rise to pyroanthocyanins.

Pyroanthocyanins are cyclic compounds that are formed by the condensation of molecules containing double bonds (acetaldehyde or pyruvic acid) with anthocyans to produce a new heterocyclic compound. These compounds are generically known as adducts. The pyroanthocyanins found in wine have a reddish-orange color and their name is derived from the formation of a fourth pyran ring at position four or five on the anthocyan molecule. They are not present in grapes but rather form during fermentation and aging of the wine.

The pyroanthocyanins have absorption peaks at between 495 and 520 nm (lower than the corresponding anthocyans) and are characterized by a peak absorption close to 420 nm, which would explain their orange tone. The inclusion in the pyran ring of the C4 from the anthocyan causes steric hindrance and makes the molecule more resistant to

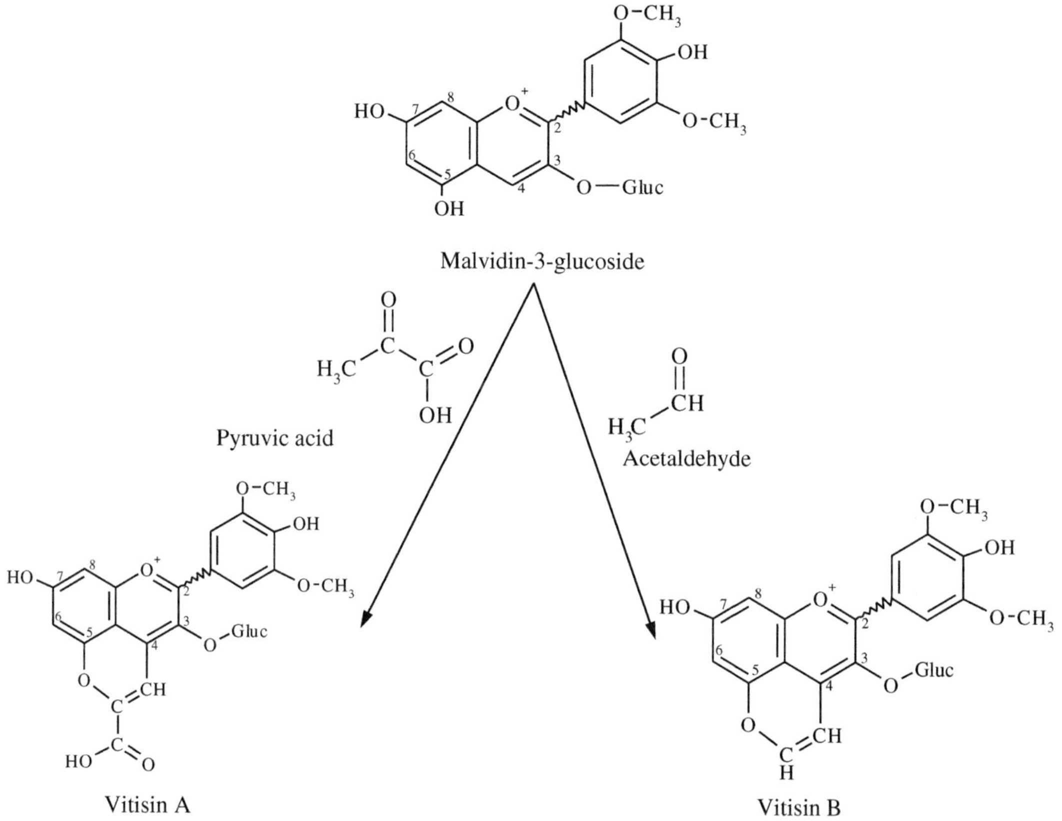

FIGURE 21.11   Formation of vitisin A and B from carbonyl compounds.

discoloration caused by SO_2, increased pH, oxidative degradation, and even temperature. Its concentration in wine is lower than that of other pigments; nevertheless, its greater stability means that almost all of the pyroanthocyanins present contribute to the color of the wine. In addition to being structurally more stable than the anthocyans, the pyroanthocyanins are not absorbed extensively by yeast cell walls, since they form at the end of alcoholic fermentation, when the cell walls are already saturated with anthocyans. The color stability observed in red wines is currently attributed to these compounds, as is the typical orange pigmentation of barrel-aged wines.

Vitisin A and B are among the main pyroanthocyanins and are derived from the addition of a molecule of pyruvic acid or acetaldehyde, respectively, to an anthocyan molecule. Both are formed during alcoholic fermentation, with vitisin A appearing earlier than vitisin B. This may be due to the fact that at the beginning of fermentation there is a greater concentration of pyruvic acid, whereas at the end, the concentration of acetaldehyde is higher and there is less combination of acetaldehyde and sulfuric anhydride. The formation of both vitisin A and B appears to follow antagonistic kinetics in which acetaldehyde competes with pyruvic acid for the anthocyan molecule.

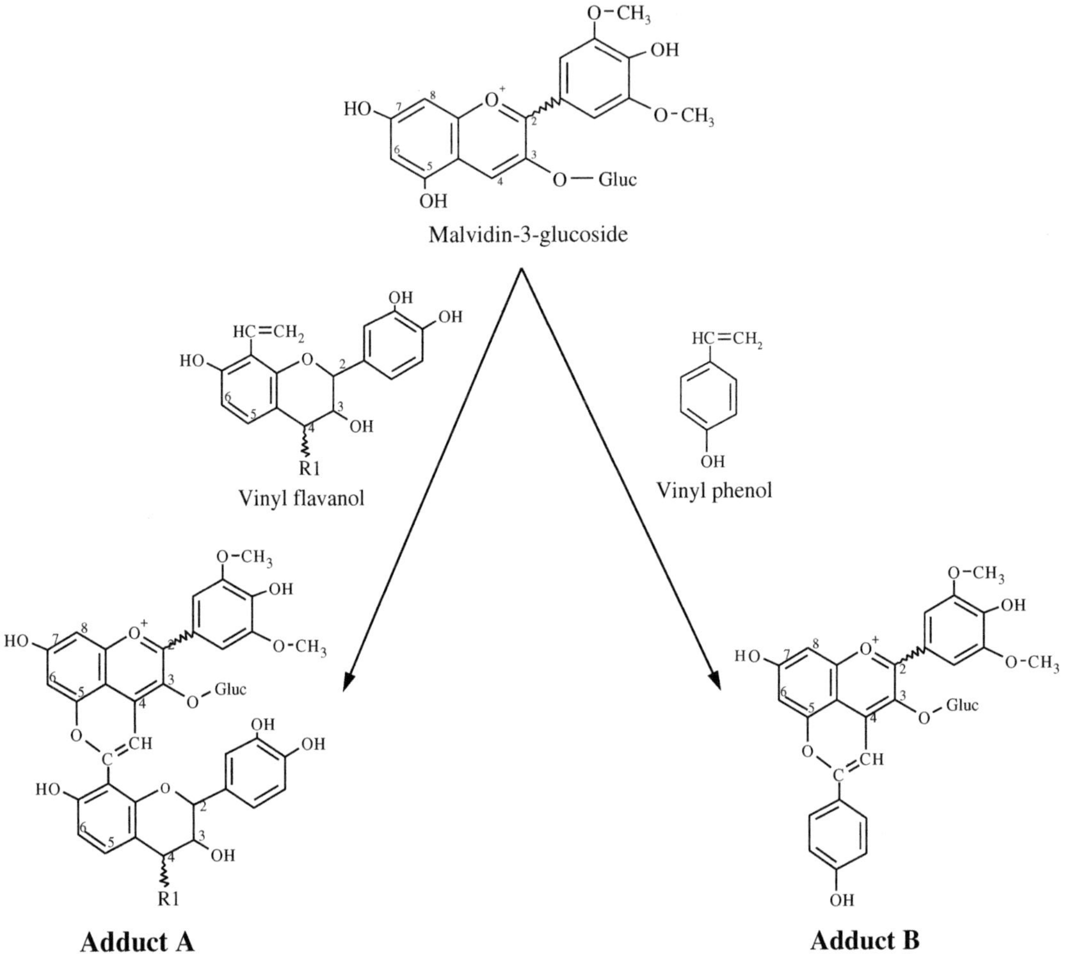

FIGURE 21.12   Formation of adducts from vinyl phenols and vinyl flavanols.

Other pyroanthocyanins are derived from the reaction between anthocyans and vinyl fla-vanols, which are in turn derived from rupture of the flavanol units or flavanol-anthocyan units mediated by acetaldehyde. This can also occur with vinyl phenols derived from decar-boxylation of hydroxycinnamic acids.

In addition to pyruvic acid and acetaldehyde, other molecules containing carbonyl groups may be able to form pyroanthocyanins: including α-ketoglutaric acid, glyoxylic acid, and even acetoin, acetone, and diacetyl. The combination of diacetyl and anthocyans gives rise to castavinols, which can act as a reserve of coloring matter.

In summary, oxygen participates in a wide range of reactions between anthocyans and fla-vanols that result in stabilization of the coloring matter and reduction of astringency. Although not all of the reactions involving oxygen that occur in wine nor their effects on the organoleptic properties of the wine are known, experience shows that both the time in the barrel and the assimilation of oxygen that this provides transforms the phenolic structure of the wine and leads to significantly improved organoleptic properties. Using these insights, alternative techniques to traditional aging have been sought in an effort to stabilize color and reduce astringency. The most notable is micro-oxygenation, which involves administering an appropriate dose of oxygen to the wine in order to control or even accelerate the natural process that takes place in oak barrels.

5. FACTORS THAT INFLUENCE THE COMPOSITION OF THE WOOD

As discussed, when wine is placed in contact with wood, numerous compounds that have a considerable effect on its organoleptic characteristics are extracted. Similarly, the diffusion of oxygen through the pores of the wood leads to a series of reactions that influence color stability and reduce astringency.

Nevertheless, a number of factors linked to the wood itself also influence the type and concentration of the substances released during aging.

5.1. Type of Oak

The contribution made to the wine by the oak barrel can be influenced by the geographic origin of the wood. For instance, American oak is richer in vanillin and lactones and therefore contributes more aromas of vanilla and coconut, respectively, than does French oak. Oak from the forests of Tronçais is rich in eugenol, whereas that from Limousin is better adapted to aging of distillates, which require stronger oxidation than wines. This is explained by the high levels of tannins and the greater porosity of Limousin oak, and because it contains higher levels of hydroxymethylfurfural, which contributes caramel and toasted almond notes.

The current trend is to combine different types of oak in the same winery and even use different types of wood (cherry, chestnut, etc.) in order to obtain wines with a unique personality.

5.2. Drying System

Prior drying of the wood has a substantial influence on the final quality of the barrel. There are two types of drying: natural drying, which requires large sheds and a wait of at least 2 or 3 years, and artificial drying, which is undertaken in drying ovens and takes a few weeks. In general terms, natural drying contributes more aromatic substances and a lower quantity of bitter and astringent compounds.

5.3. Degree of Toasting

During barrel making, the inner surfaces of the staves are burnt and, according to the depth that this reaches, the toasting is classified as light (3 to 5 mm), medium (up to 5 mm), and strong (charring to a depth of more than 5 mm). In principle, a more toasted oak will contribute more notes of vanilla, clove (due to greater generation of eugenol), almond (the toasting generates 5-methylfurfural and hydroxymethylfurfural from the hemi-celluloses in the oak), coconut (due to the presence of lactones), smoke (due to higher concentrations of volatile phenols), etc. In general, the increase in the substances that give aromatic complexity to wine aged in wooden barrels is proportional to the degree of toasting.

5.4. Number of Uses

The number of times a barrel can be used is limited. The contribution of volatile and nonvolatile compounds reduces considerably as the barrel is used. Some studies have shown that the contribution of lactones and ellagitannins is practically nil after 5 years of use. The number of uses also has a determining influence on the efficiency of transfer of extractable compounds from the wood, and the diffusion of oxygen through the wood pores, since these gradually become blocked by precipitated coloring matter and poorly soluble salts. Reduced efficiency leads to reduced diffusion of oxygen and therefore slowing of the reactions in which it is involved.

1. INTRODUCTION

When wine is deposited in an inert container such as a glass bottle, its composition changes over time, as a result of chemical reactions that occur naturally between its components. Bottle aging is normally carried out under conditions of constant temperature and humidity, and efforts are made to isolate the wine from external agents, particularly oxygen. This process is intended to facilitate the development of the natural physical and chemical properties of the wine.

Bottle aging is usually carried out under reducing conditions and is preceded by an oxidative aging process. This is particularly true of red wines. The purpose is to refine the characteristics of color and tannicity that the wine acquired during aging in wooden barrels and, ultimately, improve its organoleptic properties through transformations that are only achieved under conditions with a low oxidation potential. The time required for the wine to achieve optimal quality varies according to its characteristics at the beginning of the process, and ranges from years to decades.

The chemical transformations that occur during bottle aging mainly affect the volatile compounds responsible for aroma and the phenolic compounds responsible for color, bitterness, and astringency.

2. EFFECTS ON VOLATILE COMPOUNDS

Bottle aging causes transformations of volatile compounds that fall in the following main categories:

1. Esters
2. Aldehydes and alcohols
3. Monoterpene compounds
4. Norisoprenoid compounds
5. Volatile phenols
6. Sulfur compounds

2.1. Esters

Wine is an aqueous alcohol mixture with a high organic acid content and an acidic pH. It is therefore not surprising that esterification reactions occur within it. The general esterification reaction is represented by the following:

$$R_1\text{-}COOH + R_2\text{-}CH_2OH \rightleftarrows R_1\text{-}COOCH_2\text{-}R_2 + H_2O$$
$$\text{Acid} \qquad \text{Alcohol} \qquad\quad \text{Ester} \qquad \text{Water}$$

The equilibrium constant for the esterification reaction between an organic acid and ethanol corresponds to the constant for the formation of the corresponding ethyl ester. This constant has almost no dependence on the temperature or the nature of the acids, and it has a value of close to four:

$$K_e = \frac{[R\text{-}COOCH_2CH_3]\,[H_2O]}{[RCOOH]\,[CH_3\text{-}CH_2OH]} \approx 4$$

By using this value of four, certain considerations can be established for a mole of acid and a mole of alcohol:

$$R\text{-}COOH + CH_3\text{-}CH_2OH \rightleftarrows R\text{-}COOCH_2CH_3 + H_2O$$

At Equilibrium: $\quad (1-x) \qquad (1-x) \qquad\qquad x \qquad\qquad x$

Resolving the equation for K_e:

$$4 = \frac{x^2}{(1-x)(1-x)}; \quad x_1 = \frac{2}{3}; \quad x_2 = 2. \text{ Only solution } x_1 \text{ is valid.}$$

In other words, starting from a solution with a 1 M concentration of acid and 1 M concentration of ethyl alcohol, at equilibrium, two thirds of the acid will be esterified, leaving one third in a free state.

Conditions and Mechanism of the Esterification Reaction

Various conditions influence the equilibrium position of the esterification reaction:

1. Only undissociated acids are esterified.
2. The equilibrium is independent of the temperature, but the rate of the esterification reaction and, therefore, the rate at which the equilibrium is established, is temperature dependent. In the cold, the reaction proceeds so slowly that a synthetic solution of acid and alcohol takes more than 3 years to reach equilibrium, whereas at 100°C the same solution reaches equilibrium within 48 hours.
3. The rate of the esterification reaction depends on the nature of the acid.
4. The reaction occurs in acid media.

The mechanism of esterification in an acid medium involves four main steps:

1. Electrophilic attack by the hydrogen ion on the carboxylic oxygen of the acid:

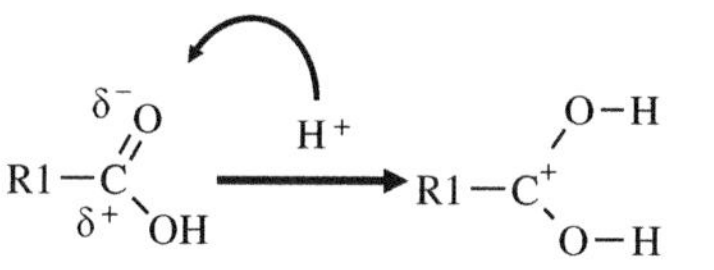

As a consequence the carbon of the carboxyl group becomes positively charged and is therefore easily attacked by the hydroxyl group of the alcohol.
2. Nucleophilic attack by the alcohol:

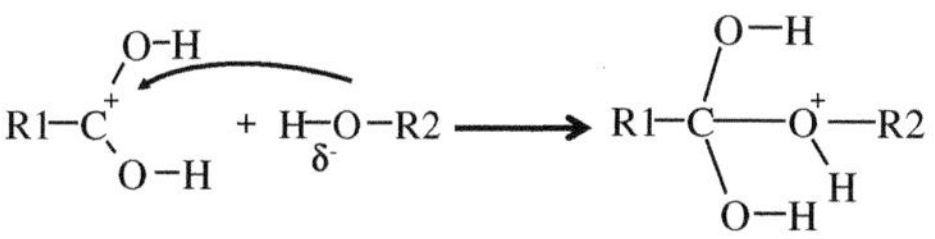

3. Change in the position of the proton from the hydroxyl group of the entering alcohol:

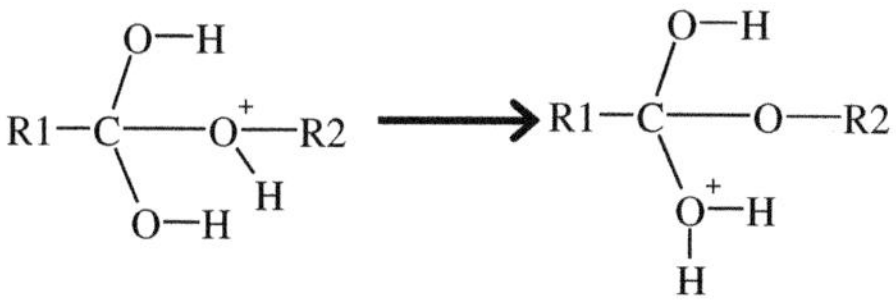

4. Formation of a molecule of water and recovery of the catalyzing proton:

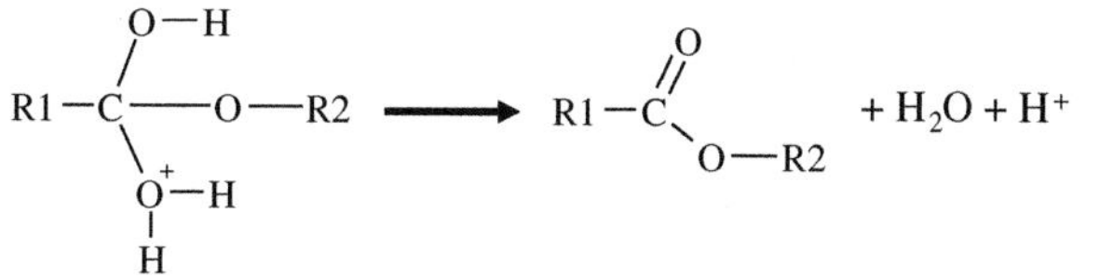

A molecule of water is formed from the −H in the hydroxyl group of the alcohol and the −OH from the carboxyl group of the acid.

The order of reactivity of the alcohols is as follows:

$$\text{Primary} > \text{secondary} > \text{tertiary alcohols}$$

The same order of reactivity is observed for the acids:

$$\text{Primary} > \text{secondary} > \text{tertiary acids}$$

In other words, the reaction occurs more easily between primary acids and alcohols than between secondary or tertiary compounds. This order of reactivity is due to steric hindrance, which keeps the substituents (alcohol and acid groups) apart, particularly in step two of the reaction.

Calculation of the Percentage of Initial Free Acid that can be Esterified

In 1863, Berthelot established an empirical formula to calculate the percentage of the initially present acid that can be esterified over time, based on the quantity of ethanol:

$$Y = 0.9\,A + 3.5$$

$Y =$ esterified acid as a percentage of the initial acid content
$A =$ grams of alcohol in 100 g of wine without dry extract

Using this formula for the example of a wine with the following characteristics:

$\%(\text{vol/vol})$ ethanol $= 10\%.$ Dry extract $= 20\,\text{g/L}$
Volumetric mass in g/mL: wine $= 1.00;$ pure ethanol $= 0.791$

Using the appropriate calculations, we obtain:

$$10\%\ \text{vol/vol} \Rightarrow 100\,\frac{\text{mL ethanol}}{\text{L wine}} \Rightarrow 100\,\frac{\text{mL ethanol} \times 0.791\,\frac{\text{g}}{\text{mL}}}{\text{L wine}} = 79.1\,\frac{\text{g ethanol}}{\text{L wine}}$$

$$79.1\,\frac{\text{g ethanol}}{\text{L wine}} \Rightarrow 79.1\,\frac{\text{mg ethanol}}{\text{mL wine} \times 1\frac{\text{g}}{\text{mL}}} \Rightarrow 79.1\,\frac{\text{mg ethanol}}{\text{g wine}} \equiv 79.1\,\frac{\text{g ethanol}}{1000\ \text{g wine}}$$

$$79.1\,\frac{\text{g ethanol}}{1000\ \text{g wine} - 20\ \text{g extract}} = 0.0807\,\frac{\text{g ethanol}}{\text{g wine without extract}}$$

According to Berthelot, $A = 100 \times 0.0807 = 8.07.$ $Y = 0.9 \times 8{:}07 + 3.5 = 10.76\%.$

This equation is now of little relevance, however, since it has been replaced by the application of the concept of equilibrium and the value of the esterification constant. In other words, an empirical rule has been replaced by a calculation procedure based on scientific data.

Applying the esterification constant to the previous example in order to calculate the percentage of esterification of a free acid in an aqueous alcohol solution, we obtain:

% (vol/vol) ethanol = 10%. Dry extract = 20 g/L
Volumetric mass in g/mL: wine = 1.00; pure ethanol = 0.791

$$79.1 \frac{\text{g ethanol}}{1000 \text{ g wine} - 20 \text{ g extract}} = 79.1 \frac{\text{g ethanol}}{980 \text{ g (water } + \text{ ethanol)}}$$

$$\frac{\text{Moles ethanol}}{\text{L wine}} = \frac{79.1 \frac{\text{g ethanol}}{\text{L wine}}}{46 \frac{\text{g}}{\text{mol}}} = 1.72 \frac{\text{mol}}{\text{L}}; \quad \frac{\text{Moles of water}}{\text{L wine}} = \frac{(980 - 79.1) \frac{\text{g}}{\text{L wine}}}{18 \frac{\text{g}}{\text{mol}}}$$

$$= 55.05$$

The equilibrium can then be written as:

$$\text{R-COOH} + \text{CH}_3\text{-CH}_2\text{OH} \rightleftarrows \text{R-COOCH}_2\text{CH}_3 + \text{H}_2\text{O}$$

At Equilibrium:　　$N_i - x$　　　　$1.72 - x$　　　　　　x　　　　　$50.05 + x$

Using the value for the equilibrium constant, the proportion of ester formed once equilibrium is achieved can be calculated:

$$K_e = 4 = \frac{(50.05 + x)x}{(N_i - x)(1.72 - x)}$$

$$\frac{x}{(N_i - x)} = \frac{1.72}{50.05} \times 4 = 0.1375$$

In this approximation, x is considered very small compared to 1.72 and also compared to 50.05, since in the best-case scenario, the concentration of esters in wine does not exceed 0.20 g/L. (This concentration when expressed as mol/L of the ester ethyl acetate corresponds to 0.0026 mol/L, which is clearly negligible compared to 1.72 and to 50.05 moles.) In other words, at equilibrium, only 13.75% of the acids present are esterified.

To compare this value with the value obtained by Berthelot, the percentage should be converted to a percentage of the free form and the esterified form in the equilibrium and then compared with the initial acid content:

$$\text{R-COOH} + \text{CH}_3\text{-CH}_2\text{OH} \rightleftarrows \text{R-COOCH}_2\text{CH}_3 + \text{H}_2\text{O}$$

Equilibrium:　　100 free　　　　　　　　　　　　13.75 esterified
Initial:　　　　　100 + 13.75　　　　　　　　　　　　0

Here, 13.75 represents the following percentage of the initial acid content:

$$\% \text{ initial esterified acids} = \frac{13.75}{100 + 13.75} \times 100 = 12.09 \%$$

TABLE 22.1   Maximum Percentage of Ethyl Esters Formed in Solutions With Different Ethanol Concentrations

% (vol/vol) Ethanol	According to the Equilibrium Constant	According to Berthelot
7	8.51	8.55
9	10.89	10.02
10	12.09	10.76
12	14.34	12.20
14	16.62	13.67
16	18.84	15.31

According to Berthelot, the percentage of acid that can be esterified in a wine containing 10% ethanol is 10.76%, and according to the equilibrium constant it is 12.09%.

Using the equilibrium constant, we can calculate the maximum percentage of acids that can be esterified in wines with a specific alcohol content.

Ribéreau-Gayon and Peynaud subjected the various acids found in wine to esterification in a 10% ethanol solution at pH values of 3 and 4 and a temperature of 100°C for 24 hours or 30 days. The results revealed even under extreme conditions (pH = 3 and temperature = 100°C for 30 days) the maximum levels of esterification predicted by the equilibrium constant for this alcohol concentration are not achieved, as shown in Table 22.2.

In summary, the value of the equilibrium constant for esterification can be applied to each of the acids found in wine to calculate the maximum possible percentage of esterification. It has also been observed that when the esterification reaction occurs only via a chemical process in wines, the maximum theoretical values are never achieved.

TABLE 22.2   Percentages of Esterification at 100°C for a 10% (vol/vol) Ethanol Solution

Acid	pH = 3		pH = 4	
	24 Hours	30 Days	24 Hours	30 Days
Succinic	8.4	10.2	3.9	9.3
Malic	9.0	10.2	3.8	9.1
Lactic	8.5	9.8	3.0	8.8
Tartaric	5.0	9.3	1.5	8.6
Citric	4.4	6.7	3.0	6.3
Acetic	2.7	8.7	0.8	7.5
Propanoic	2.4	9.0	1.2	7.7
Butanoic	1.4	8.7	0.7	7.6

Esterification Reactions in Wine

Given the large number of different acids and alcohols found in wine, many different esters can be formed. Ethyl esters are nevertheless the most abundant due firstly to the large quantity of ethanol present in wine, and secondly because of the greater reactivity of primary alcohols in this medium. Grapes contain minimal quantities of esters, and therefore the origin of esters in wine is essentially from the enzymatic reactions produced by yeast during fermentation, and to chemical esterification reactions during aging and storage.

ETHYL ACETATE

The most abundant ester in wine is ethyl acetate. It is only formed in small quantities by yeast, and any large concentrations are due to the presence of acetic acid bacteria during storage and/or barrel aging. It is responsible for the aromatic characteristics of pricked or vinegared wines. Its perception threshold is 160 mg/L, but even below this level it can have a serious effect on wine aroma. It is estimated that at concentrations of between 50 and 80 mg/L it has a favorable effect on the aromatic complexity of the wine. Red wines can tolerate higher concentrations of ethyl acetate than can white wines, explaining why this ester does not negatively affect the organoleptic characteristics of the wine.

ETHYL ESTERS OF FATTY ACIDS AND ACETATES OF HIGHER ALCOHOLS

The ethyl esters of fatty acids, mainly hexanoic, octanoic, and decanoic esters, are synthesized by yeasts during alcoholic fermentation. The concentrations that are present in young wines are incompatible with the value of the esterification constant and are therefore hydrolyzed over time. They have very pleasant aromas (unlike the corresponding acids) and participate in the complexity and aromatic finesse of young red and white wines.

Other esters to take into consideration are those formed from acetic acid and higher alcohols (chains containing more than two carbon atoms). These include isoamyl acetate and 2-phenylethyl acetate, both of which have very pleasant aromas (banana and rose). As with ethyl esters, their concentrations in recently fermented wines are higher than those permitted by the value of the equilibrium constant, and they are therefore hydrolyzed to generate the acid and the alcohol, leading to a reduction in the fruity aroma of young wines.

Chemical Origin of Esters

The numerous organic acids present in the wine alongside the high ethanol content explains the formation of ethyl esters during storage and aging, although as we have seen, the theoretical concentration is never achieved in practice. Consequently, it is of interest to establish a relationship between the ester content and the age of the wine. When the ester content is analytically quantified and related to the maximum theoretical content that can be obtained based on the esterification constant, we find that this relationship approaches unity (0.73–0.798) as the aging time increases.

The percentage of chemical esterification increases with the age of the wine, although very slowly, particularly after 2 to 3 years (Table 22.3). The concentration of esters in wine, both those produced enzymatically and those produced through purely chemical reactions, is related to the ethanol concentration, the acid concentration, and the age of the wine. Young wines have ester concentrations of between 2 and 3 meq/L, whereas those aged for a long

TABLE 22.3   Values for the Ratio of Total Percentage
Esters to Maximum Percentage Esterification

Age of the Wine (years)	Range	Mean Values
22 − 43	0.73 − 0.79	0.75
6 − 10	0.57 − 0.71	0.66
4 − 5	0.59 − 0.73	0.64
3	0.49 − 0.67	0.62
2	0.50 − 0.65	0.56
8 months	0.28 − 0.38	0.34

period in barrels have concentrations of between 9 and 10 meq/L. In barrel-aged wines, around 10% of the acids are esterified.

In the case of acids with more than one carboxyl group, the formation of neutral esters or acid esters depends on the pH, with low pH values favoring the formation of neutral esters. At the pH normally found in wine (3−3.8), the quantity of neutral esters formed is very small, and wines contain almost no neutral ethyl esters of malate, tartrate, succinate, or citrate.

ETHYL LACTATE

The formation of ethyl lactate is linked to the development of malolactic fermentation, since during this type of fermentation considerable quantities of lactic acid are produced and this facilitates lactic acid esterification. Although the existence of a bacterial esterase has not been ruled out, the formation of the lactate ester is basically chemical. The presence of ethyl lactate is linked to aromas such as sour milk, yoghurt, and butter.

2.2. Aldehydes and Alcohols

Aldehydes are highly reactive compounds that produce hemiacetals and acetals upon reaction with alcohols. Wines aged for a long period of time contain abundant acetals formed from acetaldehyde with ethanol and other abundant alcohols such as glycerol.

2.3. Monoterpene Compounds

Both free and bound forms of monoterpene alcohols, which are released slowly from their glycosides, undergo hydrolysis reactions and transform into other monoterpenes.

Hydrolysis of Glycosidically Bound Monoterpenes

Slow hydrolysis reactions occurring at the pH of wine release free volatile monoterpenes. The rate of the hydrolysis reaction is greater for terpenes that carry an alcohol functional group on a primary carbon (such as α-terpineol, geraniol, and nerol) than monoterpenes that carry this group on tertiary carbons close to a double bond (linalool), due to the stability of the carbocations formed.

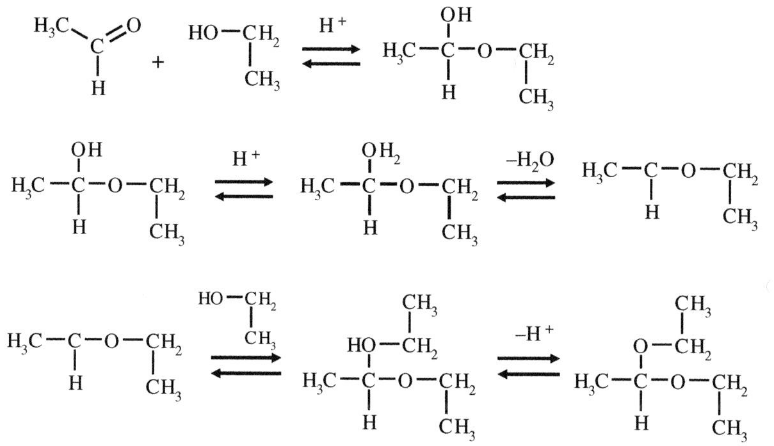

H₃C–C(=O)–H + HO–CH₂–CH₃ ⇌ (H⁺) H₃C–C(OH)(H)–O–CH₂–CH₃

H₃C–C(OH)(H)–O–CH₂–CH₃ ⇌ (H⁺) H₃C–C(OH₂)(H)–O–CH₂–CH₃ ⇌ (–H₂O) H₃C–C(H)–O–CH₂–CH₃

H₃C–C(H)–O–CH₂–CH₃ ⇌ (HO–CH₂–CH₃) H₃C–C(H)(O–CH₂–CH₃)–O–CH₂–CH₃ ⇌ (–H⁺) H₃C–C(H)(O–CH₂–CH₃)–O–CH₂–CH₃

FIGURE 22.1 Acetylation reaction between acetaldehyde and ethanol.

Transformations in Acidic Media

In an acidic medium, monoterpene alcohols undergo isomerization, cyclization, hydration, dehydration, and oxidation reactions in the presence of powerful oxidizing agents. The most common effect observed during aging is a reduction in open-chain monoterpene alcohols and an increase in cyclic forms or the most hydroxylated derivatives. Open-chain monoterpene alcohols are abundant in young wines and have a lower perception threshold than their cyclic equivalents. This accounts for the reduction in the typical aroma of muscat wines during storage or aging.

2.4. Norisoprenoid Compounds (C₁₃)

Norisoprenoids are related to the monoterpenes. They are derived from grapes and during aging react in a manner similar to that described for monoterpene alcohols in acidic media. Some norisoprenoid compounds increase their concentration and others appear as a result of conversion of other norisoprenoids.

Isomers of vitispirane (aromas of camphor and eucalyptus), teaspiran (aroma of tea), β-damascone (aromas of exotic flowers such as the damascene rose) and 1,1,6-trimethyl-1,2-dihydronaphthalene (petrol aromas) have a strong impact on aroma.

1,1,6-Trimethyl-1,2-dihydronaphthalene appears during the aging of Riesling wines and is found at levels above its perception threshold (20 ppb). However, when its concentration exceeds this threshold by five- to ten-fold, it has notably negative effects on the organoleptic quality of the wine. This often occurs in hot winegrowing regions, since exposure of the grapes to sunlight and greater ripeness increase the synthesis of its precursors. High storage temperatures also increase the formation of this compound.

2.5. Volatile Phenolic Compounds

Vinyl phenols such as 4-vinylphenol and 4-vinyl guaiacol are formed during vinification from the tartaric esters of *p*-coumaric and ferulic acid, which give rise to cinammic acids through reactions catalyzed by cinnamyl esterases. These acids produce vinyl phenols by

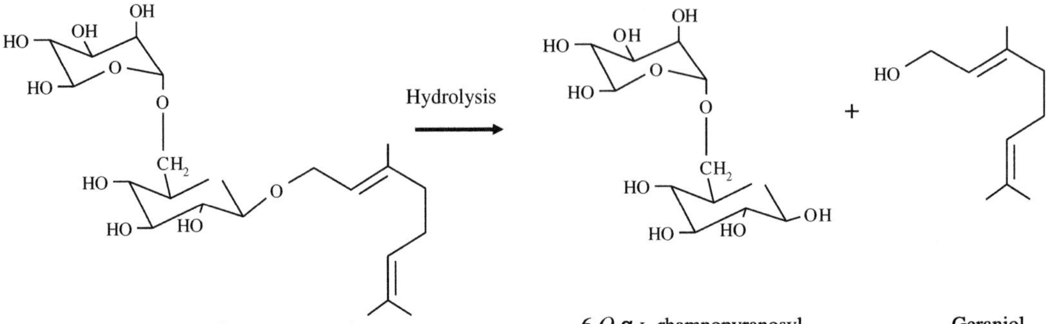

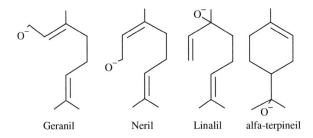

Geraniol 6-O-α-L-rhamnopyranosyl-
ß-D-glucopyranoside

Hydrolysis

6-O-α-L-rhamnopyranosyl-
ß-D-glucopyranoside

Geraniol

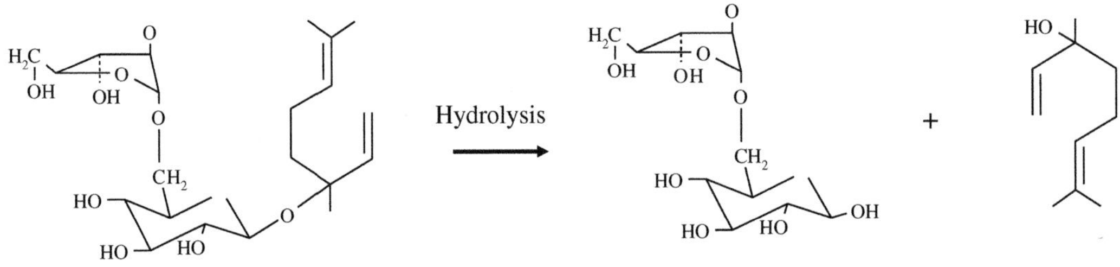

Linalool 6-O-α-arabinofuranosyl-
ß-D-glucopyranoside

Hydrolysis

6-O-α-arabinofuranosyl-
ß-D-glucopyranoside

Linalool

FIGURE 22.2   **Hydrolysis of monoterpene glycosides.**

decarboxylation. Vinyl phenols have unpleasant aromas (associated with chemist's shop, phenol, medicines) when the sum of their concentrations exceeds the perception threshold of around 725 ppb. The aroma disappears during storage of the wine due to a progressive reduction in the concentration of vinyl phenols over time.

This reduction in the concentration of vinyl phenols is due to the high reactivity of the vinyl group, which reacts by addition of ethanol to the double bond in an acidic medium, forming ethoxyethyl phenols. Other addition reactions can occur involving other nucleophilic agents found in the wine, and vinyl oligomers can even form during aging or storage.

Geranil   Neril   Linalil   alfa-terpineil

FIGURE 22.3   **Carbocations generated by hydrolysis of monoterpene glycosides.**

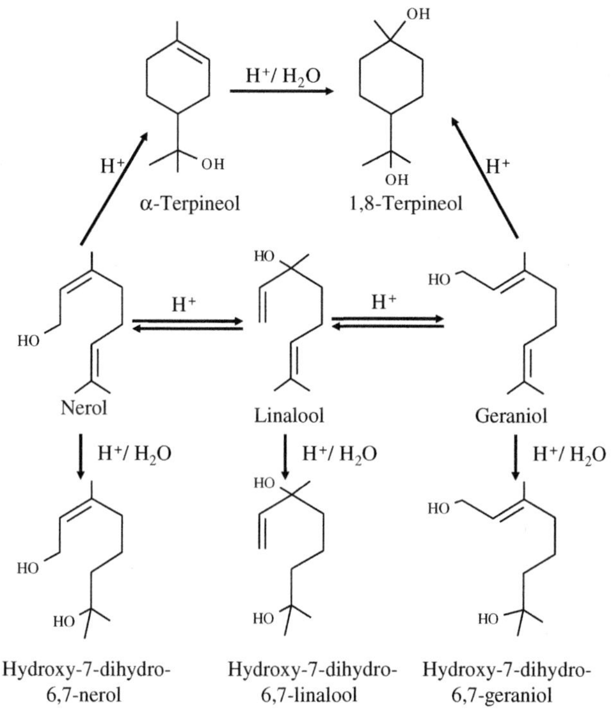

FIGURE 22.4 Conversion reactions of terpene compounds in acidic media.

Ethyl phenols can attain excessive levels during aging and storage of wine, and this has a negative effect on quality due to the unpleasant aromas of leather, animals, stables, wells, or canals that they produce. Their appearance is linked to the presence of contaminating yeasts of the genus *Brettanomyces/Dekkera*, and concentrations above 425 ppb for the mixture

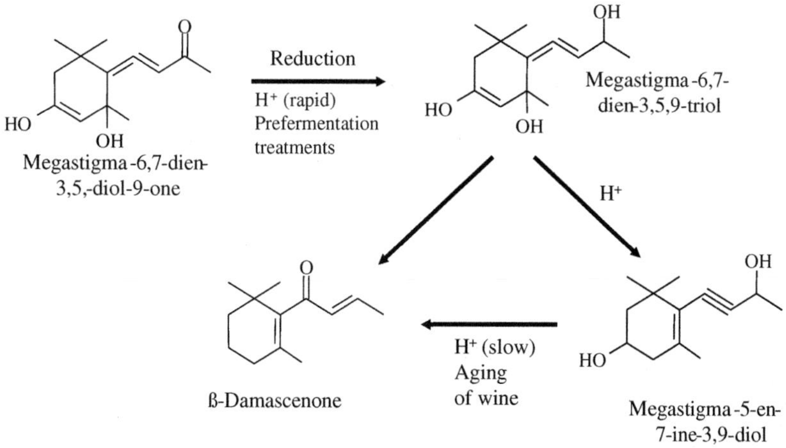

FIGURE 22.5 Formation of β-damascenone during the different phases of vinification.

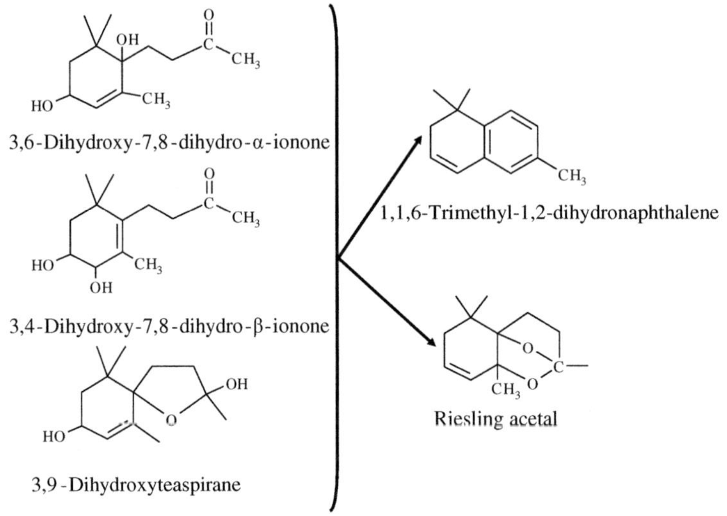

3,6-Dihydroxy-7,8-dihydro-α-ionone

3,4-Dihydroxy-7,8-dihydro-β-ionone

3,9-Dihydroxyteaspirane

1,1,6-Trimethyl-1,2-dihydronaphthalene

Riesling acetal

FIGURE 22.6 Relationship between 1,1,6-trimethyl-1,2-dihydronaphthalene and acetal in a Riesling containing different norisoprenoids.

of 4-ethylphenol and 4-ethyl guaiacol at a ratio of 1:10 (in Bordeaux wines) are responsible for the aroma of leather and animals present in some affected wines. The compounds are stable and do not undergo modification over time. Other volatile phenols that have been described are derivatives of guaiacol with methanol and butanol derived from hydrolysis of the glycosides present in the grape; their contribution to the aroma of wines stored under a cushion of air has not been demonstrated, however.

2.6. Sulfur Compounds

Volatile sulfur compounds generally contribute unpleasant aromas to wine. At the right concentration, however, they can contribute a distinctive note that characterizes certain wines, as in the case of dimethylsulfide $((CH_3)_2S)$ in bottle-aged white Chenin, Colombard, or Riesling wines.

The perception threshold for dimethylsulfide is 25 ppb in white wines and 60 ppb in red wines. Its aroma is described using terms such as asparagus, cabbage, or molasses. Whereas concentrations of 40 to 60 ppb are acceptable in Riesling wines, the same is not true of other wines, with concentrations of 23 ppb proving to be unpleasant in red wines. The mechanism of its formation in wine during aging and storage in the bottle has not been elucidated.

Another sulfur compound, ethanethiol, is believed to be responsible for the aroma of onions in some wines. Its perception threshold is 1.1 ppb in white wines, and its concentration exceeds this limit in wines with sulfur aromas. Ethanethiol, as well as methanethiol acetate and 2- and 3-methylthiophene have been identified in aged champagnes.

Sulfur compounds such as dimethylsulfide, dimethyl disulfide, and hydrogen sulfide are responsible for the flavor of gunpowder and form in champagne wines from sulfated amino

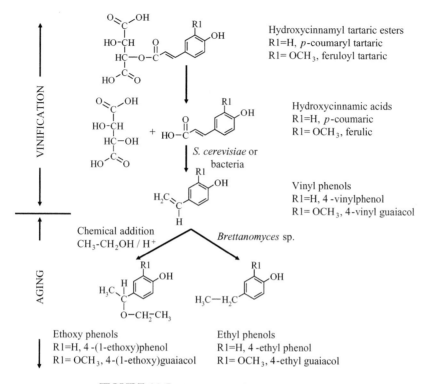

FIGURE 22.7 Formation of volatile phenols.

acids via degradative photo-oxidation. Wines aged on lees may also become enriched in volatile sulfur compounds and their precursors without involvement of light.

3. EFFECT ON NONVOLATILE PHENOLIC COMPOUNDS

One of the most visible consequences of aging in white wines is the formation of brown pigments that lead to browning of the wine and the appearance of reddish-yellow tones. These pigments mainly form as a result of reactions involving polymerization of flavanols.

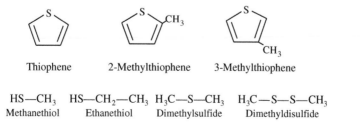

FIGURE 22.8 Structures of the sulfur compounds most commonly found in wine.

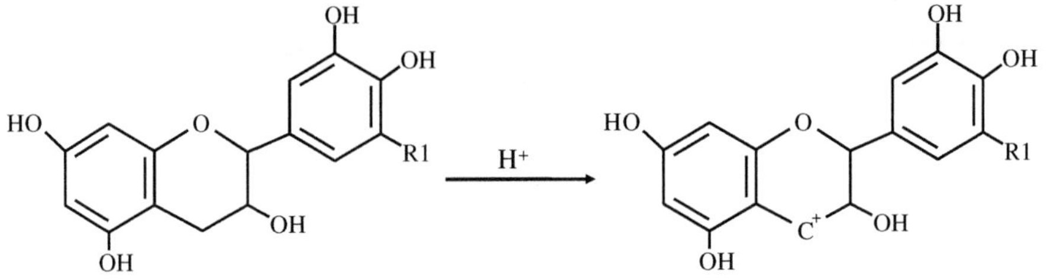

FIGURE 22.9　Formation of the carbocation at position 4 from the corresponding flavanol.

Flavanols are also known as leukoanthocyanidins and correspond structurally to 3- and 4-hydroxyflavanols and 3,4-hydroxyflavanols. In acid media they can give rise to a carbocation at position 4. The polymerization of these compounds at positions 4 and 8 forms tannins in white grapes. These are known as procyanidols, since they produce anthocyanidols and flavanols with a hydroxyl group at position 3 and 3,4 by acid hydrolysis.

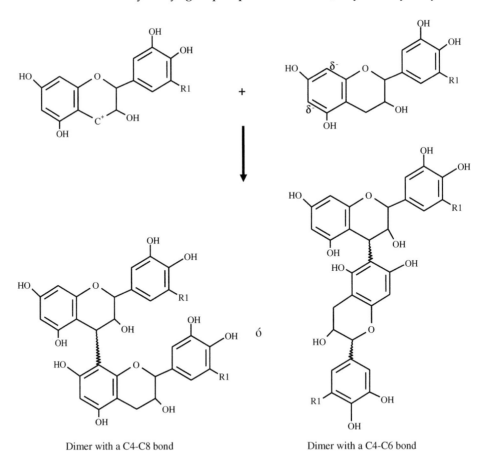

Dimer with a C4-C8 bond

Dimer with a C4-C6 bond

FIGURE 22.10　Formation of a C4-C6 and C4-C8 dimer from two molecules of flavanol.

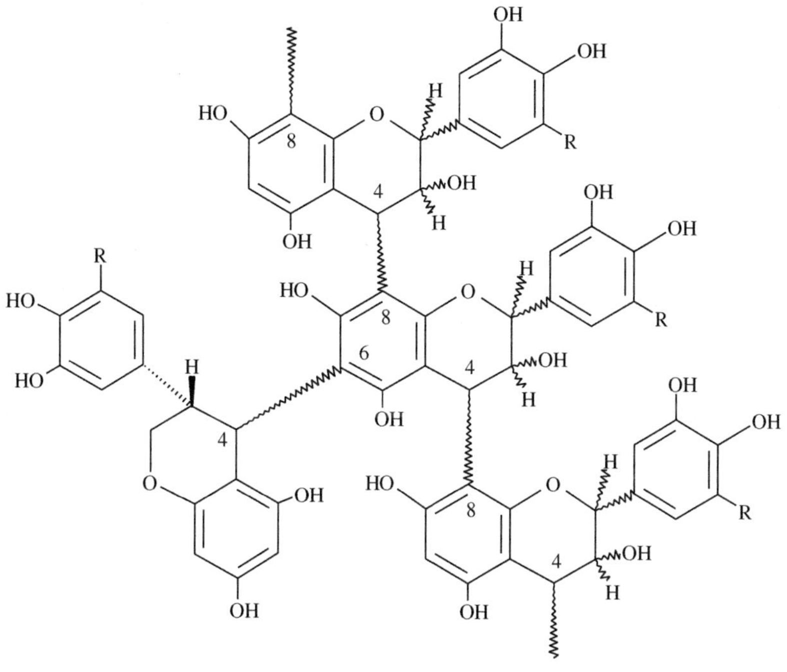

FIGURE 22.11 Structure of tannins.

The occurrence of two hydroxyl groups at positions 5 and 7 of the flavanols leads to the appearance of a negative residual charge at positions 8 and 6. Consequently, these carbons can act as nucleophilic agents that add to the carbocation at position 4 of the pyran ring of another flavanol. This reaction leads to the formation of a flavanol dimer with C4-C6 or C4-C8 bonds.

The formation of another carbocation at free position 4 or in another flavanol molecule, and the addition of the dimer, leads to formation of a trimer. Successive additions then generate oligomers and polymers with an increasingly high molecular mass (tannins). As the number of monomers grows, the molecular mass of the polymer increases until the factors that stabilize it are insufficient to maintain the macromolecules in suspension, and they flocculate and form sediment on the bottom of the vessel. As discussed in Chapter 5, the number of flavanol units that make up the polymer influences the astringency and bitterness.

In the case of red wines, polymerization continues to occur in reactions involving the tannins from the wine and those contributed by the oak of the barrels. As a result, bottle-aged wines have a lower astringency and bitterness than those taken immediately from the barrel.

Free anthocyanins ultimately disappear and other pigments come to predominate. These include pyroanthocyanins, which are responsible for the development of orange tones. If the aging process is extended for a long period of time in the bottle, precipitated coloring matter appears as a result of the progressive polymerization of the pigments responsible for the color of the wine.

Biological Aging

1. INTRODUCTION

Unlike oxidative aging, which takes place in oak barrels and is characterized exclusively by chemical phenomena, biological aging is performed by *Saccharomyces* yeasts that are capable of growing on the surface of wine, where they form a thin layer, or biofilm, known as a *flor*.

Biological aging starts spontaneously every spring in recently fermented wines in the Montilla-Moriles and Jerez regions of Spain and in other wine-producing regions around the world. The first phase involves the growth of yeast colonies on the surface of the wine. These colonies, which resemble small white or ivory-colored flowers, grow until they cover the entire surface of the wine. This thin biofilm gradually acquires a rough, creamy appearance and can reach a thickness of up to 5 mm. The yeasts are known as *flor* yeasts (*flor* means flower in Spanish) to distinguish them from *Candida* yeasts, which are harmful to wine.

Certain varieties of *Saccharomyces cerevisiae* yeasts are capable of switching from a fermentative metabolism to an oxidative (respiratory) metabolism at the end of alcoholic fermentation and of forming a *flor* film on the surface of the wine. *Flor* yeasts consume dissolved oxygen, create a reducing environment, and alter the composition of the wine, conferring specific organoleptic properties that shape the nature of the final product, known as *fino* wine. The aroma of this very distinctive white wine is determined more by the activity of the *flor* yeasts than by the variety of grape used to make the wine. Most of the varieties used are neutral or lightly aromatic.

The traditional method of producing *fino* wine involves a dynamic aging system known as *criaderas* and *solera*, which essentially consists of periodically blending wines of different ages to produce a uniform final product. The main purpose of this system is to ensure a supply of nutrients to the yeasts to keep the *flor* active for as long as possible. The terms *year* or *vintage* therefore are not used for *fino* wines in the same sense as they are for red wines as the final product is a blend of wine from various years. The system also produces a highly uniform product in terms of composition and organoleptic quality. The main disadvantages of biological aging are that it is a complex, labor-intensive process and that large volumes of wine are placed on hold for at least 4 years, which is the minimum time required to achieve the desired quality in terms of analytical and organoleptic profiles.

FIGURE 23.1   *Flor* biofilm on the surface of wine.

2. BIOLOGICAL AGING: THE PROCESS

The biological aging method is used in Sardinia and Sicily (Italy), Jura (France), Tokaj (Hungary), California (USA), and in different regions of South Africa and Australia. However, the most famous — and most studied — biologically aged wines are those produced in the south of Spain, mainly in the regions of Jerez and Montilla-Moriles, using the traditional method shown in Figure 23.2. As mentioned earlier, the main aim of biological aging is to ensure the formation and maintenance of an active film of yeast on the surface of the wine for as long as possible.

Several months after the completion of alcoholic fermentation, the wines are racked off their lees. In Jerez, the wines are fortified with grape spirit to an alcoholic strength of between 15.0 and 15.5% (vol/vol). Fortification is not necessary in Montilla-Moriles wines as, thanks to the climate of this region and the unique characteristics of the predominant grape variety, Pedro Ximénez, the wines naturally acquire an ethanol content of over 15% (vol/vol).

Once alcoholic fermentation is complete, the wine is stored in conical cement vats until it is ready to be transferred to the dynamic, biological aging system. During this time, the wine undergoes malolactic fermentation and often develops a thin layer of yeast on its surface. This wine is known by the term *sobretablas*.

Biological aging starts when the *sobretablas* wine is transferred to American oak barrels that have been used for a long period of time and whose size varies with their location within the *criaderas* system. The barrels are filled to approximately four-fifths of their capacity to allow room for the *flor* to grow on the surface of the wine. The barrels are then stacked on top of each other in three or four numbered rows known as *criaderas*. The wine within each row is of the same type and age.

The bottommost row is called the *solera* and contains the oldest wine. This is the wine that is bottled. No more than 40% of the content of each barrel on this row can be bottled per year. This wine is normally taken out three or four times a year and is replaced by an identical volume of wine from the row above, known as the first *criadera*. The barrels in the first *criadera*, in turn, are replenished with wine from the second *criadera*, and so on up to the topmost *criadera*, which contains the youngest wine in the pyramid (Figure 23.2). This wine is replenished with the *sobretablas* wine from the cement vats. There are generally four to six rows, and

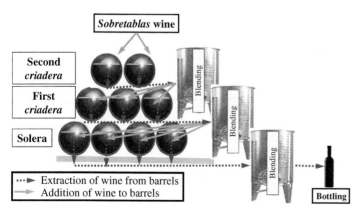

FIGURE 23.2 Traditional *criaderas* and *solera* biological aging system.

the quality of wine produced in a six-row system is generally superior to that produced in a four-row system.

The transfer of a younger wine to a barrel containing an older wine is known as *rocio*. This generally involves a series of operations designed to homogenize the wine within the same *criadera*, before transferring it to the row below. The ultimate goal is to ensure that the wine in each row has undergone the same amount of aging. The wine taken from barrels in the same row is blended in stainless steel containers before being transferred to barrels below. Because no more than 40% of the content of a barrel can be removed in a single year, in a winery in which the wine is removed four times a year, it can be assumed that during each *rocio*, 10% of the wine from any given row is blended with 90% of wine in the row below. Particular caution is required during these operations to ensure that the *flor* is kept intact.

The system ensures that the *solera* wine is a complex, uniform blend of wines from different years and means that the bottled products all have similar organoleptic properties, regardless of the harvest. The *rocio* method also brings younger wine into contact with older wine, with the former providing the latter with the nutrients it needs to keep its *flor* active. It also aerates the wine, which has many benefits for both the wine and the *flor* yeasts. Because the barrels used are very old, their purpose is simply to store the wine (i.e., there is practically no transfer of compounds from the wood to the wine).

In other parts of the world, biological aging is performed using a static system. This is the case for the *vins jaunes* (yellow wines) of the Jura region in France. Other countries use a semi-dynamic system, which is shorter — and less expensive — than the traditional system used in Jerez and Montilla-Moriles. The *vins jaunes* from Jura are made from the grape variety Savagnin Blanc (similar to Traminer), which yields wine with an alcohol content of 12% (vol/vol). After malolactic fermentation, the wine is transferred to old barrels with a capacity of 228 liters. A space of 5 or 6 liters is left unfilled and the barrels are sealed hermetically and stored for 6 years and 3 months in cellars with a temperature of between 7°C (winter) and 17°C (summer) and a relative humidity of 60 to 80%. These conditions trigger the spontaneous development of *flor* yeasts on the surface of the wine. These yeasts alter the characteristics of the wine, giving it a characteristic dark yellow-gold color and an acetaldehyde content of 600 to 700 mg/L.

3. CHANGES INDUCED BY *FLOR* YEASTS

Most of the changes that affect the composition of biologically aged wine are linked to the metabolism of *flor* yeasts, although the wine also undergoes changes common to all wines that are aged, for example, crystalline precipitation, chemical reactions between the components of the wine, and the extraction of compounds from the wood.

The main carbon sources used by *flor* yeasts are ethanol, glycerol, and acetic acid, and the amino acid L-proline is used as a source of nitrogen. Yeast autolysis results in the release of amino acids, peptides, nucleotides, mannoproteins, esters, alcohols, aldehydes, acids, and lactones. All of these compounds make an important contribution to the organoleptic properties of biologically aged wines.

The concentration of *flor* yeast-derived metabolites depends on the specific conditions under which the wine is aged, such as the number of *criaderas* and *rocios*, the physical and

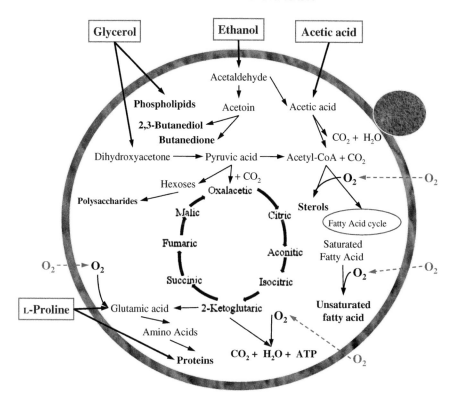

FIGURE 23.3   Metabolic activity of *flor* yeast.

environmental conditions, the chemical composition of the wine, and the dominant species or variety in the yeast *flor*.

The ethanol, glycerol, and acetaldehyde contents are most affected by the metabolism of *flor* yeasts, but other compounds such as organic acids, nitrogen compounds, higher alcohols, esters, lactones, and polyphenols are also affected.

3.1. Ethanol

Ethanol is a source of carbon and energy for yeasts. A proportion of the ethanol is converted to other compounds such as acetaldehyde, acetic acid, butanediol, diacetyl, and acetoin, while another part is metabolized via the tricarboxylic acid cycle and incorporated into cell material in the form of carbohydrates, fats, and proteins. The result is a decrease in ethanol content. The extent of this decrease depends on the dominant yeast race in the *flor*, the growth phase of the *flor*, and the position of the wine in the *criaderas* and *solera* system. More ethanol is consumed in the *flor*-formation phase than in the maintenance phase, and the consumption of ethanol is highest in the rows containing the youngest wine. By the end of biological aging, up to 1% (vol/vol) of ethanol may have been lost. Laboratory experiments have shown a consumption of close to 4% (vol/vol).

3.2. Glycerol

Glycerol is the third most abundant component of wine, surpassed only by water and ethanol. It is used by yeast as a source of carbon and its concentration decreases with aging. The glycerol content can be used to indicate the age of a wine, with levels rarely exceeding 1 g/L at the end of aging.

3.3. Acetaldehyde

Acetaldehyde is the main product of biological aging. It has a major impact on organoleptic properties and is responsible for the intense aroma that characterizes wines aged using this method. The acetaldehyde content of *fino* wines at the end of aging generally lies between 350 and 450 mg/L. This aldehyde is mostly derived from the enzymatic oxidation of ethanol by alcohol dehydrogenase II (Figure 23.4).

The acetaldehyde produced by yeast during aging is highly reactive and participates in numerous reactions. One of the most relevant reactions is its combination with ethanol to form diethyl acetal, also known as 1,1-diethoxyethane. The levels of this acetal depend on the ethanol and acetaldehyde content, and can on occasions exceed 100 mg/L, particularly in wines that have been aged for over 5 years. Diethyl acetal confers an intense nutty, licorice aroma to the wine.

Acetaldehyde also participates in other reactions:

- It combines with SO_2, thereby increasing the proportion of bound SO_2 in the wine.
- It acts as a bridging molecule to form addition compounds with certain polyphenols, such as tannins and procyanidins.
- Its chemical oxidation gives rise to the formation of acetic acid, albeit in quantities that do not significantly affect the volatile acidity or the quality of the wine.
- It participates in the formation of sotolon, a powerful odorant that confers nutty, curry aromas.

3.4. Nitrogen Compounds

Amino acids, and L-proline in particular, are the main source of nitrogen for *flor* yeasts. Wine is rich in L-proline as this amino acid is not metabolized by yeast during alcoholic fermentation due to the lack of molecular oxygen. The transformation of L-proline to glutamic acid allows *flor* yeast to synthesize the other amino acids they need for growth.

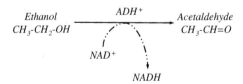

FIGURE 23.4   Enzymatic transformation of ethanol into acetaldehyde by *flor* yeasts. NAD = nicotinamide adenine dinucleotide.

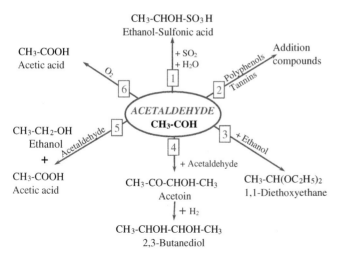

FIGURE 23.5 Overview of reactions involving acetaldehyde.

3.5. Organic Acids

Because the *sobretablas* wine that is added to the *criaderas* and *solera* system normally undergoes malolactic fermentation, it does not contain large quantities of malic acid. On the other hand, the reduction in tartaric acid content that occurs during biological aging is linked to crystalline precipitation.

Acetic acid is produced by yeast during fermentation and its levels do not generally exceed 0.3 g/L in a well-conducted fermentation. This acid is metabolized by *flor* yeasts during biological aging, but it is not advisable to biologically age wines with concentrations of over 0.5 g/L due to the risk of microbial contamination.

3.6. Higher Alcohols and Esters

Higher alcohols make a very important contribution to the aroma of *fino* wines. While the concentration of specific alcohols changes during aging, the overall change is not significant. Yeast autolysis increases the concentration of certain higher alcohols such as propanol, isobutanol, and isoamyl alcohols.

The concentration of higher alcohol acetates generally decreases in the first few months of aging due to hydrolysis; that of the ethyl esters of organic acids, in contrast, increases, and particularly so in the case of lactic and succinic acid.

3.7. Lactones

Soleron (4-acetyl-γ-butyrolactone) is the most important lactone in wines that develop under a *flor* biofilm; it is abundant in *fino* wines and Jura *vins jaunes*, and for years was considered one of the most powerful contributors to the aroma of these wines. In recent years, however, it has been shown that sotolon (3-hydroxy-4,5-dimethyl-2(5H)-furanone) has a greater impact on aroma in biologically aged wines. It possesses a highly characteristic nutty or curry odor with a very low perception threshold (10 μg/L). Other lactones that have been detected in these wines are γ-butyrolactone and pantolactone (2,4-dihydroxy-3,3-dimethylbutyric acid γ-lactone), which is characteristic of sherry-like wines. Biologically aged wines contain few wood-derived lactones as the barrels used are very old and are treated to ensure a minimum transfer of compounds.

3.8. Polyphenols

Flor yeasts exert a protective effect against air because they consume oxygen and the cell walls retain brown compounds, thereby preventing the browning of the wine underneath. The total polyphenol content is low and decreases further with aging. Absorbance values at 420 and 520 nm are particularly low in biologically aged wines and decrease with time, partly explaining the pale-yellow color that characterizes *fino* wines.

Biological aging affects the chemical composition of wine in numerous ways, the most important of which are summarized below:

- Decrease in ethanol content due to consumption by yeast and partial evaporation through pores of the wood (in damaged barrels).
- Decrease in glycerol content due to yeast consumption.
- Decrease in volatile acidity and titratable acidity (titratable acidity decreases with volatile acidity and potassium bitartrate precipitation).
- No change in pH.
- Increase in acetaldehyde content.
- Increase in volatile compounds such as alcohols, acetals, and esters.
- Decrease in total polyphenol levels and absorbance levels at 420 and 520 nm.
- Decrease in amino acid content, and in L-proline content in particular.

4. FACTORS THAT AFFECT BIOLOGICAL AGING

The main characteristic of biological aging is the formation of a *flor* layer on the surface of the wine. The rate at which the *flor* forms, together with its thickness, appearance (roughness and color), and activity, is related to microbiological and chemical factors, as well as physical factors such as the environmental conditions in which the *flor* grows. The main factors that influence the formation and maintenance of the *flor* are listed below and then discussed in more detail:

1. Yeast species or variety
   The most common species of yeast responsible for *flors* in Montilla-Moriles wines are:
   - *Saccharomyces cerevisiae* G-1
   - *Saccharomyces bayanus*
2. Composition of wine
   - Ethanol and dissolved oxygen content
   - SO_2 content and pH
   - Sugar, polyphenol, and vitamin content
3. Physical and environmental conditions of the cellar
   - Temperature
   - Relative humidity
   - Ratio of *flor* surface area to volume of wine

4.1. Yeast Species or Variety

Flor yeasts exhibit diauxic growth as they can adapt their metabolism to an anaerobic medium (alcoholic fermentation) or an aerobic medium (*flor*). This adaptive capacity is related to the *HSP12* gene and is externally manifested as a change in the shape, size, and hydrophobicity of the cell membrane. The increase in the hydrophobic nature of the membrane explains, in part, why the yeasts join together to form the *flor* and it also explains their increased ability to float. The majority of *flor* yeasts that have been isolated in the Montilla-Moriles region are *Saccharomyces cerevisiae* variety *capensis G-1* and *Saccharomyces bayanus F-12*. Figure 23.6 shows the appearance of a *flor* formed by two of these yeasts.

4.2. Composition of Wine

Ethanol

The *flor* forms in wines with an ethanol content of between 10 and 16% (vol/vol), with optimal levels lying between 13 and 15% (vol/vol). *Flor* films have, however, also been observed in wines with 17% (vol/vol) ethanol. *Flor* yeasts use both glycerol and ethanol as a source of carbon, with a preferential consumption of glycerol in the initial *flor* formation stages. The lower the glycerol content, the more ethanol will be consumed.

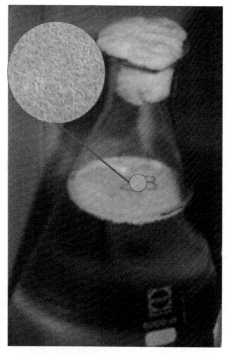

FIGURE 23.6 *Flor* formed by two species of yeast.

Oxygen

The presence of dissolved oxygen in wine favors cell viability and growth of *flor* yeasts. The yeasts consume dissolved oxygen during the formation of the *flor*; the duration of this phase varies from one variety or species to the next. Once the *flor* is formed, the yeasts consume any added oxygen until the wine becomes semi-saturated (this takes just a few hours).

SO₂

The resistance of the *flor* to SO_2 varies according to the yeast variety involved in its formation and also according to the development stage of the *flor*. *Flor* formation is delayed with doses of 100 mg/L but completely inhibited when levels exceed 200 mg/L.

pH

Flors will form at the pH of healthy wine (2.7 to 4.1), but the literature describes optimal values of between 3.1 and 3.4. At values of less than 3, the *flor* will grow very slowly. The relationship between pH and the proportion of free SO_2 is very important in winemaking and must always be taken into account.

Tannins, Sugars, and Vitamins

Tannins stimulate *flor* growth at concentrations of between 0.01 and 0.05 mg/L; the rate of growth, however, is reduced at higher concentrations. Tannic acid, ferulic acid, vanillin, and oak extract with 15% tannin content all have a negative effect on *flor* growth. Flor biofilms have also been seen in red wines.

The concentration of reducing sugars in wines to be aged is generally less than 1 g/L. Nonetheless, it has been seen that glucose induces faster *flor* growth than fructose, and that yeast adaptation to the *flor* phase is considerably faster at higher concentrations of glucose (up to 20 g/L).

Flor yeasts need biotin to form a biofilm, with greater growth occurring at higher concentrations of this vitamin. While biotin favors the growth of *flor* yeasts, calcium pantothenate is the only essential vitamin as *flor* formation occurs in the absence of biotin, thiamine, inositol, and pyridoxine.

4.3. Physical and Environmental Conditions of the Cellar

Temperature

Flors form at a temperature of between 15 and 20°C, depending on the ethanol content of the medium. A drop in temperature in winter months slows cell division and causes the *flor* to sink. The combined action of ethanol and temperature is another important factor, as ethanol is a known mutagen for yeast and limits their growth, and these effects are potentiated at higher temperature. Critical values of 23°C and 15% (vol/vol) ethanol content have been established for biological aging. Above these values, the *flor* will deteriorate due to the increased number of mutant yeasts that are incapable of aerobic metabolism.

Relative Humidity

The relative humidity of the cellar should be close to 80%, to prevent volume losses due to the evaporation of water or ethanol and to control the ethanol concentration of the wine.

Different types of humidification systems can be installed, ranging from traditional systems involving spraying a chalky sand floor renowned for its ability to retain moisture, to automatic humidifiers that spray droplets of water into the air at predefined intervals.

Ratio of Flor Surface Area to Volume of Wine

In the traditional biological aging system, the oak barrels are only filled to four-fifths of their capacity to leave room for a wine-air interface on which the *flor* can grow. The larger the surface on which the *flor* forms, the larger the population of yeast will be and the faster the desired transformations will occur. In the traditional system used in Montilla-Moriles, the surface area to volume ratio is approximately 17 cm^2 to 1 liter of wine.

Index

Note: Page numbers followed by *f* indicate figures, *t* indicate tables and *b* indicate boxes.

C

Printed and bound by CPI Group (UK) Ltd, Croydon, CR0 4YY

08/05/2025

01864843-0002